Communication Technology and Social Change

Theory and Implications

Carolyn A. Lin and David J. Atkin
University of Connecticut

LEA's COMMUNICATION SERIES

Jennings Bryant and Dolf Zillmann, General Editors

Selected titles include

Berger • *Planning Strategic Interaction: Attaining Goals Through Communicative Action*

Bryant/Zillmann • *Media Effects: Advances and Theory in Research, Second Edition*

Ellis • *Crafting Society: Ethnicity, Class, and Communication Theory*

Fortunato • *Making Media Content: The Influence of Constituency Groups on Mass Media*

Greene • *Message Production: Advances in Communication Theory*

Reichert/Lambiase • *Sex in Advertising: Perspectives on the Erotic Appeal*

Roskos-Ewoldsen/Monahan • *Communication and Social Cognition: Theories and Methods*

Singhal/Rogers • *Entertainment Education: A Communication Strategy for Social Change*

Zillmann/Vorderer • *Media Entertainment: The Psychology of Its Appeal*

For a complete list of titles in LEA's Communication Series, please contact Lawrence Erlbaum Associates, Publishers at www.erlbaum.com

Communication Technology and Social Change

Theory and Implications

Edited by

Carolyn A. Lin and David J. Atkin
University of Connecticut

LAWRENCE ERLBAUM ASSOCIATES, PUBLISHERS
2007 Mahwah, New Jersey London

Lawrence Erlbaum Associates, Inc., Publishers
10 Industrial Avenue
Mahwah, New Jersey 07430
www.erlbaum.com

Cover design by Tomai Maridou

CIP information for this volume may be obtained by contacting the Library of Congress

Communication technology and social change : theory and implications / Carolyn A. Lin and David J. Atkin
p. cm.

Includes bibliographical references and index.

ISBN 978-0-8058-5613-2 — 0-8058-5613-7 (cloth)
ISBN 978-0-8058-5614-9 — 0-8058-5614-5 (pbk.)
ISBN 978-1-4106-1541-1 — 1-4106-1541-3 (e book)

Books published by Lawrence Erlbaum Associates are printed on acid-free paper, and their bindings are chosen for strength and durability.

Printed in the United States of America
10 9 8 7 6 5 4 3 2 1

I dedicate this book to my parents who have been and always will be the inspirational source that carries me through life.

Carolyn A. Lin

To the memory of Jane Sutherland Atkin (1922–2005), in appreciation for her encouragement and support.

David J. Atkin

Contents

Preface ix
About the Contributors xi

I. Introduction 1

1 Communication Technology and Social Change 3
Carolyn A. Lin

2 Communication Technology and Global Change 17
Christine Ogan

II. Individual and Social Setting 37

3 Computer-Mediated Technology and Social Interaction 39
Scott E. Caplan, Elizabeth M. Perse, and Janice E. Gennaria

4 Computer-Mediated Technology and Children 59
Marina Krcmar and Yuliya Strizhakova

III. Work and Organizational Setting 77

5 Information Technology and Organizational Telework 79
David J. Atkin and T. Y. Lau

6 Information Technology: Analyzing Paper and Electronic Desktop Artifacts 101
Ronald E. Rice and Sara Schneider

IV. Surveillance Setting 123

7 Media Technology and Civic Life 125
Leo W. Jeffres

8 Media Technology and the 24-Hour News Cycle 143
Eric P. Bucy, Walter Gantz, and Zheng Wang

V. Entertainment Setting 165

9 Video Technology and Home Entertainment 167
August E. Grant

10 Online Technology, Edutainment, and Infotainment 183
Joey Reagan and Moon J. Lee

VI. Consumer Setting 201

11 Interactive Media Technology and Electronic Shopping 203
Carolyn A. Lin

12 Interactive Media Technology and Telemedicine 223
Pamela Whitten

VII. Legal and Regulatory Setting 241

13 Digital Media Technology and Fair Use 243
Jeremy Lipschultz

14 Digital Media Technology and Individual Privacy 257
Laurie Thomas Lee

VIII. Summary 281

15 An Integrated Communication Technology and Social Change Typology 283
Carolyn A. Lin

Author Index 309
Subject Index 325

Preface

The present volume revisits and extends a topic that the editors addressed in their previous book, *Communication Technology & Society: Audience Adoption and Uses* (Hampton Press, 2002). As the most significant outcomes of technology adoption and uses remain the manner in which these technologies impact our lives, the current volume explicates how communication and information technologies facilitate social change. The primary goal of this book is to help enhance the intellectual understanding of these social change outcomes in its readers—including scholars, students, and practitioners alike—from the perspective of theory and effects in various social systems domains.

Specifically, the chapters in this book describe, analyze, interpret, and synthesize how communication and information technologies influence our lives in several domains, including: how we interact with each other, seek, access, process, and share information; work and informat organizational tasks; play and pursue leisure; engage in teleshopping and telehealth; and protect our rights to free speech and privacy. In doing so, our conceptual scheme also aims to expand and complement the scope of other academic books published on this subject that focus on: (a) a single or narrowly formulated topical area, (b) a set of macro- or microcritical analyses, (c) a collection of individualized or stand-alone empirical studies, or (d) an applied approach with a practitioner focus.

The social contexts that serve as the unit of analysis for this book—outlined in the Table of Contents—examine how communication and information technologies influence individuals, groups, or institutions in various social settings. Each chapter is paired into one of the following specific topical sections: Introduction, Individual and Social Setting, Work and Organizational Setting, Surveillance Setting, Entertainment Setting, Consumer Marketing Setting, and Legal and Regulatory Setting.

Each of the chapters is presented with current theoretical, empirical, and/or legal analyses, organized with the following standard headings within a particular social communication domain: (a) Background—an introduction of relevant communication technology that outlines its technical capabilities, diffusion, and uses; (b) Theory—a discussion of relevant theories used to study the social impacts of the communication technology in question; (c) Empirical Findings—an analysis of recent academic and relevant practical work that explains the impact of the communication technology on social change; and (d) Social Change Implications—a summary of the real-world implications for social change that stem from synthesizing the relevant theories and empirical findings presented earlier.

This book should be most useful and valuable for the following types of readers anywhere in the world:

1. Scholars, researchers, and professionals who wish to enhance their theory, research, and practical knowledge background in this subject area.
2. Students in any upper division undergraduate- and graduate-level courses from the different branches of the communication discipline, whose coursework focuses on or encompasses the topic of communication and information technology.
3. Students from other academic disciplines such as psychology, sociology, family studies, political science, government and policy studies, allied health, public health, marketing, business management, and law, whose coursework encompasses the topics of communication and information technologies.

As Einstein once observed, "there is nothing so practical as a good theory." Using theory to help inform their analyses, the contributors of this book collectively paint a picture of a vibrant information society advancing toward a digital age. This digital age is evolving in a dynamic technological landscape, one that represents an innovative and powerful agent for fluid and intriguing social change.

About the Contributors

Carolyn A. Lin is Professor in the Department of Communication Sciences at University of Connecticut. Her research interests focus on the content, uses, and effects of new media technologies, advertising, health communication marketing, and international communication. She is a coauthor and coeditor of several books, including *Patterns of Teletext Use in England; Communication Technology and Society: Audience Adoption and Uses;* and *International Communication: Concepts and Cases.* Her current research projects include two federally funded studies that utilize interactive media technology to implement health intervention programs.

David J. Atkin is Professor in the Department of Communication Sciences at University of Connecticut. His research interests include the diffusion of new media and program formats, media economics, and telecommunication policy. He is a coauthor and coeditor of two books, including *The Televiewing Audience* and *Communication Technology and Society: Audience Adoption and Uses.* Atkin is a past winner of AEJMC's *Kreighbaum Under 40* award for excellence in teaching, research, and public service.

Erik P. Bucy is Associate Professor in the Department of Telecommunications and Adjunct Associate Professor in the School of Informatics at Indiana University, Bloomington. His research focuses on the social-political impact of new communication technologies, psychological responses to political news, and normative theories of media and democracy.

Scott E. Caplan is Associate Professor in the Department of Communication at the University of Delaware. His research interests include online social interaction and other issues related to computer-mediated interpersonal processes.

Walter Gantz is Professor and Chair of the Department of Telecommunications at Indiana University, Bloomington. His research interests include mediated sports and sports fanship, advertising content, and news diffusion processes.

Janice K. (Recchiuti) Gennaria completed her M.A. in Communication from the University of Delaware in 2004. She is currently working as a consultant for a market research firm providing insight and analysis to consumer package goods companies.

August E. (Augie) Grant is Associate Professor in the College of Mass Communications and Information Studies at the University of South Carolina. Grant is editor of *Communication Technology Update,* and his research deals with the interplay of organizations and individuals in the adoption and use of communication technologies.

Leo W. Jeffres (PhD, University of Minnesota, 1976), Professor of Communication at Cleveland State University, focuses his research on urban communication, neighborhood newspapers, technology, and media effects.

Marina Krcmar (PhD, University of Wisconsin) is Associate Professor at Wake Forest University. Her research interests include the effects of TV and video games on children, adolescents, and families; and the role of media in adolescent risk taking.

Tuen-yu Lau is Director of the Master of Communication in Digital Media Program at the University of Washington. His major research interests include digital media, media management, and global media.

Laurie Thomas Lee is Associate Professor in the College of Journalism and Mass Communications at the University of Nebraska–Lincoln. Her research interests are in the areas of telecommunications law, privacy, media economics, and new technologies.

Moon J. Lee is Assistant Professor in the Edward R. Murrow School of Communication at Washington State University. She teaches Principles of Public Relations, International Public Relations, and Public Relations Campaigns and Management, and does research on the effects of new media technologies on human cognition, health information campaigns, adolescents' information-processing and decision-making processes.

Jeremy Harris Lipschultz is Robert Reilly Diamond Professor and School of Communication Director in the College of Communication, Fine Arts and Media, University of Nebraska at Omaha. Dr. Lipschultz (PhD, Southern Illinois University, 1990) has research interests in social aspects of media law and regulation, media and information literacy, gender and elderly concerns, computer-mediated communication, media indecency law, political communication, and pedagogy of technology.

Christine Ogan is Professor of journalism and informatics at Indiana University. She conducts research on international (in Europe and the Middle East) and domestic use of media and Internet technologies, as well as issues related to gender in information technology higher education.

Elizabeth M. Perse is Professor and Chair in the Department of Communication at the University of Delaware. Her research interests are the uses and effects of new mass communication technology.

Joey Reagan is Professor in the Edward R. Murrow School of Communication at Washington State University. He teaches communication technologies and research methods and does research about technology adoption.

Ronald E. Rice is the Arthur N. Rupe Professor in the Social Effects of Mass Communication in the Department of Communication at the University of California, Santa Barbara. His research interests include computer-mediated communication, diffusion of innovations, public communication campaigns, and communication networks.

Sara Schneider worked on this project while completing her degree in Communication at Rutgers University, The State University of New Jersey. She currently works for a major consulting/accounting firm.

Yuliya Strizhakova (PhD, University of Connecticut) is Assistant Professor in the Communication and Journalism Department at Suffolk University. Her research interests include effects of new technologies on society, marketing applications of new technologies, and broader motives of leisure consumption.

Zheng Wang is a doctoral candidate in the Department of Telecommunications at Indiana University, Bloomington. Her research interests focus on the cognitive processing of mediated messages.

Pamela Whitten is Assistant Dean and Professor for the College of Communication Arts and Sciences at Michigan State University. She conducts research, often referred to as telemedicine or telehealth, that examines the application of communication technologies to facilitate the delivery of health care services.

I

INTRODUCTION

1

Communication Technology and Social Change

Carolyn A. Lin

Throughout human history, a confluence of cultural, economic, political, and technical forces has triggered profound and enduring social changes. Although some of these changes were abrupt and eruptive, others were more gradual and evolutionary. Because revolutions often result from traumatic political or economic transitions (e.g., depressed economies), evolution generally accompanies the progressive adoption of an innovative idea or invention (e.g., public education, automobiles; see Rogers, 2003). Communication technology innovations have thus catalyzed far-reaching social changes since the 19th century (e.g., Dizard, 2000).

Dating to the invention of the telegraph—which enables people to communicate across great distances from different locales around the world—the advent of communication technologies has gone as far as one's imagination can extend. As millions of radio signals travel across the globe to their respective destinations each day, communication technologies have allowed us to share our experiences and cultures wherever these linkages exist. Even though national borders survive today, they're essentially political boundaries maintained by nation-states (e.g., Margolis & Resnick, 2000; Pelton, 2003). Social changes facilitated by the diffusion of communication technologies in one society may no longer be confined by these artificial boundaries.

The main purpose of this book is to introduce the social changes resulting from communication technology diffusion in society. Specifically, this book provides a portrayal of the social changes that have occurred due to people's adoption and uses of new communication technologies in various social contexts, including home, workplace, surveillance, entertainment, consumer, and legal settings. Each chapter contains a technology primer that explains (a) the technical functions of each communication technology under consideration, (b) a review of significant theories used to study associated social changes, (c) empirical research evidence from the field, and (d) a synthesis of the implications of these social changes to society. The section to follow provides a succinct introduction to the contents of each chapter.

A CHANGING COMMUNICATION INFRASTRUCTURE

As we consider the social influence of communication technology, an analysis of technology diffusion trends is in order. Contrary to Orwell's dystopian prophecies about life beyond 1984, telematic "technologies of freedom" (Poole, 1982) now challenge Big-Brother totalitarianism. One need look no further than the popularity of Western satellite and Internet fare in emerging democracies—particularly former Communist block nations—to find a case in point (Anokwa et al., 2003).

In the early 1980s, almost every American was served by the AT&T telephone monopoly, a "shared monopoly" of three broadcast television networks and a local newspaper. Two-way communication was largely relegated to the telephone, while electronic media typically delivered one-way, least-objectionable materials on a rigid air-time schedule. Today, two-thirds of Americans enjoy Internet access, 85% of TV homes receive multichannel TV service, and two thirds use wireless carriers, which are now competing with land lines for telephone usage. Electronic media outlays are thus subsuming larger portions of our work and leisure allocations and—despite problems with the consolidation of communication and media ownership consolidation, losses due to piracy, and so on—end users generally enjoy greater choice, immediacy, and control over their mediated communication. Indeed, the United States stands among the most "wired," computer-literate nations in the world.

So great have been the economic implications of this shift that in April 2004 the Dow Jones stock exchange removed AT&T from its list of 500 leading indicators of industrial performance, only to replace it with Verizon, a Regional Bell progeny that had almost doubled the $34 billion revenue of its parent. Within a year, AT&T was acquired by another of its progeny, SBC. Yet the growth in mediated communication modalities does not necessarily imply increases in user access to greater content diversity, due to continued industry ownership integration, both horizontal (e.g., Hewlett Packard's merger with Compac computers) and vertical (e.g., Disney's merger with ABC television). And, despite the fact that the millennial generation represents the first one to grow up in a mutichannel communication environment replete with cable, the Internet, and so on, more work still needs to address the social impact of these technologies in domestic, workplace, and societal settings (Kundanis, 2003). In outlining the direction of our book, this chapter surveys the influence of these communication technologies on the discourse structure of society from a family, work, and sociocultural perspective.

AN OVERVIEW OF THIS VOLUME

Consider the following scenario. The U.S. military conducted their "shock and awe" campaign in Iraq by launching tens of thousands of bombs and missiles in the spring of 2003. This entire event was captured by the embedded journalists and observed by people from around the world in real time. Every manner of telecommunication device was employed to gather, transmit, and receive the news from the Iraq war by the news media as well as citizens across the globe. The Iraq war was thus the first war in history to be reported live from the battlefield (e.g., Joslyn, 2003). A full range of lower and medium-orbit satellites, in addition to other mobile and fixed satellite services operating in the geosynchronous orbit, have been utilized to keep this uninterrupted 24-hour news cycle running (see Pelton, 2003).

In particular, the Internet medium has been used as a rallying point for those parties who don't have formal access to regular media outlets (e.g., Johnson, Braima, & Sothirajah, 1999). Most notably, these parties include the various anti-American and terrorist groups that have been threatening a worldwide jihad against the American infidels. Journalists, on the other hand, came to rely on such mobile communication technologies as satellite telephones to conduct live coverage. Political and power elites from different sides of the Iraq war controversies have utilized their media platforms to channel their hegemonic beliefs and agenda-setting tactics to manipulate public opinion (Joslyn, 2003). From the viewpoint of social change, the communication technology advances have helped shape the world's views toward the Iraq war—from one living room to another—across the globe.

Chapter 2 by Ogan examines the phenomenon of how communication technologies—including the satellite and videocassette recorder (VCR)—allow citizens worldwide to observe and react to the same events, wherever they occur. These events could be related to issues that impact the world's health (e.g., the transmission of AIDs or SARs), economic growth (e.g., the World Trade Organization meeting or the North American Free Trade Agreement[NAFTA]), unexpected disasters (e.g., the 9/11 attack), geopolitical conflict (e.g., the Iraq war), cultural clashes (e.g., Western media dominance of the youth culture), and so on. The crucial questions that arise from this interconnected world via a vast set of communication technology networks (Wood & Smith, 2001) include, but are not limited to, the following: Which technologies have been most effectively utilized to establish the communication channels between nations and cultures? Are members of the public becoming better informed about the world around them as a result of communication technology usage? Has technology helped enhance communication effectiveness between peoples of different countries? Have national cultures become more interconnected due to the technical breakdown of communication barriers?

The best starting point for discovering any real changes in communication patterns between people and cultures will be at the individual level, or within a network of friends and family. When the image of Neil Armstrong's moon landing was transmitted live via satellite in 1969, the impact of this salient event was felt most strongly by those viewers who saw it unfolding with their friends and families around the world. After the successful moon landing, anything seemed possible in the new era of the emergent home technology environments. First came the VCR for home consumers in 1975, which was followed by digital video disc players. While cable TV systems started to spring up around suburbs and cities, large satellite receiving dishes completed the multichannel broadband transmission landscape in the rural and remote areas of the country. The decade that followed witnessed the growth of personal computers, cellular telephones, and the Internet among other technologies.

Chapter 3 by Caplan, Perse, and Gennaria investigates how the various communication technologies are utilized at personal (one-on-one) as well as societal levels (e.g., Dholakia, Mundorf, & Dholakia, 1996). In particular, what are the primary and secondary functions of different communication technologies as perceived by different users in a family or social setting? What are the dynamics of interpersonal communication patterns and social interactions, as facilitated by the uses of these technology units? Does the Internet serve as the major source of sharing information, images, and social support that would otherwise not be possible, particularly between individuals who may not belong to a family unit or social network? How do cellular phones influence interpersonal communication patterns? Does e-mail help create, maintain, or enhance interpersonal communication relations between friends and families across distance? How are mediated social interactions different from (or similar to) face-to-face interpersonal or other social interactions?

The question of when and how these types of negotiations take place in society is an important one, especially when children are involved. Considering the ubiquitous media contents available at the click of a mouse, the socialization of our children in an era of abundant channel outlets becomes ever more complicated (Kundanis, 2003). Chapter 4 by Krcmar and Stirzhakova examines the topic of children's uses of media technologies at home. Key issues include the patterns of children's media technology use, children's cognitive and affective responses to media content, the social learning impacts of media on children, and the role of parental guidance and mediation in children's media technology and content use (e.g., Kaiser Family Foundation, 1999). Other examples of significant social change issues include the following: Do violent video games desensitize our youth? Are children capable of distinguishing misleading messages from legitimate information online? What is the role of parents and schools in developing digital media literacy skills?

When communication technologies revolve around the home environment, one promising application—telecommuting—reflects another emerging social change phenomenon (e.g., Kraut, Scherlis, Mukhopadhyah, Manning, & Kiesler, 1996; Stanek & Mokhtarian, 1998). Unlike home workers, telecommuters are part of the corporate structure, where certain workers, due to the independent nature of their jobs, are permitted to work away from the corporate office and remain connected via the corporate communication network through most of their workday. This "virtual workforce" is growing as corporations trim operating costs and accommodate the lifestyle choices of valued employees (U.S. Department of Commerce, 1998).

Chapter 5 by Atkin and Lau visits this home work phenomenon in relation to highly skilled professionals working in a technology-savvy environment, one that is unique to an information society (e.g., Rogers, 2004). Digital communication networks, operated via dedicated satellite networks or fiber optics networks to provide high-speed broadband connections across different corporate and off-site workplaces, are the backbone of this telecommuting phenomenon (e.g., Bucy & Newhagen, 2004; Kraut, Steinfield, Chan, Butler, & Hoag, 1998). From the perspective of organizational communication, as telecommuters primarily conduct their work away from the physical space that other employees may share, what impacts does telecommuting have on organizational communication dynamics and effectiveness? What changes in corporate culture does telecommuting create? Other social change issues, such as what new household communication and lifestyle patterns might emerge due to telecommuting, are also interesting social change topics worthy of exploration.

Even as some workers telecommute, the majority of workers still physically commute to an office environment that may revolve around a web of communication technology networks. It is a well-known fact that corporate America is usually the first to adopt new communication technology. As time is often of the essence in the business world, the ability of each technology to efficiently help accomplish organizational communication goals is of utmost importance (e.g., Kraut et al., 1998). Technology adoption at the organizational level involves an array of factors that go above and beyond the consideration of the technical functions themselves. As Monge and Fulk (1999) suggest, human factors remain the key determinants of the success (or failure) of technology applications in an organizational setting.

Chapter 6 by Rice addresses the role of information technology, mediated communication channels, and information workers in shaping the communication culture in an organizational environment. Depending on the characteristics of the organization in terms of its management culture, personnel structure, and decision-making hierarchy, the adoption and uses of information and communication technologies can vary widely. For instance, processing an information task with a network of coworkers and for a group of clients via the use of faxes, PDF files, written memos, e-mails, an electronic bulletin board (e.g., electronic blackboard), audio conferences, video conferences, or webcast conferences could result in different degrees of success for the ideas communicated, discussed, and exchanged. The potential differential effects resulting from these different means of informating (sharing information) and communicating signal the importance of choosing the right channel for the right organizational communication task (see Monge & Fulk, 1999), which is contingent on several factors. These might include, for instance, the inherent characteristics of the information and communication technology that enable a mediated communication process to transmit the highest level of isomorphism. Hence, a medium's ability to allow the users to communicate with others in the organizational infrastructure to complete their informating tasks—with fluidity and unequivocal information flow—involves significant sociotechnical factors worthy of consideration.

As communication technologies of all forms permeate homes and workplaces, the next question addresses the role these communication technologies play in a community or social setting. It is not surprising to find that community organizations of all forms have come to rely on technologies to maintain communication between their managers, organizers, and members. As some community organizations champion a social, cultural, educational, economic, or political cause, others emphasize performing different types of civic services. These community organizations and their communication purposes can range from the Little League team coaches corresponding with team parents, to a citizen action committee conducting an e-mail petition campaign to revive the community's economic base.

Chapter 7 by Jeffres can help us understand the relations between communication technologies and community organizations. Community organizations by definition are diverse in their membership, financial sources, organizational structures and goals. However, they have one thing in common—nearly all of them maintain a web site to register their presence to the larger public (see Jeffres, 2003). Some of these organizations, especially those that perform routine civic services, have also been allotted a local cable TV channel that serves as a platform for airing their agendas and activities (e.g., educational institutions, libraries, police and fire departments, municipalities, etc.). Comparatively speaking, community organizations have open access to the Internet as a source for public communication, whereas their access to a traditional media outlet such as a local cable TV system is severely limited. Nonetheless, community news items and activities that are aired on a cable TV system can be seen by those city residents who are not Internet users. As more communities utilize the Internet to rally their members on various civic actions, these localized movements can also be endorsed or adopted by distant communities. Hence, the Internet medium has enabled different communities across geographic regions with similar social and civic goals to share information, experiences, and even human resources (see Jeffres, 2003; Lievrouw & Livingstone, 2002; Wood & Smith, 2001). A number of interesting questions are thus worth explicating here. For instance, what kinds of information and communication exchanges are being delivered through the offline (e.g., cable TV) and online communication modalities? What user needs and media functions do these types of online and offline community communication structures serve? What impacts do communication technologies have on facilitating in terms of civic actions and social movements?

While communication technologies bring relevant information to and facilitate interpersonal interaction between people who are involved with a wired community, their other function as the primary information source for society remains essential in keeping these communities engaged in civic discourse. The surveillance utility of news media is one of the most valuable institutional assets of a democratic society (e.g., McChesney, 1997). Anointed as the fifth estate of the government, electronic news media have become the primary source for people to learn about current events and to monitor their environment. Television news, with its vivid transmission of visual history, is the primary storyteller of our time (see Newcomb, 2000). The 24-hour cable news channels were founded on the notion that as the world turns, it never falls asleep. This is also true of the web sites that post updated news items, as the Internet medium knows of no boundaries of time, distance, and location in its operational domain.

Chapter 8 by Bucy, Gantz, and Wang explores the 24-hour news cycle phenomenon facilitated by cable news networks and the Internet news outlets (e.g., Johnson et al., 1999). In particular, how does this instantaneous live coverage influence the ways that audiences acquire the news? Do the audiences consume more news or obtain greater knowledge gains as a result of the flourishing electronic news outlets, either online or offline? In the world of instant news updates, what institutional and human factors help ensure the credibility of a story's integrity and objectivity? Can these 24-hour news outlets come to dominate news agendas and news frames, reinforced by the abundant news opinion/talk shows (e.g., Bill O'Reilly) that relentlessly present their hosts' opinionated views? Do the 24-hour cable news channels and online news outlets compete with each other and threaten the viability of the traditional print news media? Questions such as these about news diffusion—and its impact on public opinion—are essential in advancing our understanding of how social changes may be shaped on an entire array of issues, ranging from gay marriage, stem cell research, and presidential elections to our foreign policy.

The 24-hour news outlets, disgorging a constant information flow to a sleepless audience, may have their social change influences mitigated by audience exposure to an ever-growing list of 24-hour entertainment media options, both offline and online. Even though seeking news and entertainment from mass media are the two most cited media-use motivations, more people are consuming entertainment fare than news at any given time during the day. When it comes to offline electronic entertainment media, cable and satellite transmission technologies dominate the scene. With advances in digital compression techniques, technologies such as digital video recorders (e.g., TiVo), video-on-demand (VOD) via two-way cable systems, digital cable services, and digital direct-broadcast satellite (DBS) services represent the major weavers of this entertainment content (see Grant & Meadows, 2004).

Chapter 9 by Grant explores the links between each of these offline media technologies and explains their role in generating social change. Digital video recorders (DVRs) represent a digital version of their analog counterpart, the VCR. The VCR was hailed as the medium that freed audiences from the rigid airing schedule and gave audiences control over their TV viewing (e.g., Dobrow, 1990; Levy, 1989). In essence, the VCR helped turn the TV set into a dual-function display terminal, one that transmits programs live or on a delayed basis. DVRs perform the same functions that a basic VCR does, but offer the additional ability to help select and cue the recording of a large number of programs. This DVR technology is even integrated with the major digital cable and digital DBS systems, at the cost of alienating potential advertisers. The most interesting questions pertain to what, if any, changes in audience viewing patterns have emerged and how these changes help create a new breed of leisure culture (e.g., Waldfogel, 2002). If this new self-styled home entertainment culture is indeed in place, does this imply a stronger media dependency? If a stronger media dependency occurs, then does such dependency enhance the media's ability to cultivate beliefs, attitudes, or even behavior due to social learning mechanisms? These questions are socially significant in a culture where commercially driven messages permeate our senses and dominate our perceptions of events and issues.

As media-savvy audiences take their cues from the messages produced by large media conglomerates, the younger generations—including generation-X, generation-Y, and the millenniummies—also seek out interactive infotainment (entertainment that informs) from a wide variety of online sources that are dominated by the same group of media conglomerates (McChesney, 1997). A closer examination of the online media environment manifests a vast array of attractive destinations for entertainment. For instance, one may access entertainment-oriented news and information sites, download music, videos, or films, watch sports or live webcasts of performances, or engage in interactive activities such as video games, live chat, and so on. Not only is this online environment a 24–7 service, it also caters to niche infotainment interests that are not well served by offline media entertainment services.

Chapter 10, by Reagan and Moon, helps us navigate cyberspace for the purposes of assessing the potential social changes that online media infotainment sources have instilled in our society. At its best, the online media environment offers an escape from the "mundane" offline tried-and-true media genres (Newcomb, 2000). This emerging information grid allows the distribution of more creative fare, be it commercial (e.g., e-book sales) or noncommercial (e.g., the landing of NASA's Mars probes). At its worst, the online media outlets could also pose public health threats stemming from Internet-use addiction (e.g., video games or pornography), particularly among more socially isolated individuals (LaRose, Lin, & Eastin, 2003). As the heaviest Internet users tend to be the younger segments of our society, peer cultures formed around certain patterns of Internet use have also emerged (Wood & Smith, 2001). The crucial question here is, are we risking the danger of "amusing ourselves to death" (Postman, 1986)? In other words, can these diverse online infotainment sources—created to meet our insatiable needs for stimulation—help generate positive social learning experience, while gratifying both our aesthetic and affective needs?

One aspect of this insatiable need for stimulation involves teleshopping or shopping via an electronic means. Teleshopping as a form of direct-response marketing is pervasive in our society. The recent Federal Trade Commission action to bar telemarketers from calling consumers who registered in a "no-call list"—an action backed by the Federal Communications Commission—is an indication of the intrusive nature of this marketing practice. However, other forms of "passive" teleshopping channels are still going strong, especially those associated with the Internet. The number of spam e-mails that flood Internet users' mailboxes and the number of pop-up ads that interrupt the access of the target web pages are inviting numerous complaints from the online public.

Chapter 11 by Lin takes us on a shopping tour to see what we can get from each of the teleshopping channels, why we become teleshoppers, and what kinds of gratifications and utilities we receive as teleshoppers. Teleshopping entities depend heavily on a loyal core of shoppers to sustain their operations; this speaks to shopper dependency on teleshopping channels that may indicate an obsessive or even addictive tendency (Grant & Meadows, 2004). Along the same lines, teleshopping channels such as the televised shopping networks (e.g., QVC) can in effect provide a form of pseudo-social interaction for socially immobile shoppers. Because teleshopping purchases are often considered "impulse buys," does teleshopping—as a form of cognitive and affective stimulation—trigger the need for drive reduction? The multifaceted question of what social changes our teleshopping behaviors have wrought in our daily lives is fully explored here.

Beyond the traditional paradigm of teleshopping, the use of a new form of teleshopping, namely, telemedicine, is also on the rise. As the baby boomers approach twilight years, their need for adequate medical care is pressing. And as the costs of medical care and medicines have been spiraling upward in the past decade, affordable medical care has become more scarce for average Americans. Telemedicine as an expedient, cost-effective means of dispensing medical care and medicines has been embraced by some and scorned by others. The channels of telemedicine practice typically involve the use of an electronic communication medium such as telephones and the Internet. Patients provide a description of their physical conditions and medical history via a mediated communication channel. In return, medical care professionals offer a diagnosis of treatment and medical prescriptions, depending on what is needed, without having any in-person examination or consultation.

Chapter 12 by Whitten focuses on this telemedicine phenomenon, one that carries strong implications for health communication research as well as the health of our society. Telemedicine, with all of its potential flaws, is an essential practice for those medically disenfranchised patients who are far from adequate medical facilities, physically immobile or financially disadvantaged. A successful telemedicine service will require a high level of isomorphic communication transactions among the clinicians, pharmacists, and patients. Such transactions lack the in-person exchanges, the nonverbal cues and physical examinations of the patients. This type of communication isomorphism will require the meeting of a number of cognitive and affective criteria (see Rice & Atkin, 2001).

The ability to ask the right questions and to get the right answers from the patients, for instance, is vital for the dispensation of proper care to patients. The patients, then, will need a certain level of health literacy as well as verbal skills to convey their symptoms and ailments. The potential fallacies, scorned by the critics, include the probable miscommunication and/or incomplete or erroneous descriptions that could cause harms to the patients. By contrast, the potential utilities, praised by the advocates, are the efficiency in time and costs that enables affordable medical care for needy patients. Theoretical models such as the health belief model, theory of reasoned action, social cognitive theory, elaboration likelihood model, and diffusion of innovations are all relevant conceptual dimensions that can help explain the relations between patients and telemedicine. These models are also useful in illustrating the social changes in medical care that are brought forth by a variety of different mediated communication technologies (e.g., Rice & Katz, 2003).

The ethical issues revolving around patient privacy, especially through online communication, are also pertinent to other forms of mediated communication technology. Under federal law (Privacy Rule, 45 CFR Parts 160 and 164), there are restrictions imposed on the uses and disclosures of individually identifiable health information; civil and criminal penalties are applicable for violations of such restrictions. An additional set of security standards, with the compliance date in April 2005, also applies to protect health information conveyed in electronic form. These security standards contain three areas of concern: administrative safeguards (e.g., risk analysis and management), physical safeguards (e.g., access to facilities and workstations), and technical safeguards (e.g., access controls and audits authentication and transmission security).

Chapter 13 by Lipschultz explores the different privacy zones related to different communication technologies, discussing the privacy issues of concern and the relevant laws, regulations, or rules that may or may not be available to protect our privacy (e.g., Gandy, 1993; Noam et al., 2004). Significant electronic privacy issues revolving around the use of several technologies—namely, unsolicited commercial e-mail (or spam), radiofrequency identification (RFID) technology, caller IDs, cellular phones, global positioning systems (GPS), and electronic surveillance at public and work places—all carry strong implications for the future of safeguarding individual privacy. In this advanced information technology environment, each time an individual uses a communication technology to transmit, acquire, or receive information, or to authenticate one's identity or physical location, there lies a danger of invasion of this individual's privacy. When a radio-station frequency tuned to by drivers passing by billboards can spur the display of certain types of demographically targeted advertising messages, this signals a question regarding how intrusive communication technologies should be used to market products. Such scenarios seem reminiscent of the film *Minority Report*, where a person walking along the corridors of a shopping mall can trigger the display of targeted commercial messages due to an instant match of that individual's identity (via pupil scanning). The social implications associated with a technology environment, where personal privacy becomes a commodity, are hence numerous.

Although we are concerned about how our individual privacy can be intruded upon by others, electronic piracy committed by individuals also raises a serious information access issue in our society. Ever since the production and distribution of intellectual properties became viable in electronic form—including films, audiocassette tapes, videotapes, floppy disk, computer hard drives, CD-ROM, CDs, DVDs, the Internet, or the World Wide Web—the issue of fair use versus intellectual property rights has continued to evolve. Recent court battles over the issue of peer-to-peer file sharing involving copying online music signaled a wake-up call to technology-savvy users who are finding different ways to copy a wide variety of intellectual property products, whether for commercial or personal use.

Chapter 14 by Lee helps us dissect this phenomenon by explaining the concept of electronic intellectual property as a product of communication and the copyright owners' rights to protect their electronic intellectual creations, in addition to helping us to understand the proper applications of fair use in the federal statute. As the copyright owners are entitled to protection of their intellectual properties for the duration of their life—plus 70 years (The Sony Bono Copyright Term Extension Act of 1998, 17 U.S.C.A. §§ 302–305)—noncommercial publishers and others (including libraries) face a setback in their ability to present these copyrighted materials in the public domain. Moreover, the passage of the Digital Millennium Copyright Act of 1998 and recent FCC rulemaking (MB Docket 02-230) concerning digital broadcast content are indicative of an increase in restricting fair use zones of digital information access. The most challenging questions to answer remain: When does the public's right to access digital information truly infringe upon others' intellectual property rights and when does the restriction on fair use stifle the free exchange of information, ideas, and knowledge in society?

The notion of a technology environment that will allow information, ideas, and knowledge to freely flow through all corners of a given society (or a global society) epitomizes the ideals of an information society, where the only boundary that exists between peoples is geography (Dizard, 2000). As economic gaps between the have and have-not nations are widening, ethnic frictions between national cultures intensify, and religious differences continue to divide people in the world, an ideal information society may not arrive in the near future. Yet when greater access to all forms of modern communication technologies by the populace from around the world becomes a reality, increased communication between peoples of different ethnic, cultural, and/or religious origins can help narrow these fundamental divides between nation states. But how will this bridging process occur and what roles do communication technologies play in this progressive process? In other words, what types of social changes must emerge in different societies and how could communication technologies help facilitate these changes?

Chapter 15 by Lin and Atkin provides a summary of the major theories, empirical findings, and social change implications discussed in each chapter. Based on this summary, the chapter finds conceptual linkages, theoretical bridges, and empirical connections to further illustrate the social change implications associated with the adoption and uses of various communication technologies. Working from these theoretical and empirical explications, an analysis of the sociocultural implications of social change facilitated by the current technology environment is also presented. This is followed by a discussion of the potential social changes as an outcome of emerging technology adoption and uses in the foreseeable future. A conceptual model that links the social change setting related to the adoption and uses of the different types of communication technologies is proposed to guide future research.

REFERENCES

Anokwa, K., Lin, C., & Salwen, M. (2003). *Mass media around the globe*. New York: Wadsworth.

Bucy, E.P., & Newhagen, J.E. (2004). *Media access*. Hillsdale, NJ: Lawrence Erlbaum Associates.

Dholakia, R., Mundorf, N., & Dholakia, N. (1996). *New infotainment technologies in the home: Demand side perspectives*. Hillsdale, NJ: LEA.

Dizard, W. (2000). *Old media, new media*. New York: Longman.

Dobrow, J. (1990). *Socio-cultural aspects of VCR use*. Hillsdale, NJ: Lawrence Erlbaum Associates.

Gandy, O. H. (1993). *The panoptic sort: A political economy of personal information*. New York: Perseus.

Grant, A., & Meadows, J. (2004). *Communication technology update* (6th ed.). Boston: Focal Press.

Jeffres, L. (2003). *Urban communication systems*. Cresskill, NJ: Hampton.

Johnson, T. J., Braima, M. A. M., & Sothirajah, J. (1999). Doing the traditional media sidestep: Comparing the effects of the Internet and other nontraditoinal media with traditional media in the 1996 Presidential campaign. *Journalism & Mass Communication Quarterly, 76*(1), 99–123.

Joslyn, M. R. (2003). The determinants and consequences of recall error about Gulf War preferences. *American Journal of Political Science, 47*, 147–182.

Kaiser Family Foundation. (1999, November). *Kids & Media @ the new millennium*. Menlo Park, CA: Henry J. Kaiser Family Foundation.

Kraut, R., Scherlis, W., Mukhopadhyah, T., Manning, J., & Kiesler, S. (1996). The HomeNet field trial of residential Internet services. *Communications of the ACM, 39*(1), 55–83.

Kraut, R., Stienfield, C., Chan, A., Butler, B., & Hoag, A. (1998). Coordination and virtualization: The role of electronic networks and personal relationships. *Journal of Computer Mediated Communication, 3*(4).

Kundanis, R. M. (2003). *Children, teens, families, and mass media: The millennial generation.* Mahwah, NJ: Lawrence Erlbaum Associates.

LaRose, R., Lin, C.A., & Eastin, M. (2003). Unregulated Internet usage: Addiction, habit or deficient self-regulation? *Media Psychology, 5*, 225–253.

Levy, M. (1989). *The VCR age.* Newbury Park, CA: Sage.

Lievrouw, L. A., & Livingstone, S. (Eds.). (2002). *Handbook of new media: Social shaping and consequences of ICTs.* Newbury Park, CA: Sage.

Margolis, M., & Resnick, D. (2000). *Politics as usual: The cyberspace "revolution."* Thousand Oaks, CA: Sage.

McChesney, R. W. (1997). *Corporate media and the threat to democracy.* New York: Seven Stories Press.

Monge, P., & Fulk, J. (1999). Communication technology for global network organizations. In J. Fulk & D. DeSanctis (Eds.), *Shaping organizational form: Communication, connection, and community* (pp. 71–100). Newbury Park, CA: Sage.

Newcomb, H. (2000). *Television: The critical view.* New York: Oxford.

Noam, E., Groebel, J., & Gerbarg, D. (2004). *Internet television.* Mahwah, NJ: LEA.

Pelton, J. N. (2003). The changing scope of global telecommunications. In K. Anokwa, C. Lin, & M. B. Salwen (Eds.), *International communication: Concepts and cases* (pp. 267–284). New York: Wadsworth.

Poole, I. (1982). *Technologies of freedom.* Cambridge, MA: MIT Press.

Postman, N. (1986). *Amusing ourselves to death: Public discourse in the age of show business.* New York: Penguin.

Rice, R., & Atkin, C. (2001). *Public communication campaigns.* Beverly Hills, CA: Sage.

Rice, R. E., & Katz, J. E. (2003). Comparing Internet and mobile phone usage: Digital divides of usage adoption, and dropouts. *Telecommunications Policy, 27*, 597–623.

Rogers, E. (2003). *Diffusion of innovation.* New York: The Free Press.

Stanek, D. M., & Mokhtarian, P. L. (1998). Developing models of preference for home-based telecommuting: Findings and forecasts. *Technology and Social Change, 57*, 53–74.

U.S. Department of Commerce. (1998). *The emerging digital economy.* Washington, DC: Author.

Waldfogel, J. (2002, September). *Consumer substitution among media.* Philadelphia: Federal Communications Commission Media Ownership Working Group.

Wood, A. F., & Smith, M. (2001). *Online communication: Linking technology, identity, culture.* Mahwah, NJ: LEA.

2

Communication Technology and Global Change

Christine Ogan

The impact of communication technologies on the globe can be attributed to two of their common characteristics—their ability to cross borders, and the decentralization of control of their content and use. These characteristics have enabled video/DVD technology, satellite-delivered broadcast technology, and computer-mediated communication (CMC) to offer the hope of and deliver on the promise for social change at varying levels. Concerns have been expressed about the dual ends of a process where governments might intend to exert too much control over the information and the media products imported across a nation's borders, with fear that such imports might contain information and points of view that counter the native cultural values and traditions. For instance, during the Cold War, Americans worried that the Soviet Union might be spreading communist propaganda within their borders via satellite. In subsequent decades, developing countries thought the United States would beam its political points of view from space if those countries couldn't control access to the skies over their territories. Three categories of technologies and their impact on social change in a global environment are discussed in this chapter.

BACKGROUND

Videocassette Recorders and DVDs

Consumer videocassette recorders (VCRs) first become available in 1978 when Sony introduced its Betamax format. Although Beta was purported to have somewhat better quality and was sold in smaller cases, JVC's VHS format later became the market leader rather quickly. As Everett Rogers (1988) has told the story, less than 1% of U.S. households owned a recorder in 1980, but 7 years later, about 50%

of households had purchased one. According to the Federal Communication Commission (2005), VCRs were in 91% of American households and 70% of these households also owned a digital videodisc player or DVD player in 2004. North American VCR penetration is expected to drop compared to DVD ownership over time. The Informa Media Group predicted that although 87% of North American households had VCRs in 2005, that figure would drop to 62% by 2010 (2004).

The decline is explained by the superiority of the DVD player over the VCR and the increase in purchases of the DVD recorder. Worldwide, DVD player penetration moved from 3% in 2002 to about 25% of TV households at the end of 2004, with a forecast for that figure to jump to 46% by 2010 (Informa Media Group, 2004). Many scholars have noted that video recorders diffused rapidly because they gave people control over when they watched and what content they viewed on their TV receivers (Lin, 1993; Rubin & Eyal, 2002). DVD recorders offer the same advantage, and the Informa Media Group predicted that 31% of global TV households will own them by 2010 (2004). That power was even more important to people who lived in countries where content was more limited or censored (or both). Many countries in the Middle East experienced explosive diffusion of video even before the technology caught hold in the United States.

In countries like Saudi Arabia, where media were strictly controlled for sexual and religious content, people purchased VCRs at great cost (often more than $1,000) to be able to view a range of censored material without the government's knowledge or permission. For example, the made-for-TV movie "The Death of a Princess," which dramatized the execution of an adulterous princess of the Saudi's royal family, was reportedly in wide circulation, as the video of the film was smuggled into the country (Lin, 1987).

According to Lin (1987), although VCR penetration arrived at 10.7% in the United States in 1983, that ratio stood much higher in countries such as Kuwait (92%) and India (34%). In 1987–1988, sources put the number of players in Saudi Arabia at anywhere from 1.5 to 2 million, in a nation where the population stood at about 15 million (Ogan, 1989). The VCR use grew steadily up to 2001. At the end of that year, about 388 million homes or 79% of households worldwide owned at least one VCR ("World Video Markets," 2003, p. 331). By region, North America had the highest penetration with 90% of TV households, and Asia-Pacific (with the exception of Japan, Australia, and New Zealand) had the lowest with only 10% ("World Video Markets," 2003, p. 333).

In 2003, *Screen Digest* reported that DVD had become the preferred way to watch video, overtaking VHS for the first time ("World Video Markets," 2003, p. 329). The DVD player went on the U.S. market in 1997 and has been growing annually at over 180% per year, making its adoption rate the fastest in consumer electronics hardware ("World Video Markets," 2003). While the DVD video player and recorder penetration rate was nearly doubling from 5.7% in 2001 to about 11% in 2002, VCR penetration was dropping to under 40% of the world's 970 million TV households ("World Video Markets," 2003).

Although initially time shifting was the primary use of VCRs in the United States, most other countries used it for watching films, many of which were pirated. The piracy business was so good that cassettes loaded with illegally appropriated films and TV programs were duplicated to be distributed around the globe. Often films would appear in Asia, Europe, or Africa concurrently with their theater release in the United States. The situation has repeated itself with the DVD, an even more portable format.

The Motion Picture Association of America (now called the Motion Picture Association to suit its global image) worked to stamp out piracy in the United States and abroad because its sales of films on video in 1986 exceeded its box office earnings ("Jolly Rogers," 1997). As part of the World Trade Organization's (WTO's) treaties, countries such as the People's Republic of China have agreed to enforce antipiracy rules and laws to protect the intellectual property rights of media products of foreign nations such as the United States (see http://www.wto.org for information on TRIPS [trade related aspects of intellectual property rights] for more information). A new bilateral Sino-U.S. organization to specifically address the runaway film and home video piracy was established in the summer of 2005, following the 2004 confiscation of 40 million pirated films and CDs (Coonan, 2005).

Satellite Technology

The race to space begun in 1957 with the Soviet Union's launch of Sputnik, the first man-made satellite. This quickly led to the establishment of the U.S. National Aeronautics and Space Administration (NASA) in 1958 and a resolve to beat the Soviet Union in developing and launching communication satellites. In 1965, Early Bird was the first international communications satellite launched by the international consortium Intelsat, which was dominated by the United States. During that same year, the Soviet Union launched a domestic satellite for disseminating domestic TV. But it took 4 more years to mark the first worldwide broadcast of a televised event, Neil Armstrong's walk on the moon in July 1969.

Due to the cost-prohibitive nature of building and launching satellites, most nations relied on the satellite fleets of large international satellite communication organizations such as Intelsat (largely dominated by the U.S.' Hughes Aircraft) to transmit and receive audio, video, and data communication signals across the globe. Consequently, most of the satellites launched into the orbit are controlled by the United States. This phenomenon raised the concern of how to guarantee equitable allocation of orbital slots and frequency spectrum to all sovereign nations—one of the most contentious issues in the history of telecommunications. This issue was finally resolved in 1995, when the 189 member nations of the International Telecommunications Union (a signatory organization that sets international regulations and technical standards for telecommunications systems) adopted a resolution that balances equitable access and actual needs for orbital resource use of all sovereign nations (Lin, 2004).

Nonetheless, diffusion of programs across borders took place without the consent of their producers, as it did with programs on videocassettes, due to signal spillovers linked to satellites' large footprints. Recent development of national direct-broadcast satellite (DBS) systems, however, has altered the dynamics of these cultural and legal concerns. DBS has allowed both domestic and international viewers to access a wider range of foreign-language programming.

For instance, DirecTV and Dish TV, the two major suppliers of DBS in the United States, offer at least 18 different foreign-language programs and sell them in packages based on a world region or language category. DirecTV began broadcasting to ethnic subscribers in the United States in 1999 with Para Todos, a collection of 45 Spanish-language networks. Then Dish TV followed in 2000 with 60 channels of foreign-language broadcasting (Applebaum, 2004). Now it is also possible to subscribe to other language services from GlobeCast WorldTV or other U.S. services from a variety of countries.

Although there are no definitive penetration figures on DBS diffusion around the world, such systems can be found across Europe, in Asia and Africa, and in the Americas. In Western Europe, *Screen Digest* estimates that there were 76.5 million pay-TV subscribers at the end of 2004 and that at least 25% of those were direct to home customers at the end of 2003. This figure is expected to rise to one third of subscribers by 2008 (*Screen Digest*, 2005). In the United States, in less than a decade, DBS penetration reached 25% of the American TV households, compared with 72% that subscribed to a cable TV service (Federal Communications Commission, 2005).

Internet Communication

Compared with VCRs and satellites, the capabilities of the Internet to cross borders and circumvent government and market control of information are much greater. We have seen accounts of blogging by Iranian youth, Saudi users who get around government controls by getting a dial-up connection to an ISP in a neighboring country, and China's attempts to block Internet news and information sites through the building of firewalls that are circumvented through the use of proxy servers. These cases illustrate the strength of the Internet to get messages to and from individuals and groups that seek it despite a range of authoritarian controls and technical barriers.

The Internet's ability to bypass network-control mechanisms is due to its decentralized structure, as it is a network of networks that has no direct control by any one person, group, or government. As first conceived in 1969, the precursor of the Internet—the ARPAnet—was designed to withstand a nuclear attack during the Cold War. In 1982, the Internet was born when TCP/IP protocol established a more efficient system than used by ARPAnet to allow numerous networks across the world to interconnect to become one network, or an Internet (Hafner & Lyons, 1996)

As with many other technologies, it is incredibly difficult to determine what the current Internet diffusion rate might be (e.g., Rogers, 2002). Even if we could determine how many different computers are connected to the Internet (or how many have an IP [Internet protocol] address), many computers have multiple users—particularly in poorer developing countries, where the Internet is often accessed at cyber cafés. That said, a good source for an estimate of worldwide users of the Internet, *Computer Industry Almanac*, reported that the figure will top 1 billion in mid-2005 (Worldwide Internet, 2004). Although the growth has slowed substantially in developed countries, strong growth is reported in countries such as China. The United States now has about 20% of the global users, and China takes second place with about 11% (Worldwide Internet, 2004).

THEORETICAL EXPLANATIONS

VCR Technology

The to-date unproven theoretical complex of cultural/media imperialism has often been discussed in conjunction with VCR technology. Although less prominent in the current era, media imperialism theory loomed large in the heyday of the VCR. The rampant spread of pirated video alongside the legally produced films that circulated around the world—at relatively low cost to the consumers—was a concern for those scholars who viewed the spread of Hollywood films and U.S. TV programs as cultural domination. Work by Varis (1984), Varis and Nordenstreng (1974), and Schiller (1976) demonstrated the economic impact of the world's media production, but as Tomlinson (1991) wrote, little research proved the cultural domination these primarily Western media products over people across the world.

Shifting away from the imperialism paradigm, an alternative theoretical perspective—one that explains the transborder information flow in terms of cultural regionalism—has underscored the widespread distribution of media content among neighboring countries that share cultural proximity or linguistic commonality (Straubhaar, Burch, Duarte, & Sheffer, 2002). Hence, the flow of cultural content is becoming more diversified when it comes to the content sources. For example, the development of online services such as Netflix and Blockbuster made it possible for viewers in the United States to select from a range of India-originated Bollywood (musical comedy films), Hong Kong-produced kung fu, Latin American telenovelas, and many other films made for national and international markets (e.g., Anokwa et al., 2003). Before the availability of these online services, U.S. viewers interested in foreign films were at the mercy of their brick-and-mortar video rental shops that did not offer a good selection of foreign film titles.

Another aspect of VCR's cultural impact across societies rests in its role in spreading political messages across borders and also in spreading educational messages to urban and rural Third World poor. In El Salvador, during the 1980s, the so-called video guerrillas bought VCRs and video cameras along the U.S. border and distributed videos with propaganda messages that told the Faribundo Marti para Liberacion Nacional (FMLN) story to potential supporters inside and outside the region (Epstein, 1987). Before the fall of communism in Eastern Europe, governments had chosen to allow VCRs and Western-origin videotaped content within their borders as part of a political and cultural Glasnost (or political openness). In Poland alone, 1 in 20 households were said to own a video recorder by 1988 (Diehl, 1988). Anticommunist groups in Poland and Czechoslovakia also used video to overcome censorship through content produced by underground sources (Diehl, 1988).

Other theoretical approaches, including social marketing strategies (e.g., Andreason, 1995; Kotler, 1984), examined the positive impact of educational content produced on video, particularly on poor and marginalized groups. Other scholars have been critical of such work (Melkote & Steeves, 2001), arguing that rather than using media to influence behavior, decentralized media (like audio and videocassette recorders) should be used for the empowerment of individuals and groups. These media have been part of a larger strategy of participatory communication when people decide what changes need to be made and use the smaller media to communicate messages they make to effect that change. The Self Employed Women's Association in India is such a group. This organization has used video to tell stories of women workers' conditions to bring about behavioral and policy changes (http://www.sewa.org).

Satellite Technology

As mentioned previously, issues of national sovereignty related to the allocation of slots and spectrum for deploying communication satellites in the geosynchronous orbit, where the satellite is visible from earth 24 hours a day because it circles the globe at the earth's rotation rate, were contentious between member nations of the International Telecommunications Union. As the political resolution was achieved (Lin, 2004) and digital signal-compression techniques that reduce size of satellites and spacing between satellites became viable, scholars largely withdrew from this particular discourse (e.g., Straubhaar et al., 2002).

Nonetheless, a tandem issue on national sovereignty and information privacy persists as it relates to media and cultural imperialism/hegemony. During the late 1970s, the possibility of delivering signals direct to home came about and then the furor over piracy (from the signal originators) and privacy (from the signal receivers) emerged. Developing countries feared that a range of foreign content would come streaming down from the skies into their citizens' living rooms, programming that might challenge their cultural values and system of government as well as their centralized system of control over information flow. A declaration adopted by UNESCO in 1972 specifically stressed the sovereignty of nations in determining what information would be available to their citizens when broadcast from outer space (de Sola Pool, 1979). However, as pointed out by Webster's (1984) early assessment of the development of DBS, "even allowing for cultural differences, barriers of language, and the high cost of entry into the business, [direct broadcast satellites] make possible an unregulated invasion into traditional domestic markets" (p. 1161).

As transborder satellite channels proliferated, political economy discussions shifted to the challenge that these "technologies of freedom" present to authoritarian regimes controlling broadcast content within their borders (e.g., de Sola Pool, 1983). Although the United States has always supported the free flow of information across borders in international debates, it has opposed such channels as Qatar-based Al Jazeera, which it perceives to threaten the U.S. government's point of view in the Middle East. *The Economist* has attributed great impact to the proliferation of satellite channels that gave rise to free-speech expectations for national news and information and to the standardization and acceptance of classical Arabic. "Satellite television has created a sense of belonging to, and participation in, a kind of virtual Arab metropolis. It has begun to make real a dream that 50 years of politicians' speeches and gestures have failed to achieve: Arab unity" ("The World," 2005, p. 1).

Over time, the diversity of content that flowed across national borders and the inspiration it gave to local content producers to adapt international content to the local cultural, linguistic, and political environment gave rise to the concept of glocalization in the literature on transnational media. Applied by Roland Robertson (1995) from Japanese business (Ohmae, 1990), the term *glocalization* has been used to describe the process of taking services or products that are meant for the global market and making them fit a local situation. Creeber (2004) sees the process wresting control away from public service broadcasting monopolies—the norm of European media systems (e.g., British Broadcasting Corporation [BBC])—and facilitating the reflection of the contemporary diversities of Britain and Europe. In the United States, scholars have referred to a loss of a public sphere (or a sense of national community) at one extreme and a fragmentation of audiences for important national issues at another, as immigrant populations are now able to maintain contact with their home countries by watching satellite-delivered news and entertainment (e.g., Ogan, 2001). But it remains unclear whether this prevents new arrivals from adjusting to life in their adopted country or just keeps them from getting overly homesick.

The issue of audience fragmentation derived from a proliferation of satellite-delivered channels is not unique to the United States, as it often goes to the core of what constitutes national identity. According to Mazdon (2001), when satellite and cable TV came to France, they were accompanied by a private challenge to the state broadcast monopoly and its definition of national identity. The subsequent demise of the state monopoly and the rise of the new channels have "undermined the apparent certainties of the national broadcast to a national audience carried out by a state-controlled public television" (Mazdon, 2001, p. 338). Mazdon (2001) argued that in the late 1990s no single broadcaster was capable of addressing the "diversity of identities and cultures that are now acknowledged to make up French national life" (p. 339). Creeber (2004) made a similar argument for the "cultural hegemony" of the BBC and the challenges to its "internal colonization" prior to the breaking down of the public service monopoly.

Across the world in India, a similar process was occurring when Doordarshan (or DD)—the national public broadcasting monopoly—was being challenged by satellite and cable systems that were characterized as bringing a "cultural invasion" (Sinclair & Harrison, 2004, p. 44). The foreign content that arrived on the satellite-delivered channels through Star TV (Rupert Murdoch's News Corporation service based in Hong Kong) drew a mixed response. As a result, cable services were required to carry three state DD channels, even though no banning of satellite dishes occurred as it has in Iran and Saudi Arabia. According to Sinclair and Harrison (2004), that has resulted in DD retaining dominance, despite the proliferation of new channels from Star and also from Zee TV—an Indian satellite company that provides national services, primarily in Hindi, but also exports TV programs to the Indian diaspora located throughout the world.

Internet Technology

When it comes to the Internet, some scholars have engaged in grandiose, often technologically deterministic theoretical discussions of its impact on society. Most of these systems theories have not been able to be tested in the real world. DiMaggio, et al. (2001) do an excellent job of summarizing these theories and ways they have been examined in a variety of studies. They break these up into five domains: "1) inequality (the 'digital divide'); 2) community and social capital; 3) political participation; 4) organizations and other economic institutions; and 5) cultural participation and cultural diversity" (p. 307). They note that the Internet tends to complement other media, technologies, and behaviors rather than displace them. Atton (2003) agrees. He argues that Internet use is embedded in the wider socioeconomic struggle against the internationalization of capital. (p. 4).

Dimaggio et al. (2001) say that in the 1960s (pre-Internet era), one of the big social changes that would be observed was the transformation of society from an industrial one to an information society. This change would affect everything—from the jobs people do to the way society is organized. They also cite Castells as a major theorist in this regard as he argued that the changes required to make a networked society would be as significant as the invention of the alphabet. (p. 309).

In a recent review of literature on social change and the role of the Internet, Tyler (2002) notes that, although researchers may agree that the Internet is "changing the nature of work, government, and social relationships, the key question is whether the availability of the new modality of communication represented by the Internet leads to fundamental changes in personal and social life" (p. 204).

He concludes that, although people have embraced new technologies like the Internet and its accompanying opportunities, "the fundamental issues of life that people are seeking to address remain more constant, with people seeking tools to better live their lives" (p. 205).

One area of social change based on the Internet that has been widely discussed—but undertheorized—is the change in attitude and behavior that has come about surrounding the file sharing of copyrighted information (mostly music and video; see chap. 13, this volume). Although we have no international statistics on this practice, they are not likely to be different from the domestic data. In April 2004, the Pew Internet and American Life project reported that one in seven people who said they once downloaded music files no longer do (Madden & Rainie, 2004). However, the study reported an overall increase in the number of people downloading from 2003 to 2004, from 18 million to 23 million. Those who said they no longer download music claimed that they stopped the practice because of the legal actions against file sharers by the recording Industry Association of America. But in an earlier study, Pew reported that 82% of people ages 18–29 said they had little concern for the copyright status of the music they download (Madden & Lenhart, 2003). So it appears that, at least in the United States—and also in countries where copyright is enforced less frequently—illegal file sharing is not an issue. Perhaps piracy never was important to people, never was considered stealing from the creator of the film or the music, but because it is currently so easy to access copyrighted material, young people have adopted the attitude that it should be free for the taking. More empirical research needs to be conducted internationally to determine whether there really is a shift in social attitudes concerning this practice. The attitudes held by young people in the Pew survey are not shared by those in older age groups.

From its inception, the Internet was seen as a technology geared toward social change. Several scholars have focused their attention on a global digital divide based on the imbalance of content in English. According to the Aguillo, Garcia, and Arroyo (2004) estimations using Yahoo and Google in August 2004, 57.95% of existing web pages are published in English. That contrasts rather dramatically with the percentage of web pages in Chinese (3.21%), French (5.14%), and Japanese (4.96%). These percentages do not jibe well with the languages as percent of total Internet users 31.6% are English language users, 13.2% are Chinese users, 4.1% are French users, and 8.3% are Japanese users ("Internet World Stats," 2005). African nations are among the most disadvantaged when it comes to this linguistic digital divide. Research by Roycroft and Anantho (2003) revealed that one of the strongest predictors of Internet subscription was whether English was an official language of the African country.

Whether we consider the digital divide a theory or an apt description of the way new technologies are distributed throughout the globe, the concluding statement in Drori and Jang's (2003) study of the trends and causes of the process is accurate: "Developing countries, which by definition lag behind the developed countries in many other dimensions of social development, are now also lagging in IT connectivity, which stands as a marker for cutting-edge technology and with it a marker for the prospects of engagement in world affairs" (p. 157).

EMPIRICAL FINDINGS

VCR Technology

Research into international VCR use has been difficult to conduct. As Levy (1987) put it in an article discussing research problems, one of the main issues pertained to the diffusion of the technology in the early years. While VCRs were steadily diffusing, it was both "difficult and expensive" to conduct empirical research, according to Levy (1987, p. 462). Although the adoption rate remained low, survey research methods could reach many users. This problem may explain why the literature is largely devoted to descriptive case studies that focus on the empowering nature of the video as a technology that could circumvent government control of the mass media. Two books by Gladys and Oswald Ganley that detail the political power of video and other "personal media" in various countries represent this type of research (Ganley, 1992; Ganley & Ganley, 1987). Articles and books written by Boyd (1989), Boyd, Straubhaar, and Lent (1985), and Ogan (1989) were other examples of such descriptive or analytical work.

Researchers took note of the fact that VCRs diffused faster in developing nations where fewer alternative sources of entertainment were available, where government strictly controlled the content of mass media, and where dissenting political views could not be openly expressed.

In one of the few international empirical studies conducted internationally, Schoenbach and Hackforth (1987) surveyed German VCR owners and nonowners about their attitudes and behaviors related to TV and VCRs. Although owners of VCRs were more open to technological innovations than nonowners, the biggest difference between the two groups was related to entertainment. Owners of VCRs were much more likely to see the value of ownership for watching movies that would otherwise be unavailable to them and to see video viewing as a means of escaping life's problems.

Satellite Technology

A conservative position on the impact of transnational satellite broadcasting is that its effects have been overblown in the literature. The general guideline for national TV programming schedules is that audiences prefer local programs over imports. That argument was made by Stan (2003), when he noted that "the vast majority of programming is produced, aired, and remains in a single country" (p. 10).

He supports that assertion with an analysis conducted on the amount of money spent on trade in international programming in 2002 ($2.4 billion), compared with the amount spent by broadcasters to produce domestic programs (more than $70 billion).

There are not many empirical studies that look specifically at satellite-delivered programs or channels from one country to another, as foreign programs drop into and off from broadcast schedules as well transforming themselves to adapt to local viewing tastes around the world. We visit one U.S. programmer, MTV, as an example. Like all other United States-based channels that were exported to Europe, Latin America, and Asia, MTV found itself not always welcome when it appeared amid a set of subscription channels. In India, MTV faced a number of obstacles—from its English-language format to its schedule of 100% Western pop music to the fast pace and high volume of its graphics (Cullity, 2002). When a local version of MTV began broadcasting Indian music and interspersing Hindi with English in the broadcasts, MTV realized that its format was not working and left the air to regroup. When MTV returned, it "kept its Western format but indigenized (or localized) it to suit Indian middle-class tastes... the focus on youth culture is maintained but refitted to suit Indian tastes presented by Indian players" (Cullity, 2002, p. 414). By 2003, MTV had emerged as the world's largest network, reaching 380 million households around the globe (Chalaby, 2003); with its combination of local and U.S. styles and content, MTV has found the formula for success in global music.

Another interesting, culturally transforming outcome resulting from transnational satellite TV channels is the incidental democratization of broadcast environments among authoritarian regimes. Al Jazeera—the Qatar-based channel that became so prominent following 9/11 and during the Afghanistan and Iraq wars—began in 1995 as a government-owned but editorially independent satellite TV news channel viewable by all Arab countries in the region. Miles (2005) argues that Al Jazeera is the first sign of a free and independent broadcast station in the Muslim world, as it has angered most Arab leaders at one time or another and led to Egypt's denunciation of its reporters, Saudi Arabia's advertising ban on the channel, and Iraq's total ban on its signal reception. Ayesh (2002) believes that a range of privately owned satellite channels delivered to the Arab world share in the shift to what he terms a "new pan-Arab public sphere marked by varied new agendas. Previously suppressed political perspectives and orientations have become more visible on Arab world television" (p. 139). Given the political openings presented by these emerging video influences, it is useful to consider corollary influences of emerging online channels.

Internet Technology

Perhaps it is too early in the diffusion of the Internet to have empirical findings from cross-cultural or international studies of effects of the technology on social change. We are still only a little more than a decade into the use of the Internet; because it has not diffused to a majority of the population in many countries, the lack of empirical work that goes much beyond describing Internet use is not surprising. One ambitious study conducted by a group of international researchers in collaboration with the UCLA Center for the Digital Future collected data and produced comparative results for a 12-country survey. The countries represented developed or very developed economies in seven European countries, five Asian countries, and the United States. Although the research was primarily descriptive, the study's major findings included the following:

1. Internet users in all of the countries watched TV less than nonusers.
2. Respondents thought that the Information they get from the Internet is generally reliable and accurate.
3. Even in poor countries and among the poorest residents in all countries, online use is surprisingly high (from 1.6% in the lowest economic quartile in Hungary to 46.4% in Korea to 43.1% in the United States).
4. When it comes to shopping, a gender gap existed in all countries at the time of the survey, with males purchasers outnumbering female purchasers in all countries but Korea, where 34% of women used the Internet versus 28.5% of the men. German male and female users were about equal in number.
5. The survey also included Chinese people who live in urban areas. Of those respondents, 21% said that the Internet had increased contact with others who share their political interests.

Use of a media technology can also be an effect—that is, the medium is the message. The fact that people get themselves connected to the Internet to spend time communicating with others, finding information, or making purchases means that it has some impact on their daily lives. Of late, U.S. scholars have been examining the role of the Internet in people's everyday lives. They are finding increasing degrees of embeddedness in acts of communication, shopping, conducting personal business, and entertainment, among other areas (Althaus & Tewksbury, 2000; Atkin, Jeffres, & Neuendorf, 1998; Charney & Greenberg, 2002; Flanagin & Metzger, 2001). As the Internet diffuses to more people internationally, it will become embedded in more people's lives. Even in the poorest countries, Internet cafés can be found in large numbers. For example, in China, where the government controls the Internet use in these cafés, the government closed down 12,575 in late 2004. The total number operating in that country is not known ("China's Café Count," 2005). Turkey, a country of more than 70 million people, was home to 11,222 in 2004 (DGS).

Studies of international Internet use have appeared in the literature over the past several years. One recent study compared users in India and the United States (Patwardhan, 2004). That cross-sectional online survey found few differences in activity and participation between the users in the two countries. Although the author notes that the use of a convenience sample might limit any generalizations, "activities that users in both countries were most satisfied with were also ones that they were most involved in cognitively and emotionally, spent the most time on, and exposed themselves to the most frequently" (p. 429). The users were most satisfied and also had high levels of cognitive and emotional involvement in information search and communication. Ogan and Cagiltay (in press) conducted an online survey of 4,500 users of a Turkish confession web site on which users tell stories about their lives and exchange comments with other users. They found that the motivation for social interaction figured importantly into the reasons for using itiraf.com for reading others' confessions, writing confessions, and comments and for meeting people found on the site. A study of Greek high school students found few gender differences; both boys and girls use the Internet most often for finding personal information and for entertainment (Papastergiou & Solominidou, 2005). And a survey of Internet café users in Dar es Salaam, Tanzania, found that the frequenters of the cafés were mostly males who engaged in personal communication and online entertainment site visits most often (Chachage, 2001). Cross-national studies in many locations will need to be conducted to determine whether Internet use has more universal characteristics or is more culture specific.

The most interesting signs of the impact of the Internet come from countries where governments have tried to suppress the freedom of their people and control Internet use. Internet censorship practices enforced in such countries as Saudi Arabia, Iran, North Korea, and Syria have ranged from the use of filtering software via Internet café sites and controlling content through authorized Internet service providers (ISPs), to raiding other public access locations for inappropriate and unauthorized content. In China specifically, there may be as many as 30,000 Internet police working at ISPs around the country (Morrison & Nuttall, 2004). Nonetheless, people have found ways around all of these controls.

In Iran, young people are downloading banned music and participating in digital dating; hundreds of blogs have also appeared in both Farsi and English (Walt, 2005). In China, blogger Aggressive Little Snake (or Yan Wenbo in real life) writes a blog about the way dogs are treated in China; his blog was given the 2004 International Weblog Award because jury members saw the content as a metaphor for the discussion of human rights abuses in China ("Don't be Led Astray," 2005 p. 5). When China blocks Internet sites, proxy networks are created through which users can access the blocked Internet sites (Morrison & Nuttall, 2004). It seems that where there is a will, there is a way around the censors.

SOCIAL CHANGE IMPLICATIONS

This chapter began by noting that the communication technologies that were most significant in their globe-spanning capabilities are the VCR, the

International communication satellite, and the Internet, alongside the media information and content that each technology platform has delivered. It is fair to say that we have reached the era where the expectation of a global society functioning with "technologies of freedom" (de Sola Pool, 1983) has been principally met. Such freedom has allowed peoples and societies to achieve a measure of social change via the use of these technologies in both traditional and innovative manners. That change could work for the good, in opening up authoritarian governments to scrutiny and bringing about challenges to those governments or actual democratic reforms.

In particular, these media technologies share characteristics of border-crossing capabilities that have helped facilitate the decentralization of state-sponsored media monopolies, the circumvention of government control over censored information and media content, and democratization of the flow of information between nations that are rich and poor, authoritarian and libertarian. On the other hand, the technologies can also work to undermine the circulation of copyrighted materials through various means, introduce cultural values that are incongruent with a nation's religious and/or cultural traditions, or transport uninvited signals and contents that intrude on the sovereignty of a nation's border. The bottom line is that none of these communication media is able to work independently to cause social change, but all of them support or complement other activities to allow social change to occur.

As many scholars have written, the scenarios surrounding information and communication technologies related to social change have generally fallen into two camps. The first is the utopian scenario in which information and communication technologies (ICTs) are described as bringing about positive social change (e.g., making life easier and more efficient, allowing our children to be better educated and more successful, and helping to bring peace to warring factions in the world). The opposing dystopian scenario has portrayed ICTs as causing only dire consequences for society (bringing innocent children in contact with pedophiles online, encouraging individuals to become addicted to online gambling, or leading gullible people to fall for phishing schemes that appear in their e-mail inbox). The truth is that ICTs and/or the Internet are not the direct cause of any of these behaviors. The Internet is not a technology that, by itself, can cause much of any social consequence. ICTs are socially contexualized and part of a social system where they play some role in social change, but probably do not act as independent agents. Rob Kling (2000) referred to this complex environment as "sociotechnical networks."

An example of an online network that arose during the Seattle-based World Trade Organization talks in 1999 is the network of independent media centers (IMCs or indymedia). It is a "backbone of communication for the broad coalition of groups that comprise the anti-capitalism movement" around the world (Atton, 2003, p. 3). In 1995, the indymedia.org web site listed 93 separate web locations outside the United States where news, information, and opinion are being published on the Internet. Whether this network effects social change or just serves as a vehicle to communicate and organize activities to create social change is not yet known. But what is known about this network is that without the Internet such a movement would be nearly impossible.

As the power of the Internet to network and organize people rises with those who are maximizing its technical might, it also reveals the issue of the digital divide that separates those Internet-capable and media-rich populations from their Internet-incapable and media-poor counterparts. *Digital divide* is a concept that directly relates to international social change because the premise of this idea is that as wealthier, more educated, more urban, and generally more privileged members of societies gain access to computers and to the Internet, the gap between rich and poor grows larger. In the United States, the digital divide is often thought to have been closed, as men and women are about equally represented online; senior citizens have joined youth in being Internet users, and most of the nation's schools enjoy Internet access for all students. But internationally, the Internet has not diffused to the poor and disadvantaged in such proportions. One measure of a nation's relationship with the rest of the world regarding this issue is its e-readiness ranking. The Economist Intelligence Unit publishes a composite index that evaluates how "amenable a market is to Internet opportunities" in the world's 60 largest economies (2004, p. 3). In the 2004 index, the lowest countries on the list included Algeria, Pakistan, Kazakhistan, and Azerbaijan. Of course, many other African, Asian, and Latin American countries did not even make the list.

Many programs have been instituted to help address this divide at the international level, including the establishment in rural areas of telecenters where people can access the Internet at a local community center and where technicians can also instruct the visitors on how to use the computer to obtain the information needed for their work, health, or family. As most research of the divide in developing countries has shown, people need more than access to computers and the Internet—they need to be educated in ways to translate the technology into effective tools for development. Although case studies of telecenter success appear from many parts of the world, Roman has pointed out that there is little systematic research on the effects of telecenter diffusion (2003).

Manuel Castells (1997) argues in his trilogy of books on the networked society that new communication technologies are fundamental for social movements to exist, that they provide the organizational infrastructure. In the conclusion of his discussion of global social movements, Castells writes:

> However, the second and *main agency* detected in our journey across the lands inhabited by social movements, is a *networking, decentered form of organization and intervention, characteristic of the new social movements*, mirroring and counteracting, the networking logic of domination in the informational society.... These networks do more than organizing activity and sharing information. *They are the actual producers, and distributors, of cultural codes*. (p. 362; italics original)

Howard Rheingold has continued to support the use of the Internet and ICTs for social change in his latest book, *Smart Mobs* (2003). But he also provides a caution in his discussion of the capabilities of new media.

> Technologies and methodologies of cooperation are embryonic today, and the emergence of democratic, convivial, intelligent new social forms depends on how people appropriate, adopt, transform, and reshape the new media once they are out of the hands of engineers—as people always do. (pp. 214–215)

Both Rheingold and Castells note that power does not just disappear because the decentralized media like the Internet exist for others to spread their messages of change. The powerful seek to exploit these new media to maintain their control. The challenge is for the less powerful to use the technologies to create power shifts to achieve real and positive social change.

REFERENCES

Aguillo, I. F., Garcia, I., & Arroyo, N. (2004). *Regional and linguistic patterns in positioning*. Madrid: Centro de Información y Documentatión Científica. Retrieved September 23, 2005, at www.csi.ensmp.fr/csi/4S/download_paper/download_paper.php?paper=aguillo_garcia_arroyo.pdf

Althaus, S. L., & Tewksbury, D. (2000). Patterns of Internet and traditional news media use in a networked community. *Political Communication*, *17*, 21–45.

Anokwa, K., Lin, C. A., & Salwen, M. B. (Eds.), (2003). *International communication: concepts and cases*. Belmont, CA: Thomson/Wadsworth.

Andreason, A. R. (1995). *Marketing for social change.* San Francisco: Jossey-Bass.

Applebaum, S. (2004, July 5–18). Mano a mano! Cable takes the ethnic fight to the skies [Electronic version]. *CableWORLD.* Retrieved March 9, 2005 from http://www.cable-accessintel.com/cgi/cw/show_msg.cgi?pub=cw&mon=070504&file=manoamano.htm

Atkin, D. J., Jeffres, L. W., & Neuendorf, K. A. (1998). Understanding Internet adoption as telecommunications behavior. *Journal of Broadcasting & Electronic Media*, *42*, 475–490.

Atton, C. (2003). Reshaping social movement media for a new millennium. *Social Movement Studies, 2*(1), 3–15.

Ayesh, M. I. (2002). Political communication on Arab world television: Evolving patterns. *Political Communication, 19*, 137–154.

Boyd, D. (1989). The videocassette recorder in the USSR and Soviet-bloc countries. In M. Levy (Ed.), *The VCR age: Home video and mass communication* (pp. 252–270). Newbury Park, CA: Sage.

Boyd, D., Straubhaar, J., & Lent, J. A. (1989). *Videocassette recorders in the third world.* White Plains, NY: Longman.

Castells, M. (1997). *The information age: Economy, society and culture: Vol. II. The power of identity.* Malden, MA: Blackwell.

Chachage, B. L. (2001). Internet cafés in Tanzania: A study of the knowledge and skills of end-users. *Information Development, 17*(4), 226–233.

Chalaby, J. K. (2003). Television for a new global order: Transnational television networks and formation of global systems. *Gazette, 65*(6), 457–472.

Charney, T., & Greenbearg, B.S. (2002). Uses and gratifications of the Internet. In C. Lin & D. Atkin (Eds.), *Communication technology and society: Audience adoption and uses* (pp. 379–407). Cresskill, NJ: Hampton.

China's café count. (2005). *Communications of the ACM, 48*(4), 9–10.

Coonan, C. (2005, September 1). org planning China antipiracy push. *Variety.com.* Retrieved September 16, 2005, from http://www.variety.com/article/VR1117928427?categoryid=1338&cs=1&s=h&p=0

Creeber, G. (2004). "Hideously white" British television, glocalization, and national identity. *Television & New Media, 5*(1), 27–39.

Cullity, J. (2002). The global desi: Cultural nationalism on MTV India. *Journal of Communication Inquiry, 26*(4), 408–425.

De Sola Pool, I. (1979). Direct broadcast satellites and the integrity of cultures. In K. Nordenstreng & H. I. Schiller (Eds.), *National sovereignty and international communication* (pp. 120–153). Norwood, NJ: Ablex.

De Sola Pool, I. D. (1983). *Technologies of freedom*. Cambridge, MA. MIT Press.

Diehl, J. (1988, April 17). VCRs on fast forward in Eastern Europe. *The Washington Post,* p. A1.

DiMaggio, P. E., Hargittai, Neuman, W. R., & Robinson, J. P. (2001). Social implications of the Internet. *Annual Review of Sociology, 27*, 307–336.

Directorate General of Security (DGS). (2004). *Turkish Internet Café statistics, September 2004*. Ankara, Turkey: Author.

Don't be led astray. (2005, January 17). *South China Morning Post*, p. 5.

Drori, G. S., & Jang, Y.S. (2003). The global digital divide: A sociological assessment of trends and causes. *Social Science Computer Review, 21*(2), 144–161.

Economist Intelligence Unit. (2004). *The 2004 e-readiness rankings*. Retrieved March 8, 2005, from http://graphics.eiu.com/files/ad_pdfs/ERR2004.pdf

Epstein, N. (1987, August 6). Salvadorean rebels videotape their cause, then smuggle cassettes to US. *The Christian Science Monitor,* p. 3.

Federal Communications Commission. (2005, February 4). *FCC media bureau releases 11th annual mvpd competition report* (FCC, MB Docket No. 04-227). Retrieved March 4, 2005, from http://www.fcc.gov/mb/csrptpg.html

Flanagin, A. J., & Metzger, M. J. (2000). Perceptions of Internet information credibility. *Journalism & Mass Communication Quarterly*, *77*, 515–540.

Ganley, G. D. (1992). *The exploding political power of personal media.* Cambridge, MA: Program on Information Resources Policy, Harvard University.

Ganley, G. D., & Ganley, O. H. (1987). *Global political fallout: The VCR's first decade.* Cambridge, MA: Program on Information Resources Policy, Harvard University.

Hafner, K., & Lyon, M. (1996). *Where wizards stay up late.* New York: Simon & Schuster.

Informa Media Group. (2004, July). *Global DVD & video forecasts*. Retrieved September 16, 2005, from http://shop.telecoms.com

Internet world stats: Usage and population statistics. (2005, July 23). The Internet Coaching Library. Retrieved September 23, 2005, from http://www.internetworldstats.com/stats7.htm

Jolly Rogers flying high. (1987, October 31). *The Economist*, p. 65.

Kling, R. (2000). Learning about information technologies and social change: The contribution of social informatics. *The Information Society, 16*, 217–232.

Kotler, P. (1984). Social marketing of health behavior. In L. W. Frederiksen, L. J. Solomon, & K. A. Brehony (Eds.), *Marketing health behavior* (pp. 23–39). New York: Plenum.

Levy, M. (1987). Some problems of VCR research. *American Behavioral Scientist, 30*(5), 461–470.

Lin, C. (1987). A quantitative analysis of worldwide VCR penetration. *European Journal of Communications, 13,* 131–159.

Lin, C. A. (1993). Exploring the role of VCR use in the emerging home entertainment culture. *Journalism Quarterly, 70,* 833–842.

Lin, C. A. (2004). Satellite communication trends and issues. In A. E. Grant & J. H. Meadows (Eds.), *Communication technology update* (pp. 299–311). Boston: Focal.

Madden, M., & Lenhart, A. (2003, April). *Music downloading, file sharing and copyright: A Pew Internet Project memo.* Retrieved March 9, 2005, from http://www.pewinternet.org/PPF/r/96/report_display.asp

Madden, M., & Rainie, L. (2004, April 25). *14% of Internet users say they no longer download music files: Data memo from PIP and comScore Media Metrix.* Retrieved March 9, 2005, from http://www.pewinternet.org/PPF/r/124/report_display.asp

Mazdon, L. (2001). Contemporary French television, the nation, and the family. *Television & New Media, 2*(4), 335–349.

Melkote, S., & Steeves, H. L. (2001). *Communication for development in the third world: Theory and practice for empowerment.* Thousand Oaks, CA: Sage.

Miles, H. (2005). *Al-Jazeera: The inside story of the Arab news channel that is challenging the West.* New York: Grove.

Morrison, S., & Nuttall, C. (2004, December 8). Police blocks on the information highway. *The Financial Times,* p. 15.

Ogan, C. (1989). The worldwide cultural and economic impact of video. In M. Levy (Ed.), *The VCR age* (pp. 230–231). Newbury Park, CA: Sage.

Ogan, C. (2001). *Communication and identity in the diaspora: Turkish migrants in Amsterdam and their use of media.* Lanham, MD: Lexington Books.

Ogan, C., & Cagiltay, K. (in press). Confession, revelation and storytelling: Patterns of use on a Turkish web site. *New Media & Society.*

Ohmae, K. (1990). *The borderless world: Power and strategy in the interlinked economy.* New York: Harper Business.

Papastergiou, M., & Solomonidou, C. (2005). Gender issues in Internet access and favourite Internet activities among Greek high school pupils inside and outside school. *Computers & Education, 44*(4), 377–393.

Patwardhan, P. (2004). Exposure, involvement and satisfaction with online activities. *Gazette, 66*(5), 411–436.

Rheingold, H. (1994). *The virtual community.* London: Secker & Warburg.

Robertson, R. (1995). Glocalization: Time-space homogeneity-heterogeneity. In M. Featherstone, S. Lash, & R. Robertson (Eds.), *Global modernities* (pp. 27–44). London: Sage.

Rogers, E. (1988, Winter). Video is here to stay. *Media & Values, 42*. Retrieved July 31, 2006, from http://www.medialit.org/reading_room/media_and_values_5.html#mv42

Rogers, E. (2002). The information society in the new millennium: Captain's log, 2001. In C. Lin & D. Atkin (Eds.), *Communication technology: Audience adoption and uses* (pp. 43–64). Cresskill, NJ: Hampton.

Roman, R. (2003). Diffusion of innovations as a theoretical framework for telecenters. *Information Technologies and International Development, 1*(2), 53–66.

Roycroft, T. R., & Anantho, S. (2003). *Telecommunications Policy, 27*(1/2), 61–75.

Rubin, A., & Eyal, K. (2002). The videocassette recorder in the home media environment. In C. Lin & D. Atkin (Eds.), *Communication technology: Audience adoption and uses* (pp. 329–349). Cresskill, NJ: Hampton.

Schiller, H. (1976), *Communication and cultural dominance.* New York: International Arts and Science Press.

Schoenbach, K., & Hackforth, J. (1987). Video in West German households. *American Behavioral Scientist, 30*(5), 533–544.

Screen Digest. (2003). November World video markets shuffle: World spending on DVD overtakes VHS for first time, pp. 329–336.

Screen Digest. (2005). DTH to capture one third of pay TV markct. Accessed March 4, 2005, from http://www.screendigest.com/ezine/#tv

Sinclair, J., & Harrison. M. (2004). Globalization, nation, and television in Asia. *Television & New Media, 5*(1), 41–54.

Stan, F. (2003). On the future of "global television": An economic and historical approach to understanding the basics and trajectory of world television. *Ad Astra, 2*(1), 1–11.

Straubhaar, J., Burch, E., Duarte, L. G., & Sheffer, P. (2002). International satellite television networks: Gazing at the global village or looking for "home" video? In C. A. Lin & D. Atkin (Eds.), *Communication technology: Audience adoption and uses* (pp. 307–327). Cresskill, NJ: Hampton.

The world through their eyes. (2005, February 24). *The Economist.* Retrieved March 10, 2005, from http://www.economist.com/displayStory.cfm?story_id=3690442

Tomlinson, J. (1991). *Cultural imperialism.* Baltimore, MD: Johns Hopkins University Press.

Tyler, T.R. (2002). Is the Internet changing social life? It seems the more things change, the more they stay the same. *Society for the Psychological Study of Social Issues, 58*(1), 195–205.

Varis, T. (1984). The international flow of television programs. *Journal of Communication, 34*(1), 143–152.

Varis, T., & Nordenstreng, K. (1974). *Television traffic: A one-way street?* Paris: UNESCO.

Walt, V. (2005, February 20). Iran's young lose faith in reform. *Sunday Times,* p. 24.

Webster, D. (1984, Summer). Direct broadcast satellites: proximity, sovereignty and national identity. *Foreign Affairs,* pp. 1161–1174.

Worldwide Internet users will top 1 billion in 2005: USA remains #1 with 185m internet users. (2004, September 3). *Computer Industry Almanac, Inc.* Retrieved March 7, 2005, from http://www.c-i-a.com/pr0904.htm

II

INDIVIDUAL AND SOCIAL SETTING

3

Computer-Mediated Technology and Social Interaction

Scott E. Caplan, Elizabeth M. Perse, and Janice E. Gennaria

The person has become the portal. (Wellman, 2002, p. 15)

This chapter presents an exploration of computer-mediated communication (CMC) technologies by examining how and why people use instant messaging (IM), e-mail, and chat rooms in a social context. We also examine the positive and negative impacts such technologies have on relational processes and on the individuals who use them. After a brief historical background of these technologies and their diffusion patterns, the remaining sections of this chapter explore the social change implications based on the differences between CMC and face-to-face (FtF) communication with an emphasis on the benefits (e.g., relational development and social support) and risks (e.g., problematic Internet use) of online social interaction.[1]

BACKGROUND

Although people have used text-based e-mail and synchronous CMC chat technologies (e.g., IRC, ICQ, MUDS, and MOOS; see Herring, 2004; Parks & Roberts, 1998; Turkle, 1995; Wallace, 1999) for decades, new forms of these applications such as IM and SMS (short message service or text messaging) have become ubiquitous social tools for many people, especially those in younger generations (Herring, 2004; Thurlow, 2003). In fact, Thurlow (2003) dubs the current generation of adolescents and young adults "Generation Text." Herring (2004)

[1]Space constraints demand that we offer a representative, rather than comprehensive, review of the vast literature on interpersonal communication technologies.

notes that IM is, thus far, largely an American phenomenon, whereas SMS is more popular outside of the United States (see also Thurlow, 2003).

Although early networked interpersonal technologies were used, primarily, by members of military and academic communities, the growing presence of broadband Internet connections into American households has further expanded the variety and sophistication of available technologies to ordinary people, especially teens and young adults (see Herring, 2004; Horrigan & Rainie, 2002; Madden & Rainie, 2003).

According to a 2003 study by the Pew Internet and American Life Project, "the rapid growth of broadband use at home has been perhaps the most striking development in the Internet population in the past 4 years" (Madden & Rainie, 2003, p. 67). Most recently, another Pew study (Rainie, 2005) reported that as of November 2004, 60 million American homes had broadband connections. Moreover, wireless technology has enabled people to use cellular phones and other devices to further expand their repertoire of CMC activities with their friends and family.

E-mail

E-mail marked the beginning of computer-based interpersonal communication. The original Internet, ARPAnet, was a computer network that connected military and academic researchers; the first e-mail was sent on that system in 1971 (Zakon, 2005). In 1979, Usenet was established; it allows users to form topic-specific newsgroups and to post discussions of technical, scientific, and other "fun" topics on an electronic bulletin. By 1986, there were 2,500 sites connected to the network, posting about 500 messages a day. In 1981, BITNET was developed. Similar to Usenet, BITNET supported scientific and technical discussions among academic and nonacademic researchers through the use of LISTSERV, which e-mailed all contributions to subscribers instead of posting them on an electronic bulletin.

E-mail use grew slowly in its early years, but as personal computers entered the home, e-mail use increased. By 1997, 45% of U.S. households owned a computer and 15% of all Americans used e-mail (Klopfenstein, 2002; Rice & Webster, 2002). A later study tracking changes in e-mail use found that in 2003, 54% of respondents used e-mail on a typical day, an increase of 20% from 2000 (Madden & Rainie, 2003). According to a more recent estimate, at the end of 2004, approximately 58 million (or 28% of) American adults reported using e-mail on a daily basis (Rainie, 2005). Baym (2002) argues that e-mail is the primary reason that many people began to use the Internet. Herring (2004) recounts that e-mail was the "default mode of CMC for many people" (p. 27) in the early 1990s. Not surprisingly, to date, much of the research on people's online social behavior has focused on e-mail (e.g., Dimmick, Kline, & Stafford, 2000; Golden, Beauclair, & Sussman, 1992; Hill & Monk, 2000; Romm & Pliskin, 1999;

Stafford, Kline, & Dimmick, 1999), in addition to CMC/Internet usage as a whole (Dainton & Aylor, 2002; Flaherty, Pearce, & Rubin, 1998; McKenna, Green, & Gleason, 2002; Papacharissi & Rubin, 2000).

Instant Messaging

Instant messaging (IM) is an interpersonal CMC technology in which users are able to communicate with each other in one-on-one synchronous conversations. Many users participate in multiple IM sessions simultaneously, with each conversation appearing in a different window on the user's screen. Some forms of IM have been around since the 1980s. In 1996, AOL introduced its "Buddy List" feature, which popularized IM and made it easier to use. That same year, an Israeli company, Mirablis, developed ICQ ("I seek you"), another IM program. In 1998, AOL acquired ICQ, becoming the dominant IM provider. Although there are various different IM programs (e.g., AIM [AOL's], ICQ, MSN Messenger, and Yahoo), most of these are proprietary (i.e., one can only communicate with others who are using the same software). The popularity of IM has taken hold among teenagers and young adults (see Jones, 2002; Rideout, Roberts, & Foehr, 2005; Shiu & Lenhart, 2004).

A recent Pew research study (Shiu & Lenhart, 2004) estimated that 42% of adult American Internet users (53 million) use IM; this reflects a growth rate of about 29% from 41 million users since 2000. The same study surveyed the IM users and found that "21% of IM-ers in each of the Gen Y (18–27) and Gen X (28–39) age groups log onto IM several times a day, followed by 17% of Trailing Boomers (40–49), 15% of Leading Boomers (50–59), 10% of Matures (60–68), and a mere 9% of the After Work (69 and older) age group" (Shiu & Lenhart, 2004, p. iii). There were also differences among IM users within the youngest generations: 42% of Gen Y members reported using IM more frequently than e-mail, while only 18% of Gen X-ers (28–39 years) used IM more often than e-mail.

Online Chat Rooms

Although IM enables participants to engage in multiple one-on-one conversations, chat rooms allow multiple users to converse with one another in groups (see Herring, 2004; Kendall, 1996; Parks & Roberts, 1998; Suler, 1996; Werry, 1996). Chat rooms have dual origins: in Internet Relay Chat (IRC), begun in the late 1980s, and in multi-user dungeons (MUDs). Both of these allowed real-time text-based online conversations among participants. Online chat rooms appear to be an activity in which many young adult Internet users take part as well. Toward the mid-1990s, many chat systems began to incorporate graphics and multimedia elements into what had previously been text-only communication environments (e.g., The Palace; see Suler, 1996). Recent commercial examples of fully animated three-dimensional graphical social environments include Habbo Hotel (www.

habbohotel.com), There (www.there.com), and Second Life (secondlife. com; see Järvinen , Heliö, & Mäyrä, 2002).

According to the estimate of one recent study (Madden & Rainie, 2003), about 29 million people used chat rooms online in 2002 and about 5 million of them participated in a chat room on a typical day. The same study also indicated that, in general, chat-room participation appears to be less popular than IM. Specifically, it reported that "Internet users are almost twice as likely to use IM as they are to take part in a chat room. Furthermore, they are three times as likely to use IM on a typical day compared to chatting" (p. 11). Consistent with IM use, the study also found that teens and young adults create most of the traffic to chat rooms and online discussions.

THEORETICAL EXPLANATIONS

The rapid and widespread adoption of e-mail, IM, and (to a lesser extent) chat rooms suggests that people find these social interaction-capable technologies useful. Several major communication perspectives that explain the appeal of new communication technologies are reviewed here. The first perspective, focusing on the motives and uses of those technologies, reviews uses and gratifications theory (e.g., Katz, Blumler, & Gurevitch, 1974). By contrast, the second perspective, focusing on the attributes of the technologies, discusses theoretical explanations including the diffusion of innovations tradition (Rogers, 2003). The third perspective examines theoretical issues that focus on the role of technologies in maintaining social relations by reviewing such theories as social identification model (Culnan & Markus, 1987).

Motives and Uses

The uses and gratifications perspective focuses on the reasons people use different communication channels (e.g., Katz et al., 1974; Rubin & Rubin, 1985) and is an audience-centered functional approach to communication. At first, the framework was applied primarily to the study of mass communication systems and later to interpersonal processes (Rubin & Rubin, 1985). Researchers now recognize that it is especially useful for exploring the adoption and uses of Internet-based technologies (e.g., Morris & Ogan, 1996).

Uses and gratifications theory suggests that people use the media to satisfy certain cognitive (e.g., information learning) and affective (e.g., escape) needs. Computer-based personal communication technologies are used for a range of different personal, educational, or business reasons, as well as for information gathering, social connections, and entertainment reasons. Research has already noted the positive and negative consequences of CMC use (e.g., Caplan, 2002, 2003; Caplan & Turner, in press; Kraut et al., 1998; McKenna & Bargh, 1998;

Wright, 2002; Young & Rogers, 1998). Uses and gratifications theory posits that knowledge of the motivational reasons that people use communication technologies is key to understanding more fully their use of these technologies and the consequences of that use.

Palmgreen (1984) pointed out that new communication technologies are most likely associated with new motives for use. The literature on computer-mediated social interaction technologies shows that these technologies are motivated by new and different reasons for use. Research by Dimmick (Dimmick et al., 2000) and Recchiuti (2003) points out that, similar to FtF communication, these technologies have social/interpersonal uses that focus on relationship development and maintenance. But these technologies also are motivated by reasons that are similar to some of the traditional mass communication motives: entertainment, escape, and information seeking.

Another way to understand the appeal and use of online social interaction technologies is to focus on their attributes in terms of how those attributes contribute to the ways people adopt and use these technologies, in addition to the effects of such adoption and use. Moreover, it is also useful to examine how mediated communication differs from other forms of nonmediated communication in a social interaction context.

Technology Attributes

Diffusion of innovations (Rogers, 2003) is a theoretical perspective that focuses on the adoption of new ideas, practices, and technologies. According to Rogers, perceptions about the attributes of the innovation can affect whether and how quickly a new communication technology is adopted. Specifically, innovations are more likely to be adopted (a) if people believe that they offer advantages over what already exists; (b) if they are compatible with potential users' existing values, experiences, and needs; (c) if they are perceived as being fairly easy to understand and use; (d) if users can try them before they commit to purchase; and (e) if users can easily observe their use by others.

Perceived attributes of an innovation is also a concept that can be explicated by the theory of social presence. In contrast with the diffusion of innovations theory, the concepts of media richness (e.g., Daft & Lengel, 1984) and social presence (Short, Williams, & Christie, 1976) focus specifically on the attributes of *communication channels* that lead them to be adopted and used. Media richness is an attribute of a channel based on the channel's bandwidth capacity to transmit different types of communication cues. Communication channels are considered "rich" when they are more interactive and convey many different types of information such as visual, oral, and nonverbal. Social presence is the ability to "feel" the other person's presence during the mediated communication interactions. Social presence can be influenced by channel attributes, but it is primarily defined by users' perceptions. For instance, desktop conferencing over webcams may be perceived as projecting a greater degree of social presence of conference participants due to the participants' ability to see each other's image, compared with SMS or IM.

Social Relations

Researchers have sought to understand and explain the core differences between computer-mediated and FtF communication processes (for reviews, see Caplan, 2001, 2003; Hancock & Dunham, 2001; Ramirez, Walther, Burgoon, & Sunnafrank, 2002; Riva, 2002; Walther, 1996, 2004; Walther, Anderson, & Park, 1994; Walther & Parks, 2002). Among the earliest theories to emerge was the cues-filtered-out perspective (Culnan & Markus, 1987; also see Walther & Parks, 2002), which suggested that some forms of CMC are less personal than FtF activity because of the reduced number of contextual and nonverbal cues available in text-based online social interaction. The cues-filtered-out perspective asserts that the diminished available cues available in CMC create a heightened sense of anonymity, which leads to a more impersonal communication exchange than is present in FtF interaction. As Ramirez and Burgoon (2004) note, however, researchers have moved away from early perspectives focusing solely on cue deficits, toward more sophisticated theories that consider the cognitive and behavioral mechanisms people use to compensate for the lack of cues available in text-based CMC.

One particularly influential theoretical perspective that describes how CMC and FtF processes differ is Walther's (1996) hyperpersonal communication perspective. According to Walther, interpersonal CMC can become *hyperpersonal* because it affords message senders a host of communicative advantages over traditional FtF interaction. Compared to ordinary FtF situations, due to the reduced number of available nonverbal cues, a hyperpersonal message sender has a greater ability to strategically develop and edit self-presentation, enabling a selective and optimized presentation of oneself to others (Walther, 1996, Walther & Burgoon, 1992). This process then allows senders to selectively control the quantity, quality, and even validity of personal information available to other participants (e.g., age, race, physical appearance, sex), to form idealized impressions of their partners and, consequently, engage in more intimate exchanges than people in FtF situations (Tidwell & Walther; 2002; Walther, 1993, 1996; Walther & Burgoon, 1992). For example, in IM, e-mail, and chat rooms, one can choose any sort of screen name to identify oneself. These screen names are forms of self-presentation and are also often the only information available to others about the sender, at least within initial interactions.

It is clear that CMC technologies are being used to develop and maintain interpersonal and social relationships (Flanagin & Metzger, 2001; McKenna et al., 2002; Parks & Floyd 1996; Parks & Roberts, 1998; Rabby & Walther, 2003; Ramirez & Burgoon, 2004; Tidwell & Walther, 2002; Walther & Parks, 2002). Rabby and Walther (2003) note that many people use CMC to supplement their FtF relational communication. To date, researchers have offered several theories to explain relational CMC processes.

The social identification model of deindividuation effects (SIDE) proposes that, despite offering fewer interpersonal cues (e.g., Culnan & Markus, 1987), CMC is not necessarily impersonal; rather, impression formation online results in more socially categorical, rather than personal, impressions of others (Lea & Spears, 1992; Reicher, Spears, & Postmes, 1995; Spears & Lea, 1992, 1994; Spears, Postmes, & Lea, 2002). Similarly, social information-processing (SIP) theory (Walther, 1992, 1993; Walther & Burgoon, 1992; Walther, Loh, & Granka, 2005) also takes issue with the notion that CMC is necessarily impersonal; instead, SIP theory suggests that online interpersonal relationship development might require *more time* to develop than traditional FtF relationships.

In addition to research on CMC and relational communication, in general, other studies have examined therapeutic relational communication online. There is a good deal of evidence suggesting that online support and therapeutic discussion groups are an important positive aspect of the Internet (e.g., see Caplan & Turner, in press; Walther & Parks, 2002; Wright, 1999, 2000, 2002; Wright & Bell, 2003). To date, researchers have not firmly established whether participation in online emotional support has therapeutic value that is less than, equivalent to, or beyond that obtained via FtF support (Finfgeld, 2000; Owen, Yarbrough, Varga, & Tucker, 2003; Walther & Boyd, 2002). The few studies that have compared computer-mediated and FtF psychotherapy sessions have reported that participants in both groups exhibited relatively equivalent outcomes (e.g., Cohen & Kerr, 1998; Day & Schneider, 2002; for a review, see Rochlen, Zack, & Speyer, 2004). To advance understanding of online emotional support, Walther and Parks (2002) recommend that researchers begin to develop explanations for why CMC might be particularly effective as a support medium.

EMPIRICAL FINDINGS

User Motivations

Recchiuti (2003) reports an in-depth study of the reasons underlying the use of e-mail, IM, and chat rooms. She focused her research on college students because at the time of data collection (2003), IM was not widely used outside the youthful audience. Based on her sample of 446 college students (168 males and 278 females who ranged in age from 18 to 33, with an average age of 19.80 years), Recchiuti (2003) identified three common motives—information seeking, interpersonal utility, and entertainment—to all three forms of CMC studied. Separate motives found for using e-mail included: (a) convenience, or because it is easy and comfortable; (b) pass time, or when they have nothing better to do; and (c) escape, or to get away from work and other pressures. Three motives were unique to IM use, including (a) escape; (b) companionship, or to overcome feelings of loneliness; and (c) anonymity, out of a desire to be anonymous. Two additional motives were identified for online chat room use, namely, (a) pass time and (b) benefits, which included companionship and anonymity reasons. Additional findings suggest the following: (a) E-mail more than IM and online chat rooms was used for information seeking and task-related communication, reflecting a more instrumental form of CMC; and (b) IM more than e-mail and online chat rooms was utilized for entertainment, interpersonal communication, and socially related communication, indicating more socially oriented forms of CMC.

Earlier research on e-mail adoption in the workplace found three main motives for its use: to fill free time, to take breaks from work, and for entertainment (Rice & Steinfield, 1994). Papacharissi and Rubin (2000) found that college student e-mail use was predicted by greater entertainment motives and lower levels of information seeking. Other uses and gratifications research focused on wider types of uses and found that people are using e-mail as an effective way to sustain relationships (Stafford et al., 1999).

Dimmick and his colleagues (2000) compared gratifications of e-mail and telephone usage. They found that e-mail's ability to allow for asynchronous communication was an advantage it held over the telephone. E-mail allowed people to communicate with friends and family who lived far away, or in different time zones, and with those with whom they did not have time to stay in touch in person. These researchers found that one gratification aspect that e-mail fulfilled was the ability to communicate and maintain relationships with people across geographic distance and schedule conflicts. The gratification opportunities that e-mail presents are part of the reason that people are motivated to use it as a means of sustaining relationships. Some of these gratification opportunities include that it is faster than postal mail, cheaper than the telephone, and a convenient way to share opinions, information, and ideas with friends and family. This can be done in either an asynchronous or synchronous manner across distance and time zones (Stafford et al., 1999).

Research addressing IM indicates that IM is widely used by a younger population (e.g., Lenhart, Rainie, & Lewis, 2001; Rideout et al., 2005; Shiu & Lenhart, 2004). Teenagers, in particular, use it to ask each other out, to break up with each other, and to make plans with friends (Lenhart et al., 2001). Hardaf Segerstad and Ljungstrand (2002) found that college students use WebWho, a web-based IM program, to collaborate on assignments and coordinate social activities. This finding suggests that IM is used for both task and social activities, reflecting instrumental and ritual functions of CMC use, respectively. Leung (2001) found that students' motives for ICQ differed depending on whether they were heavy or light users. Heavy use of ICQ was motivated by the need to express affection and sociability; light ICQ use, in contrast, was motivated by keeping up with fashion trends. Leung (2001) also found that males used ICQ to fill time between classes; females, by comparison, used ICQ to show or seek affection and to socialize with friends. One advantage that IM holds over the telephone is its ability to enable users to multitask. IM allows users to have multiple synchronous conversations at once (Lenhart et al., 2001).

Some limited research has looked at people's motives for using online chat rooms. Motivations for online chat rooms include seeking conversation and initiating relationships with others (Peris et al., 2002). Whitty (2002) found that 63% of respondents sought emotional support from the use of chat rooms. Stone and Pennebaker (2002) found that some people use online chat rooms as a way of coping with trauma.

Technology Attributes

Several researchers have analyzed new communication channels, applied these criteria, and found that new channels have attributes that foster adoption (e.g., Rogers, 2003; see also Lin, 1999, 2001). According to this approach, innovations with superior attributes will displace older, less desirable products, ideas, and practices. It is clear that e-mail, IM, and chat rooms offer several advantages over traditional FtF and telephone conversations that have led to their widespread adoption. Early writings about electronic mail (Rice & Bair, 1984), for example, listed many advantages to "electronic messaging." IM and chat rooms also offer many advantages, such as speed, convenience, relationship maintenance, and ease of use (Recchiuti, 2003). Additionally, as we discuss later in this chapter, online social interaction differs in important respects from typical FtF communication, particularly in ways that may be attractive to some users.

There has been little research exploring people's perceptions of the social presence of e-mail, IM, and chat rooms. Perse and Courtright (2004), however, found that asynchronous Internet use, including web surfing and e-mail, was rated second to interpersonal channels (e.g., cell phones and FtF conversations) in social presence. IM and chat rooms, along with synchronous CMC, were rated quite low in social presence. These results, however, were based on adult, noncollege student samples and do not reflect IM's popularity among the young in the general population. Yet the results do suggest that perceptions about the social qualities of communication technologies are related to adoption; the most positively rated CMC technology, e-mail, was also rated the highest in social presence.

Social Relations

As noted earlier in this chapter, many of the new personal CMC applications employ graphical elements (ranging from small user icons to fully immersive three-dimensional virtual environments with animated avatars) to allow users a wider variety of tools and strategies for self-presentation than earlier text-based systems. Although one might choose a photograph of oneself to use as an avatar or icon, one could just as easily use any image (Caplan, 2001). Suler (1996) notes that some people have collections of different avatars at their disposal that they might implement and choose from in order to purposefully influence recipients' perceptions of the sender (i.e., one might change from a "happy" or "pleasant" image to a "sad" or "threatening" one).

For instance, in one study comparing FtF to CMC romantic relationships, Cornwell and Lundgren (2001) found that CMC partners engaged in greater misrepresentation during self-presentation than their FtF counterparts. They attributed the difference in levels of misrepresentation to a lower level of relational involvement among CMC romantic partners, compared to those using an FtF channel. In another study, Joinson (2001) reported that levels of spontaneous self-disclosure were greater in CMC exchanges than in FtF interactions when there was a heightened sense of private self-awareness and a lower sense of public self-awareness associated with CMC exchange.

Other researchers have reported that, compared to FtF interactions, CMC exchanges include more direct and more intimate uncertainty reduction strategies (e.g., greater proportions of direct questions and self-disclosing statements; Tidwell & Walther, 2002), along with less detailed and more intense impressions of communication partners (Hancock & Dunham, 2001). As Rabby and Walther (2003) explain, "The development of relationships online may simply be temporally retarded in comparison to FtF relationship development" (p. 148). Empirical evidence supports this hypothesis. In one study of CMC and impression formation, Walther (1993) found that members of FtF groups developed impressions of one another more quickly than their CMC counterparts, but after a 6-week period, the CMC groups formed impressions that were as well developed as those exhibited by the FtF participants.

In terms of communicating social support, Burleson and Goldsmith (1998) argue that the type of conversational environment most conducive to effective comforting requires reducing the distressed other's self-presentational anxiety. Caplan and Turner (in press) propose that establishing such an environment might be easier and more effective if the conversation is computer-mediated. They further assert that computer-mediated social support interactions might be especially helpful at creating a conversational context that is less socially risky than its FtF counterpart. For example, Walther and Boyd (2002) contend that computer-mediated discussions of stigmatized topics are likely to be perceived as less threatening than their FtF counterparts due to their increased anonymity and increased social distance, which facilitate better stigma management. These findings that reflect how CMC social support affords its users reduced social stigma and increased anonymity were further validated by other similar studies (Gustafson et al., 1999; McKenna & Bargh, 1998; White & Dorman, 2001; Wright, 2002). Online health applications are discussed further in chapter 12 by Whitten (this volume).

Caplan and Turner (in press) also point out that computer-mediated emotional support allows support seekers who have limited mobility to participate in groups that they would be less willing, if at all able, to attend if offered in an FtF format (Braithwaite, Waldron, & Finn., 1999; White & Dorman, 2001; Wright, 2002).

Along a similar line, online conversation partners are not bound by proximity and geographical barriers; individuals can communicate with a seemingly limitless number of diverse people who would be difficult or impossible to locate in most FtF cases (Barrera, Glasgow, McKay, Boles, & Feil, 2002; Braithwaite et al., 1999; Finfgeld, 2000; Finn, 1999; Sharf, 1997; Walther & Boyd, 2002; White & Dorman, 2001; Wizelberg, 1997; Wright, 2002).

Taken together, the empirical evidence reviewed here identifies a number of features of some CMC applications that might be particularly useful to people who perceive themselves as being low in social competence. First, with regard to self-disclosure, CMC interaction allows individuals greater flexibility in self-presentation by way of omitting and reshaping personal information that they perceive to be negative or harmful, as there is a greater opportunity to fabricate, exaggerate, or intensify more positive aspects of oneself to others online. Thus, for some, the Internet represents a place where they can exercise greater control over the impressions that others form of them. Second, as Tidwell and Walther (2002) found, participants in FtF conversations exhibit a smaller proportion of direct question-asking, self-disclosing statements, and politeness strategies than those in CMC interactions, suggesting that effective FtF communication demands greater communicative flexibility and creativity than CMC interaction. Third, CMC can offer an opportunity for people to receive social support in an anonymous fashion that is less socially risky—and in a manner that is not limited by time and geographic boundaries—as compared to FtF social support opportunities.

SOCIAL CHANGE IMPLICATIONS

Research from a variety of disciplines, including communication, reflects a growing concern with problematic Internet use and its potential ill effects (for reviews, see Beard & Wolf, 2001; Brenner, 1997; Caplan, 2002; Davis, 2001; Griffiths, 1998, 2000; Young, 1998; Young & Rogers, 1998). Problematic Internet use is a multidimensional syndrome consisting of cognitive and behavioral symptoms that result in negative social, academic, or professional consequences (Caplan, 2002, 2003, 2005; Davis, 2001; Davis, Flett, & Besser, 2002; Morahan-Martin & Schumacher, 2003). With regard to this chapter's focus on CMC technologies and social interactions, one particularly noteworthy finding to emerge from the growing literature on problematic Internet use is that individuals who report negative outcomes associated with their Internet use appear to be especially drawn to the interpersonal uses of the Internet (Caplan, 2002, 2003; McKenna & Bargh, 2000; Morahan-Martin & Schumacher, 2000; Young, 1998; Young & Rogers, 1998).

For instance, Young (1998) reported that dependent users mainly utilized the social and interactive functions of the Internet, whereas nondependents used the Internet more for information gathering. Similarly, Young and Rogers (1998) note that problematic or abusive Internet use involves primarily socially interactive uses, which also appear to be associated with low self-esteem. In another study, Morahan-Martin and Schumacher (2000) found that the social aspects of Internet use consistently differentiated those with more Internet use problems from others, as the former were more likely to use the Internet for seeking emotional support, talking with others, and playing highly socially interactive games. LaRose, Lin, and Eastin (2003), based on uses and gratifications as well as operant conditioning theory, argue that when individuals have received repeated expected gratifications (or rewards) from Internet use over time, their Internet use behavior can turn into a conditioned habit (operant conditioning). If these individuals subsequently preoccupy themselves with this habit due to deficient self-regulation of their Internet use behavior, they can go one step further and isolate themselves from society to the extent of becoming addicted to Internet use.

Similarly, based on the cognitive behavioral theory of problematic Internet use (Caplan, 2002; Davis, 2001), Caplan (2003, 2005) suggests that individuals who suffer from psychosocial problems (i.e., low social skill and loneliness) develop a preference for online social interaction as an alternative to FtF communication because they perceive it to be less threatening and perceive themselves to be more efficacious when interacting with others online. Over time, people who prefer online social interaction may engage in compulsive and excessive use of some synchronous CMC applications to the point that they suffer negative outcomes at home and at work, further exacerbating existing psychosocial problems. Such findings raise interesting questions for researchers. For example, is CMC a functional alternative for FtF communication for some people but not others—and if so, why? Do people who prefer online social interaction use CMC technologies differently than those without such a preference?

In conclusion, computer-mediated social interaction technologies have facilitated significant changes in how people relate to members of their personal and professional social networks. For example, physical distance or proximity between network members is becoming increasingly less important. Thus, as Meyerowitz (1985) observes, "*Where* one is has less and less to do with what one knows and experiences. Electronic media have altered the significance of time and space for social interaction" (p. viii). These changes in social interaction channels also create new challenges for parents. Growing concerns about children's safety online, for example, stem from the increasingly permeable physical boundaries that once separated families from the larger community. Meyerowitz (1985) notes that "the walls of the family home, for example, are no longer effective barriers that wholly isolate the family from the larger community and society. The family home is now a less bounded and unique environment" (p. viii).

As computer-mediated social interaction becomes more widespread, we can expect that physical location will become an increasingly less salient predictor of with whom we interact. Hampton and Wellman (2000) make a similar point, observing that "whatever happens, new communication technologies are driving out the traditional belief that community can only be found locally" (p. 195).

Clearly, communication scholars will need to adapt communication theories to evolving technologies and changing contexts in order to understand the uses and effects of computer-mediated social interaction technologies.

REFERENCES

Barrera, M., Glasgow, R. E., McKay, H. G., Boles, S. M., & Feil, E. G. (2002). Do Internet-based support interventions change perceptions of social support?: An experimental trial of approaches for supporting diabetes self-management. *American Journal of Community Psychology, 30*(5), 637–654.

Baym, N. K. (2002). Interpersonal life online. In L. A. Lievrouw & S. Livingstone (Eds.), *Handbook of new media: Social shaping and consequences of ICTs* (pp. 62–76). London: Sage.

Beard, K. W., & Wolf, E. M. (2001). Modification in the proposed diagnostic criteria for Internet addiction. *Cyberpsychology and Behavior, 4,* 377–383.

Braithwaite, D. O., Waldron, V. R., & Finn, J. (1999). Communication of social support in computer-mediated groups for people with disabilities. *Health Communication, 11*, 123–151.

Brenner, V. (1997). Psychology of computer use: XLVII. Parameters of Internet use, abuse and addiction: The first 90 days of the Internet Usage Survey. *Psychological Reports, 80,* 879–882.

Burleson, B. R., & Goldsmith, D. J. (1998). How the comforting process works: Alleviating emotional distress through conversationally induced reappraisals. In P. A. Anderson & L. K. Guerrero (Eds.), *Handbook of communication and emotion: Theory, research, application, and contexts* (pp. 245–280). San Diego, CA: Academic Press.

Caplan, S. E. (2001). Challenging the mass-interpersonal communication dichotomy: Are we witnessing the emergence of an entirely new communication system? *Electronic Journal of Communication, 11.* Retrieved April 6, 2005, from http://www.cios.org/get file%5CCaplan_v11n101

Caplan, S. E. (2002). Problematic Internet use and psychosocial well-being: Development of a theory-based cognitive-behavioral measure. *Computers in Human Behavior, 18*, 533–575.

Caplan, S. E. (2003). Preference for online social interaction: A theory of problematic Internet use and psychosocial well-being. *Communication Research, 30*, 625–648.

Caplan, S. E. (2005). A social skill account of problematic Internet use. *Journal of Communication*, *55*(4), 721–736.

Caplan, S. E., & Turner, J. (in press). Online social support: Bringing theory to research on computer-mediated supportive and comforting communication. *Computers in Human Behavior.*

Cohen, G. E., & Kerr, B. A. (1998). Computer-mediated counseling: An empirical study of a new mental health treatment. *Computers in Human Services*, *15*, 13–26.

Cornwell, B., & Lundgren, D. (2001). Love on the Internet: Involvement and misrepresentation in romantic relationships in cyberspace vs. realspace. *Computers in Human Behavior, 17*, 197–211.

Culnan, M. J., & Markus, M. L. (1987). Information technologies. In F. M. Jablin, L. L. Putnam, K. H. Roberts, & L. W. Porter (Eds.), *Handbook of organizational communication: An interdisciplinary perspective* (pp. 420–443). Newbury Park, CA: Sage.

Daft, R. L., & Lengel, R. H. (1984). Information richness: A new approach to managerial behavior and organization design. In B. M. Shaw & L. L. Cummings (Eds.), *Research in organizational behavior* (Vol. 6, pp. 191–233). Greenwich, CT: JAI.

Dainton, M., & Aylor, B. (2002). Patterns of communication channel use in the maintenance of long-distance relationships. *Communication Research Reports, 19,* 118–129.

Davis, R. A. (2001). A cognitive-behavioral model of pathological Internet use. *Computers in Human Behavior, 17,* 187–195.

Davis, R. A., Flett, G. L., & Besser, A. (2002). Validation of a new measure of problematic Internet use: Implications for pre-employment screening. *Cyberpsychology and Behavior, 5*, 331–346.

Day, S. X., & Schneider, P. L. (2002). Psychotherapy using distance technology: A comparison of face-to-face, video, and audio treatment. *Journal of Counseling Psychology, 49*, 499–503.

Dimmick, J., Kline, S. L., & Stafford, L. (2000). The gratification niches of personal e-mail and the telephone. *Communication Research, 27,* 227–248.

Finfgeld, D. L. (2000). Therapeutic groups online: The good, the bad, and the unknown. *Issues in Mental Health Nursing, 21*(3), 241–255.

Finn, J. (1999). An exploration of helping processes in an on-line self-help group focusing on issues of disability. *Health and Social Work, 24,* 220–240.

Flaherty, L. M., Pearce, K. J., & Rubin, R. B. (1998). Internet and face-to-face communication: Not functional alternatives. *Communication Quarterly, 46,* 250–268.

Flanagin, A. J., & Metzger, M. J. (2001). Internet use in the contemporary media environment. *Human Communication Research, 27,* 153–181.

Golden, P. A., Beauclair, R., & Sussman, L. (1992). Factors affecting electronic mail use. *Computers in Human Behavior, 8,* 297–311.

Griffiths, M. (1998). Internet addiction: Does it really exist? In J. E. Gackenbach (Ed.), *Psychology and the Internet: Intrapersonal, interpersonal, and transpersonal implications* (pp. 61–75). New York: Academic Press.

Griffiths, M. (2000). Does Internet and computer "addiction" exist? Some case study evidence. *Cyberpsychology and Behavior, 3*, 211–218.

Gustafson, D. H., Hawkins, R., Boberg, E., Pingree, S., Serlin, R. E., Graziano, F., & Chan, C. L. (1999). Impact of a patient-centered, computer-based health information/support system [Electronic version]. *American Journal of Preventative Medicine, 16*(1), 1–9.

Hancock. J. T., & Dunham, P. J. (2001). Impression formation in computer-mediated communication revisited: An analysis of the breadth and intensity of impressions. *Communication Research, 28,* 325–347.

Hampton, K. N., & Wellman, B. (2000). Examining community in the digital neighborhood: Early results from Canada's wired suburb. In T. Ishida & K. Isbister (Eds.), *Digital cities: Technologies, experiences and future perspectives* (pp. 194–208). Heidelberg, Germany: Springer-Verlag.

Hardaf Segerstad, Y., & Ljungstrand, P. (2002). IM with WebWho. *International Journal of Human-Computer Studies, 56,* 147–171.

Herring, S. C. (2004). Slouching toward the ordinary: Current trends in computer-mediated communication. *New Media & Society*, *6*, 26–36.

Hill, K., & Monk, A. F. (2000). Electronic mail versus printed text: The effects on recipients. *Interacting With Computers, 13*, 253–263.

Horrigan, J. B., & Rainie, L. (2002). *The broadband difference: How online Americans' behavior changes with high-speed Internet connections at home*. Washington, DC: Pew Internet & American Life Project. Retrieved April 5, 2005, from http://www.pewInternet.org/pdfs/PIP_Broadband_Report.pdf

Järvinen, A., Heliö, S., & Mäyrä, F. (2002). Communication and community in digital entertainment services: Prestudy research report. *Hypermedia Laboratory Net Series 2*. Retrieved April 7, 2005, from http://tampub.uta.fi/tup/951-44-5432-4.pdf

Joinson, A. N. (2001). Self-disclosure in computer-mediated communication: The role of self-awareness and visual anonymity. *European Journal of Social Psychology, 3*, 177–192.

Jones, S. (2002). *The Internet goes to college: How students are living in the future with today's technology*. Washington, DC: Pew Internet & American Life Project. Retrieved April 5, 2005, from http://www.pewInternet.org/pdfs/PIP_College_Report.pdf

Katz, E., Blumler, J. G., & Gurevitch, M. (1974). Utilization of mass communication by the individual. In J. G. Blumler & E. Katz (Eds.), *The uses of mass communications: Current perspectives on gratifications research* (pp. 19–32). Beverly Hills, CA: Sage.

Kendall, L. (1996). I hardly know'er!: Adventures of a feminist MUDer. In L. Cherny & E. R. Weise (Eds.), *Wired women* (pp. 207–223). Seattle, WA: Seal Press.

Klopfenstein, B. (2002). The Internet and web as communication media. In C. A. Lin & D. J. Atkin (Eds.), *Communication technology and society* (pp. 353–371). Cresskill, NJ: Hampton.

Kraut, R., Patterson, M., Lundmark, V., Kiesler, S., Mukopadhyay, T., & Scherlis, W. (1998). Internet paradox: A social technology that reduces social involvement and psychological well being? *American Psychologist, 53*, 1017–1031.

LaRose, R., Lin, C. A., & Eastin, M. S. (2003). Unregulated internet usage. Addiction, habit, or deficient self-regulation? *Media Psychology, 5*, 225–253.

Lea, M., & Spears, R. (1992). Paralanguage and social perception in computer-mediated communication. *Journal of Organizational Computing, 2*, 321–341.

Lenhart, A., Rainie, L., & Lewis, O. (2001). *Teenage life online: The rise of the instant-messaging generation and the Internet's impact on friendships and family relationships*. Washington, DC: Pew Internet & American Life Project. Retrieved April 5, 2005, from http://www.pewInternet.org/pdfs/PIP_Teens_Report.pdf

Leung, L. (2001). College student motives for chatting on ICQ. *New Media & Society, 3*, 483–500.

Lin, C. A. (1999). Predicting online service adoption likelihood among potential subscribers: A motivational approach. *Journal of Advertising Research, 39*, 79–89.

Lin, C. A. (2001). Audience attributes media supplementation, and likely online service adoption. *Mass Communication & Society, 4*, 19–38.

Madden, M., & Rainie, L. (2003). *America's online pursuits: The changing picture of whose online and what they do*. Washington, DC: Pew Internet & American Life Project. Retrieved April 5, 2005, from http://www.pewInternet.org/pdfs/PIP_Online_Pursuits_ Final.PDF

McKenna, K. Y. A., & Bargh, J. A. (1998). Coming out in the age of the Internet: Identity "demarginalization" through virtual group participation. *Journal of Personality and Social Psychology, 75*(3), 1–24.

McKenna, K. Y. A., & Bargh, J. A. (2000). Plan 9 from Cyberspace: The implications of the Internet for personality and social psychology. *Journal of Personality and Social Psychology*, *75*(3), 681–694.

McKenna, K. Y. A., Greene, A. S., & Gleason, M. E. J. (2002). Relationship formation on the Internet: What's the big attraction? *Journal of Social Issues*, *58*(1), 9–31.

Meyerowitz, J. (1985). *No sense of place: The impact of electronic media on social behavior.* New York: Oxford University Press.

Morahan-Martin, J., & Schumacher, P. (2000). Incidence and correlates of pathological Internet use among college students. *Computers in Human Behavior, 16*, 13–29.

Morahan-Martin, J., & Schumacher, P. (2003). Loneliness and social uses of the Internet. *Computers in Human Behavior, 19*(6), 659–671.

Morris, M., & Ogan, C. (1996). The Internet as mass medium. *Journal of Communication, 46*(1), 39–50.

Owen, J. E., Yarbrough, E. J., Varga, A., & Tucker, D. (2003). Investigation of the effects of gender and preparation on quality of communication in Internet support groups. *Computers in Human Behavior, 19*(3), 259–275.

Palmgreen, P. (1984). Uses and gratifications: A theoretical perspective. In R. N. Bostrom (Ed.), *Communication yearbook* (Vol. 8, pp. 20–55). Beverly Hills, CA: Sage.

Papacharissi, Z., & Rubin, A. M. (2000). Predictors of Internet use. *Journal of Broadcasting & Electronic Media, 44,* 175–196.

Parks, M. R., & Floyd, K. (1996). Making friends in cyberspace. *Journal of Communication, 46*, 80–97.

Parks, M. R., & Roberts, L. D. (1998). "Making MOOsic": The development of personal relationships on line and a comparison to their off-line counterparts. *Journal of Social and Personal Relationships, 15*, 517–537.

Peris, R., Gimeno, M. A., Pinazo, D., Ortet, G., Carrero, V., Sanchiz, M., & Ibanez, I. (2002). Online chat rooms: Virtual spaces of interaction for socially oriented people. *CyberPsychology & Behavior, 5,* 43–51.

Perse, E. M., & Courtright, J. A. (2004, November). *Functional images of communication channels: Mass and interpersonal alternatives in a fragmented media environment.* Paper presented at the Nation Communication Association annual conference, Chicago.

Rabby, M. K., & Walther, J. B. (2003). Computer-mediated communication effects on relationship formation and maintenance. In D. J. Canary & M. Daiton (Eds.), *Maintaining relations through communication: Relational, contextual, & cultural variations* (pp. 141–162). Mahwah, NJ: Lawrence Erlbaum Associates.

Rainie, L. (2005). *The mainstreaming of online life*. Washington, DC: Pew Internet & American Life Project. Retrieved August 1, 2005, from http://207.21.232.103/pdfs/Internet_Status_2005.pdf

Ramirez, A., & Burgoon, J. K. (2004). The effect of interactivity on initial interactions: The influence of information valence and modality and information richness on computer-mediated interaction. *Communication Monographs*, *71*, 422–447.

Ramirez, J. R. A., Walther, J. B., Burgoon, J. K., & Sunnefrank, M. (2002). Information-seeking strategies, uncertainty, and computer-mediated communication: Toward a conceptual model. *Human Communication Research, 28*, 213–228.

Recchiuti, J. K. (2003). *College students' uses and motives for e-mail, IM, and online chat rooms.* Thesis, University of Delaware, Department of Communication, Newark.

Reicher, S. D., Spears, R., & Postmes, T. (1995). Effects of public and private self-awareness on deindividuation and aggression. *Journal of Personality and Social Psychology, 43*, 503–513.

Rice, R. E., & Bair, J. H. (1984). New organizational media and productivity. In R. E. Rice & Associates (Eds.), *The new media: Communication, research, and technology* (pp. 185–215). Beverly Hills, CA: Sage.

Rice, R. E., & Steinfield, C. (1994). Experiences with new forms of organizational communication via electronic mail and voice messaging. In J. H. Andriessen & R. A. Roe (Eds.), *Telematics and work* (pp. 109–134). Hillsdale, NJ: Lawrence Erlbaum Associates.

Rice, R. E., & Webster, J. (2002). Adoption, diffusion, and use of new media. In C. A. Lin & D. J. Atkin (Eds.), *Communication technology and society* (pp. 191–227). Cresskill, NJ: Hampton.

Rideout, V., Roberts, D. F., & Foehr, U. G. (2005). Generation M: Media in the lives of 8–18 year olds. The Henry J. Kaiser Family Foundation. Retrieved April 5, 2005 from http://www/kff.org/entmedia/7251.cfm

Riva, G. (2002). The sociocognitive psychology of computer-mediated communication: The present and future of technology-based interactions. *CyberPsychology & Behavior, 5*(6), 581–598.

Rochlen, A. B., Zack, J. S., & Speyer, C. (2004). Online therapy: Review of relevant definitions, debates, and current empirical support. *Journal of Clinical Psychology, 60*(3), 269–283.

Rogers, E. M. (2003). *Diffusion of innovations* (5th ed.). New York: The Free Press.

Romm, C., & Pliskin, N. (1999). The role of charismatic leadership in diffusion and implementation of email. *Journal of Management Development, 18,* 273–290.

Rubin, A. M., & Rubin, R. B. (1985). Interface of personal and mediated communication: A research agenda. *Critical Studies in Mass Communication, 2,* 36–53.

Sharf, B. (1997). Communicating breast cancer on-line: Support and empowerment on the Internet. *Women & Health, 26*, 65–84.

Shiu, E., & Lenhart, A. (2004). *How Americans use IM*. Washington, DC: Pew Internet & American Life Project. Retrieved March 26, 2005, from http://www.pewInternet.org/pdfs/PIP_Instantmessage_Report.pdf

Short, J., Williams, E., & Christie, B. (1976). *The social uses of telecommunications*. New York: Wiley.

Spears, R., & Lea, M. (1992). Social influence and the influence of the "social" in computer-mediated communication. In M. Lea (Ed.), *Contexts of computer-mediated communication* (pp. 30–65). London: Harvester-Wheatsheaf.

Spears, R., & Lea, M. (1994). Panacea or panopticon?: The hidden power in computer-mediated communication. *Communication Research, 21*, 427–459.

Spears, R., Postmes, T., & Lea, M.(2002). The power of influence and the influence of power in virtual groups: A SIDE look at CMC and the Internet. *Journal of Social Issues, 58*, 91–108.

Stafford, L., Kline, S. L., & Dimmick, J. (1999). Home e-mail: Relational maintenance and gratification opportunities. *Journal of Broadcasting & Electronic Media, 43,* 659–669.

Stone, L. D., & Pennebaker, J. W. (2002). Trauma in real time: Talking and avoiding online conversations about the death of Princess Diana. *Basic and Applied Social Psychology, 24,* 173–183.

Suler, J. (1996). Life at the Palace: A cyberpsychology case study. In *The psychology of cyberspace*. Retrieved April 6, 2005, from http://www.rider.edu/~suler/psycyber/palacestudy.html

Thurlow, C. (2003). Generation txt? The sociolinguistics of young people's text-messaging. *Discourse Analysis Online*, *1*(1). Retrieved March 26, 2005, from http://www.shu.au.uk/daol/articles/v1/n1/a3/thurlow2002003.html

Tidwell, L. C., & Walther, J. B. (2002). Computer-mediated communication effects on disclosure, impressions, and interpersonal evaluations: Getting to know one another a bit at a time. *Human Communication Research, 28*, 317–348.

Turkle, S. (1995). *Life on the screen: Identity in the age of the Internet*. New York: Simon & Schuster.

Wallace, P. M. (1999). *The psychology of the Internet*. New York: Cambridge University Press.

Walther, J. B. (1992). Interpersonal effects in computer-mediated interaction: A relational perspective. *Communication Research, 19,* 52–90.

Walther, J. B. (1993). Impression development in computer-mediated interaction. *Western Journal of Communication, 57*, 381–398.

Walther, J. B. (1996). Computer-mediated communication: Impersonal, interpersonal, and hyperpersonal interaction. *Communication Research, 23*, 3–43.

Walther, J. B. (2004). Language and communication technology: Introduction to the special issue. *Journal of Language & Social Psychology. Language and Communication Technology, 23*, 384–396.

Walther, J. B., Anderson, J. F., & Park, D. W. (1994). Interpersonal effects in computer-mediated interaction: A meta-analysis of social and antisocial communication. *Communication Research, 21*, 460–487.

Walther, J. B., & Boyd, S. (2002). Attraction to computer-mediated social support. In C. A. Lin & D. Atkin (Eds.), *Communication technology and society: Audience adoption and uses* (pp. 153–188). Cresskill, NJ: Hampton.

Walther, J. B., & Burgoon, J. K. (1992). Relational communication in computer-mediated interaction. *Human Communication Research, 19*, 50–88.

Walther, J. B., Loh, T., & Granka, L. (2005). Let me count the ways: The interchange of verbal and nonverbal cues in computer-mediated and face-to-face affinity. *Journal of Language & Social Psychology, 24*(1), 36–65.

Walther, J. B., & Parks, M. R. (2002). Cues filtered out, cues filtered in: Computer-mediated communication and relationships. In M. L. Knapp & J. A. Daly (Eds.), *Handbook of interpersonal communication* (3rd ed., pp. 529–563). Thousand Oaks, CA: Sage.

Wellman, B. (2002). Little boxes, globalization, and networked individualism. In M. Tanabe, P. van den Besselaar, & T. Ishida (Eds.), *Digital cities: II. Computational and sociological approaching* (pp. 10–25). Berlin: Springer-Verlag.

Werry, C. (1996). Linguistic and interactional features of Internet Relay Chat. In S. C. Herring (Ed.), *Computer-mediated communication: Linguistic, social and cross-cultural perspectives* (pp. 47–61). Philadelphia: John Benjamins.

White M., & Dorman S. M. (2001). Receiving social support online: Implications for health education. *Health Education Research, 16* (6), 693–707.

Whitty, M. T. (2002). Liar, liar! An examination of how open, supportive and honest people are in chat rooms. *Computers in Human Behavior, 18,* 343–352.

Wizelberg, A. (1997). The analysis of an electronic support group for individuals with eating disorders. *Computers in Human Behavior, 13*, 393-407.

Wright, K. (1999). Computer-mediated support groups: An examination of relationships among social support, perceived stress, and coping strategies. *Communication Quarterly, 47*, 402–414.

Wright, K. (2000). Perceptions of on-line support providers: An examination of perceived homophily, source credibility, communication and social support within on-line support groups. *Communication Quarterly, 48*, 44–59.

Wright, K. (2002). Social support within an on-line cancer community: An assessment of emotional support, perceptions of advantages and disadvantages, and motives for using the community from a communication perspective. *Journal of Applied Communication Research, 30*, 195–209.

Wright, K., & Bell, S. B. (2003). Health-related support groups on the Internet: Linking empirical findings to social support and computer-mediated communication theory. *Journal of Health Psychology, 8*(1), 39–54.

Young, K. S. (1998). *Caught in the Net.* New York: Wiley.

Young, K. S., & Rogers, R. C. (1998). The relationship between depression and Internet addiction. *Cyberpsychology and Behavior, 1*, 25–28.

Zakon, R. H. (2005). *Hobbes' Internet timeline* v8.0. Retrieved March 28, 2005, from http://www.zakon.org/robert/Internet/timeline

4

Computer-Mediated Technology and Children

Marina Krcmar and Yulia Strizhakova

Historically, children's overall computer use and, more specifically, Internet use has been examined in the context of school and home, with use rates constantly increasing. As early as 1984, demographic trends indicated that about 27% of students from prekindergarten to college used computers at school and about 12% did so at home. In 2001, about 90% of children ages 5 to 17 used computers, with school use still being more prevalent than home use. About 81% of students reported using computers at school and about 65% at home (National Center for Education Statistics, 2004).

Recently, computer use has been examined among younger children. Calvert, Rideout, Woolard, Barr, and Strouse (2004) report an earlier onset of computer use and a constant linear increase in use among children ages 6 months to 6 years old. The researchers found that toddlers regularly use computers at a parent's lap at the age of 2.5 and move to autonomous computer and mouse use at the age of 3.5. Rideout, Vandewater, and Wartella (2003) add that around 48% of all children ages 6 and under have used a computer and around 70% of those in the 4- to 7-year-old range. Those who use a computer spend an average of just over an hour per day at the keyboard.

BACKGROUND

Internet Use

Rates of Internet use have also been rapidly increasing since the mid-1990s (Klopfenstein, 2002; U.S. Department of Commerce, 2002). The Current Population Survey reports that in 2003, about 59% of 3- to 17-year-olds used the Internet, with use rates increasing to 70% by Grades 6 to 8 and to 80% by Grades

9 to 12 (National Center for Education Statistics, 2004). Overall, home is the most popular use location. About 78% of students reported using the Internet at home, compared with 68% at school, 15 % at somebody else's home or public library, and 1% from a community center (National Center for Education Statistics, 2004). About 68% of children used the Internet in 2003, with 85% of children having a PC and 9% having an Internet connection in their bedroom (The Home Technology Monitor, 2003).

Not unlike adults, children and adolescents use the Internet to communicate with others, to find information, to have fun, and to do homework. In 2003, 67% of students agreed that they used the Internet to have fun. Playing learning games was the top Internet activity at school, and playing games in general was the top favorite activity for Internet use at home. Other popular activities included listening to music, making pictures/cards, e-mailing/instant messaging, and making music and movies (NetDay, 2004). Given a choice of six media, one third of children ages 8 to 17 agreed that the Web would be the medium they would want to have if they could not have any other, surpassing TV, telephone, and radio (*BBC Monitoring International Reports*, 2002). However, spending time with the Internet does not necessarily indicate spending less time with other media (e.g., Rideout et al., 2003). Subrahmanyam, Kraut, Greenfield, and Gross (2000) concluded that computers appear to increase an overall "screen" time, with an average of 3.40 hours per day for children ages 2 to 18 without computers, to 4.48 hours for those with computers. Overall, American children between 2 and 18 years of age spend an average of 6 hours and 32 minutes each day using all kinds of media. When simultaneous use of multiple media is accounted for, that exposure increases to 8 hours per day (American Academy of Pediatrics, 2001), with Internet use continuing to drive increases in time spent with media (National Center for Education Statistics, 2004).

Demographic Differences

Certain significant demographic and socioeconomic differences exist in computer and Internet use. Although gender differences in the rate of use have largely disappeared, family income and parents' education are positively associated with the technology use. Whites use technologies more than Blacks and Hispanics, whereas nondisabled persons use them more than the disabled. These differences are stronger in relation to home use than school use. Girls tend to use the Internet more for communication and educational purposes, whereas boys prefer information search and entertainment (Weiser, 2000).

Children also use computers extensively for game play, although videogames are played on several types of platforms: home consoles used with TV sets, computers, computers with access to the Internet, coin-operated arcade machines, and handheld devices including game systems, cell phones, and Palm Pilots (Kaiser Family Foundation, 2002). Since the birth of Pong, a black-and-white tennis videogame, in 1972, the videogame industry has come a long way, through the development of consoles, animation, voice recognition, play stations, and online gaming, to become the fastest growing entertainment segment (DFC, 2004). The videogame, computer game, and interactive entertainment hardware and software market is projected to grow from $23.2 billion in 2003 to $33.4 billion in 2008 worldwide. Just in the United States alone, sales of videogame hardware, software, and accessories grew 10% between 2001 and 2002 to reach $10.3 billion, surpassing all Hollywood revenues (RocSearch, 2004–2005). A large share of these revenues comes from children and teen audiences, who consistently report that video and computer game playing is among their top favorite leisure activities.

Videogame Playing

Burgeoning videogame industry revenues suggest that children have been spending more and more time playing the games. Funk (1993) found that approximately 36% of male seventh- and eighth-grade students played videogames at home for 1 to 2 hours per week, 29% played 3 to 6 hours, and 12% did not play at all. Approximately 42% of females played 1 to 2 hours, and 15% played 3 to 6 hours per week. Nearly 37% of females did not play any videogames. A decade later, Sherry, de Souza, Greenberg, and Lachlan (2003) reported that eighth graders spent an average of 17 hours per week, whereas fifth graders played an average of 12 hours. The rates were still higher for boys. According to Kaiser Family Foundation (2002), children ages 2 to 18 spend, on average, between 20 and 33 minutes a day playing videogames. Although boys and girls spend almost the same amount of time using their computers, boys spend an average of 31 minutes a day gaming, compared with just 8 minutes for girls. On any given day, 44% of boys report playing videogames, compared with 17% of girls. Among children ages 2 to 7, boys are 25% more likely than girls to play videogames on a regular basis, whereas male teenagers are 49% more likely to play than their female counterparts (Rideout et al., 2003). In addition to gender, the amount of playing varies by age and ethnicity, with 8- to 13-year-olds spending more time playing than 14- to 18-year olds and 2- to 7-year-olds. African Americans and Hispanics as well as lower and middle-class children play more than White children and children from high-income homes (Kaiser Family Foundation, 2002).

What kinds of games do adolescent children report liking and playing? Funk's (1993) study of seventh and eighth graders reported that the two most preferred categories were games that involved fantasy violence, preferred by almost 32% of subjects, and sports games, some of which contained violent subthemes, which were preferred by more than 29%. Nearly 20% of the students expressed a preference for games with a general entertainment theme, whereas another 17% favored games that involved human violence. Fewer than 2% of the adolescents preferred games with educational content. The Kaiser Family Foundation (2002) further reports that, among 2- to18-year-olds, the three genres that dominate kids' videogame playing are action or combat (42%), sports (41%), and adventure (36%).

Hence, children spend more time with the screen overall, using the computer for both educational and entertainment purposes and for communicating with their peers. In fact, the presence of children in the home may be a primary reason for the adoption of computer technology in the household. This phenomenon implies that computer technologies may play an important role in the children's socialization, allowing children to learn and explore new venues to which they might not have otherwise enjoyed access. Consider, for example, a recent study by Drotner (2000) which found that access to information and computer technology (ICT) is greater in homes with children than in those without. This difference allows children to forge the way in utilizing and establishing new technology cultures.

THEORETICAL EXPLANATIONS

As with any new technology, researchers often evoke comparisons to existing technologies as a starting point. When we consider the intersection of child development and computing technology, many researchers have drawn on related established literatures. First, we can examine the extensive literature on the effects of TV. Theories regarding the effects of violent TV content (e.g., social cognitive theory, Bandura, 1994; script theory, Huesmann, 1986, 1988; cognitive-neoassociation theory, Berkowitz, 1993; Berkowitz & Heimer, 1989; excitation transfer, Zillmann, 1983) offer a substructure for examining the effects of videogame violence and have been successful in explaining and predicting the effects of computer game violence on children. Theories regarding the effects of TV on children's cognitive development and learning (e.g., Salomon, 1979) also offer a genesis for theorizing about the role of computers in shaping the cognition of a new generation of children. However, in most cases, the study of computer and videogames is treated separately from the study of computer technology overall (see e.g., Anderson & Bushman, 2001; Sherry, 2001).

Therefore, a second theoretical approach that scholars have used is theory regarding the diffusion of innovations. This is done in order to understand how families have incorporated these new technologies into home and family life (Rogers, 1995) and into the classroom (Lievrouw & Livingstone, 2002). In fact, the diffusion of computers into the lives of children has created what Montgomery (2000) refers to as children's "digital media culture." As such, there are nonprofit sites, academic sites, and a host of commercial sites targeting children. Although some of these sites seek to enlighten, others seek mainly to generate sales. Last, although little research on computer technology per se has taken this perspective,[1] we might utilize theories of child development by considering children's cognitive, emotional, social,

[1]However, there has been extensive research on the role of children's development and exposure to TV. See, for example, the work of Joanne Cantor, Barbara Wilson, Cynthia Hoffner, Kristen Harrison, Ellen Wartella, and others.

and moral development as the starting point to explain and predict the relations between children and technology.

The Effects Approach

Several theoretical perspectives explain how exposure to computer games, especially violent games, can lead to imitative behavior. It is clear why these theories of TV violence might easily be applied in a gaming environment. Perhaps the most comprehensive theory to date is the general aggression model (GAM), which comprehensively integrates central elements from several earlier aggression theories. Included in the model are elements of social cognitive theory (Bandura, 1994), which focuses on the audience member's attention to the modeled behavior, retention of that behavior, ability to imitate, and motivation to imitate the behavior. Furthermore, social cognitive theory concentrates on the model, noting that admired and rewarded models are more likely to be imitated. As such, the theory focuses on both the contextual cues (e.g., whether violence is rewarded) and the cognitive structures that lead to imitation. Script theory (Huesmann, 1986) is also integrated into the general aggression model. Script theory focuses on learned and activated scripts, arguing that we might learn to respond to situations in particular ways based on situations that have been repeatedly modeled for us. Therefore, in a new situation (e.g., a conflict), we might draw on scripts observed in the media, such as those containing violence.

Also included in the general aggression model is cognitive-neoassociative priming theory (Berkowitz, 1993; Berkowitz & Heimer, 1989), which draws largely on network models of memory. Given that memory is organized through a network, ideas can prime or activate related thoughts. Berkowitz argued that exposure to media violence, especially over long periods of time, could serve to create a rich, intricate memory network of hostility and violence for heavy viewers. The result, according to priming theory, is that exposure to media violence could then readily activate hostility and aggressive thoughts. In addition to cognitive-neoassociative priming theory, Geen's affective aggression model (1990) explains that increases in aggression after exposure to media violence would result in hostility and negative affect. Furthermore, Zillmann's (1983) excitation transfer model focuses on the mechanism of physiological arousal as the cause of increases in aggression after exposure to violence.

The general aggression model also explains that exposure to videogame violence can increase aggressive behavior both in the short and long term by noting that aggression is largely based on existing knowledge structures or existing mental scripts that are created by the process of social learning (Anderson et al., 2004). That is, individuals can learn new skills and information by watching the behaviors of others, especially if those behaviors are rewarded, performed by attractive actors, or do not cause pain or suffering for the victim of aggression (i.e., sanitized violence). In the short term, both personalogical and situational input variables can lead to aggressive behavior. Personalogical variables include personality variables such as aggressive disposition, current states, beliefs, attitudes, and so on. Situational variables are found in the environment surrounding the person and include factors such as aggressive cues (e.g., playing a violent videogame), being provoked, or feeling pain. Both of these inputs can impact the present internal state of the person. For example, aggression may become more likely if an individual has an aggressive disposition and also plays an aggressive videogame. This may lead to feelings of hostility. Then, given the opportunity to retaliate against someone who has insulted the person, for example, that individual may behave more aggressively than someone without those personalogical or situational factors in place.

Parallel to research on TV effects, scholars have also asked what effect, if any, computers may have on children's cognitive and social development. As early as 1979, scholars made a distinction between the effect *of* and *with* television (Salomon, 1979). The former might be seen as the more traditional effect: what is left over after use, whether positive or negative. The latter might be thought of as those more far-reaching changes that occur due to an encounter with a novel way of absorbing, thinking about, and processing information because of its novel mode of representation. In other words, it is possible that the mode of presentation, say, print or TV or interactive videogames, may in fact affect how we think and how we process information. When a novel mode of presentation is introduced (perhaps an interactive web site to present a traditional classroom lecture on Peru), children may not only learn the material (i.e., effects of media), but may in fact learn to process differently (i.e., effects with media). Salomon has more recently extended his original argument and applied it to computer technology (Salomon, 1990). Although Salomon makes an excellent case for the effects *with* computers, little empirical research has been able to test this indistinct outcome, perhaps because those effects are difficult to measure. Instead, the more concrete cognitive effects of learning, what Salomon refers to as learning *of* (or from) computers, have occurred. These, like the social learning of violence, often are described in terms of social cognitive theory (Pulos & Fisher, 1987).

Diffusion of Innovations

The diffusion of innovation approach (Rogers, 1995) examines the spread of an innovation, in this case, a new technology, throughout society. Successful diffusion of new technology tends to follow a fairly similar pattern each time. Initial adopters, known as innovators, learn about a new technology and adopt it before all other groups in society. This venturesome group tends to be more cosmopolitan and acts as a catalyst for an innovation. Following the innovators, the next group members to adopt a new technology can be termed *early adopters*, who may act to inform others about a technology as they adopt it. This group is followed by both the early majority and late majority. Both groups tend to deliberate before adopting, although the late majority is likely more skeptical. Last are the laggards, who are typically slow to adopt innovations and trust in the traditional ways of doing things. They may only adopt a technology once the rest of society has moved forward. The diffusion of innovations typically follows an S-shaped curve, and this theoretical approach considers both the rate of adoption and, more important, the effect of technology adoption.

Despite children's label as *technology masters*—a label often touted by policy-makers (Facer, Sutherland, Furlong, & Furlong, 2001; Vryzas & Tsitouridou, 2002)—some skepticism is needed as we pursue research into children and computers. First, Vryzas, and Tsitouridou (2002) argue that there exists a disconnect between children's use of computers at home and in school. Although children do use computers at home to play games, e-mail, and visit fan web sites (Livingstone, 2001), these are not always skills that transfer readily to Internet research or other skills needed in schools (Kaiser Foundation, 2000). In fact, children are less likely to question the information they find on the Internet than are adults (Kaiser Foundation, 2000), despite their label as natural technology wizards. Second, and perhaps stemming from this former problem, parents and children are uncertain about what role the computer does and should play in their lives at home and in school (Buckingham, 2002; Kellner, 2002). Children claim to learn more about computers and the Internet from their friends than they do either from their parents or from school (Schmidbauer & Lohr, 1999). As a result, this free exploration at home may lead schools to lag behind in competence, allowing children to retain their role as the innovators, early adopters, and those with the mastery of skill. Therefore, the framework of diffusion of innovations continues to look not only at children's adoption of the technology, but at its integration into family and school life (Livingstone & Bovill, 2001).

A Developmental Approach

A large body of research has examined the effect of TV on children using child developmental theory as its central theoretical approach. Specifically, Piagetian theory has provided the groundwork for much research in this area. Piaget (1970) pioneered research in child development suggesting that children learn through several strategies, such as adaptation, or the continual use of the environment to learn and adjust to changes in the environment. In doing so, they proceed through rather systematic stages. Although it is not necessary to review each stage here, it is worthwhile to note the characteristics of younger versus older children. For example, Piaget noted that children younger than approximately age 5 (i.e., preoperational) tend to think in highly egocentric or personal ways, finding it difficult to imagine the point of view of another, tend to engage in centering, or concentrating on only one aspect of an object at a time, and have a difficult time distinguishing fantasy from reality. Between the ages of 5 and 7, children begin to transition into concrete operations and surmount some of these earlier obstacles.

For example, Cantor (1994) has used Piagetian developmental theory in order to explain and predict what images frighten children at different stages of cognitive developmental progress. Wilson and Weiss (1991) have also used Piagetian developmental theory in order to understand children's responses to news media. Krcmar and colleagues (Krcmar & Cooke, 2001; Krcmar & Valkenburg, 1999) have utilized Kohlberg's theories of moral development in order to understand how children of different ages respond to depictions of interpersonal violence in the media. Because Kohlberg argues that judgments about right and wrong are based on a different decision matrix for children of different ages, it makes sense that how children interpret violence, a potentially immoral act, may differ for younger versus older children. For example, children younger than age 5 tend to use the guidance of an authority figure in order to determine between right and wrong or may simply consider the outcome of an action in making such a judgment. Older children, in contrast, may consider the motive of the actor in order to decide whether an act was wrong (Kohlberg, 1984). In summary, child development, whether studied in the context of cognitive development, moral development, or emotional or social development, has provided a solid framework—one that focused on the child more than on the medium—to understand the responses of a group that is qualitatively different from its adult counterparts.

This review of child development literature then leads us to contemplate whether a cognitive developmental approach could help us to understand at what age children are able to take a critical stance to information found on the Internet. Obviously, children have been faced with faulty information prior to the Internet, but they now enjoy greater access, particularly to information that has gone through little, if any, review. What is the earliest age at which this critical stance can be taught? Should the Internet be used prior to this developmental threshold? Alternately, if we were to take a moral developmental approach, we might understand that younger children are less likely to consider the motivation for an act when deciding whether it was right or wrong than are older children (Kohlberg, 1986). Might a clearer, theoretically grounded, and empirically tested understanding of the role of development help parents make decisions regarding their children's videogame play? To this point, we have considered the broad theoretical approaches to children and computer technology. In the next section, we review some of the empirical research available on children's technology use.

EMPIRICAL FINDINGS

Computer Games and Videogames

The research on computer and videogames and children has received considerable research attention in recent years. The earliest work examined the simple relationship between computer game play and aggression, whereas more recent research has utilized an experimental approach in order to test issues of causality between the two. For example, early research found a correlation between overall videogame exposure and real-world aggressive behavior in children from 4th to 12th grade (Dominick, 1984; Fling et al., 1992; Lin & Lepper, 1987). Early experimental work (Cooper & Mackie, 1986; Irwin & Gross, 1995; Schutte, Malouff, Post-Gorden, & Rodasta, 1988; Silvern & Williamson, 1987) also found some support for the notion that violent videogame content can increase aggression; however, technological advances in the field of electronic gaming have rendered much of the very early research in this area all but obsolete.

What does more recent research tell us? Violent videogames can influence aggressive cognitions (Anderson et al., 2004; Anderson & Dill, 2000; Kirsh, 1998; Tamborini et al., 2001), as well as aggressive affect, leading to feelings of hostility (Anderson & Dill, 2000; Tamborini et al., 2001), and have been found in survey research to be associated with aggressive delinquent behavior, even after controlling for aggressive personality traits (Anderson & Dill, 2000). Experimental work using up-to-date games has also found that playing violent games increases aggressive behavior (Anderson & Dill, 2000; Anderson & Murphy, 2003; Anderson et al., 2004).

In their narrative review of the empirical literature, Dill and Dill (1998) concluded that short-term exposure to violent videogames increases aggression. Similarly, Bensley and Van Eenwyk (2001) conclude there is evidence that playing violent videogames can increase short-term aggression in young children. Meta-analyses conducted on the research on violent videogames have also supported an effect of game play on aggression. The first such comprehensive study was conducted by Anderson and Bushman (2001). Across all studies included in their meta-analysis, the authors found that exposure to violent videogames was positively associated with increased levels of aggression. Anderson (2004) recently updated this original meta-analysis and concluded that when only those studies with the soundest methodological approaches were used, results showed even stronger effect sizes, suggesting that methodologically weaker studies actually underestimate the true effects of exposure to violent videogames. Another meta-analysis by Sherry (2001) using 25 studies found evidence for a small effect of videogame play on aggression. However, Sherry also found that effect sizes have increased over time, with more current studies producing stronger effects, presumably due to the greater realism of today's games. Game type was also important, as games classified as human violence or fantasy violence were found to be more strongly related to aggression than sports games.

Effects on Cognitive Development

Despite the early promise of computers for educational purposes, many educators are still unsure of their value as educational tools (Buckingham, Scanlon, & Sefton-Green, 2001). However, the main strength of computers and the Internet may lie in the home, where, when the Internet is available, children ages 6 to 20 find learning via the Internet a fun activity (Kuchinskas, 1999). Unlike structured school exercises, Internet exploration at home may feel more like free play and may therefore be more appealing to children and adolescents (see play theory; Reiber, Smith, & Noah, 1998). Although web sites that contain some educational content for children abound (e.g., MaMaMedia, NickJr.com), the more interesting question may be what effect computers and the Internet have on children's cognitive skill and development overall. Although children can recall what they learn from a computer, especially when the computer praises their efforts (Bracken & Lombard, 2004), it is crucial to ask how computers, overall, have affected skill learning, education, and cognition, and how they affect the skills necessary for reasoning, problem solving, reading, and creativity.

The earliest research, a longitudinal study of home computer use, found that students who used educational software at home had higher overall grades (Rocheleau, 1995); however, it is likely that these high-performing students also came from families with higher incomes. Nonetheless, even among those with computers, those who spent more time with educational software (but not with game play) performed better academically than those who were light users. In a more conclusive experimental design, children who participated in an after-school program that used computers performed better academically than those who did not participate (Blanton, Moreman, & Hayes, 2000). Perhaps most notably, the Networked Interactive Media in Schools (NIMIS) project, funded by the European Union, found that computers in the classroom encouraged motivation and learning in children—but only when the technology was properly implemented and supported. Interestingly, the greatest progress was made not only by classrooms with access to computers, but by those classrooms with both access and high-quality peer and teacher relationships. In other words, computers can enhance education in the classroom when properly integrated into a successful classroom.

Computers also appear to have an effect on visual intelligence. For example, playing Marble Madness, a game requiring children to visualize and manipulate objects in their minds, was found to enhance spatial skill (Subrahmanyam & Greenfield, 1994), although playing a game that requires the visual tracking of an object on a screen also improved this type of spatial skill outside of the computer environment (Greenfield, De Winstanley, & Kilpatrick, 1994); playing the computer game Concentration—a learning game—improved iconic representation skills (Greenfield et al., 1996). Therefore, there is some evidence that computers, even computer game play, can improve visual skills outside of the computer environment. However, for long-term positive effects to occur in the classroom, computers must be integrated into the classroom and not simply placed in it. Similarly, much research is needed on the use of computers in the home in terms of their influence on cognitive and academic growth in children.

Effects on Social and Emotional Development

Although early critical work on children and computers expressed concern over possible social isolation resulting from the use of computers, the research results on children have been mixed. For example, 80% of parents with children under age 18 reported going online with their children (America Online/ Roper Starch, 1999). Although there is reason to doubt this figure due to the fact that parents may inflate their reports in order to sound like "good" parents, academic research suggests that young children enjoy going online with their peers (Rhee & Bhavnagri, 1991). More recent research suggests that two thirds of students agreed that they used the Internet to have fun, including playing learning and entertainment games, e-mailing, and instant messaging (NetDay, 2004). At least the latter of these computer use activities can be seen as social activities.

Other social objectives are achieved through computer use as well. For example, a recent study by Facer et al. (2001) surveyed 855 children on their home computer use. Although children did use the computer for solitary pursuits, much of the use occurred for the purposes of contstructing and integrating oneself and being included in a group identity (i.e., to be part of a social group). However, the introduction of computers into the home has been associated with a decline in family interaction (Kraut et al., 1998). Quite simply, when families adopt yet another solitary medium into their home, time spent with the computer takes away from time spent with other family members. Therefore, computer use can have a negative impact on social interaction in the family, but may increase children's feelings of belongingness with a peer group that is defined by media.

Perhaps the greatest social concern about children and computers is the growing incidence of unwanted or inappropriate sexual or pornographic material encountered by children. Although the occurrence of adult sex offenders meeting children online is on the rise (Arnaldo, 2001), little systematic research has examined the effects of this contact other than the obvious criminal implications. Pornography, too, is a concern. One in five children has encountered porno graphic material online (Media-Awareness, 2000), and one in three teens has found pornography online (Kaiser Foundation, 2000). Although young children find this material that is accidentally encountered to be embarrassing or upsetting (Von Feilitzen & Carlson, 2000), at least some of the pornography encountered by teens is sought out by them. For ethical reasons, no research has examined the effect of pornography exposure on children and teens; however, research on the effects of exposure to televised sexual situations suggests both cultivation and possible effects on sexual behavioral intentions.

SOCIAL IMPLICATIONS AND CONCLUSIONS

The Macro Perspective

It is unclear, as yet, what the long-term social implications of computer-mediated communication technology use will be. We do know that children frequently use these media technologies—downloading music from the Internet, blogging, and instant messaging (IMing) friends. They spend more time in front of the screen to access a seemingly limitless supply of information than their counterparts of a decade ago. What, if any, are the social implications that we might surmise from this? We might first consider some of the inevitable relationships that arise from children's computer use—relationships that have been there all along, but are somehow hastened and more readily accessed through computer use. The first of these is the child–peer interaction. With the development and adoption of computers and the Internet, we see that children and adolescents do spend somewhat less time with their families, but have not become social isolates as traditionalists feared. They feel included in a culture of their peers, of which technology is a part. They e-mail and chat and IM. The computer has become one kind of intermediary in the child–other relationship. Here, we might say that the child–peer relationship has been facilitated by computers, and we might continue to explore the ways in which computers affect relationships between children and others.

The second of these relationships is the child–world relationship. Through the Internet, the child has access to information, both harmful and beneficial. With creative and engaged parents and teachers, the Internet indeed can become a vital source, an example of all that is good in technology, opening new doors and encouraging curiosity. Of course there is probably as much misinformation, myth, and problematic content on the Internet as there is of the enlightening kind. But this has been true of books and poorly informed peers long before computers. With greater and simpler access, the child–world relationship has become potentially more enriching as well as more damaging. For those adults who want to control the flow of information, the challenge is now greater and more daunting. In this particular scenario, the computer is not the mediator, as in the child–other relationship, but the door. Here we might also say that the child–world relationship has been facilitated by the computer, requiring more mediation and vigilance, but potentially offering more benefit than before.

Last, the relationship that we might consider is the child–self. It is here that the computer is perhaps the greatest mystery. Perhaps it affects children's development. The very way we process information may be affected. Here, we question the interaction between the child and a computer or not the computer as a kind of mediator between parties or the computer as quick and easy library or but computer itself as another. There may be, as Salomon (1990) suggests, effects *with* computers. Although the former two relationships may be more readily explored by the empiricists, it is not entirely clear how the third relationship, child–self, may be explored in regard to computers. How does this interaction, which many of us engage in *daily*, affect the world of childhood? What is the effect of the computer on the child–self relationship?

The Micro Perspective

As with any innovation, we can only look at the effects retrospectively and then use those assessments to predict later outcomes. We speculate about, first, where we will take the technology, intentionally, and then where the technology will take us, unintentionally. In the world of childhood, we are just beginning to see how computer technology has affected children.

One way that computers have affected children is that they have had an impact on children's free time. Children spend more time with media now and less time with their families. Is the computer to be blamed for this? Or would children simply fill their time otherwise without the computer, in ways that deviate from our idealized vision of families sprawled around the fireplace playing Scrabble together? At least some evidence exists that the integration of computers into family life is just one technology in an already complex media environment. However, with greater complexity, the computer more than earlier home media technologies has permitted children to be the experts, in some cases helping their parents adapt to the technology. Children may act as early adopters, encouraging their late-adopting parents, or may act as teachers, showing their sometimes resistant or struggling parents the ins and outs of a new software product (Ribak, 2001). Therefore, one social impact of the computer has been to displace family time; however, it is unclear how this "time" would have been spent otherwise.

Computer gaming has also had an impact on the culture of childhood, and indeed on the marketplace. Children are the primary consumers of computer games and spend more time than adults with the interactive gaming technology than other groups in society. Although the negative effect of violent computer games on children is well documented (e.g., Sherry, 2001), other games can positively affect visuospatial skills (e.g., Greenfield et al., 1994). Therefore, like the TV that came before it, the quality of the game matters, leaving room for either largely positive or alternately negative effects from children's game play.

Another effect of computers has been in the classroom. The best example of that may be the extensive study of computers in the classroom (Cooper & Brna, 2002). After assessing the outcome of "classrooms of the future," the authors concluded that computers in the classroom could truly aid learning outcomes only when the computers were set up in an ecologically useful way—with proper technological support, teacher training, and inventive teachers, who previously had good emotional relationships with their students. If all of these things were in order, then true progress could be made by the students. However, computers could also be the high-tech version of the old-fashioned worksheet, filling students' time in uncreative and ultimately unhelpful ways.

Perhaps one of the more important threads to be gleaned from all of this work can be described as follows. First, some children begin using the computer as early as 6 months and regularly on a parent's lap at 2. Second, computers can be used effectively in the classroom, and by extension in the home, if they are used thoughtfully and in the context of a larger positive learning environment. Third, as children begin to use computers at such an early age, children can use them inventively and creatively because youngsters have the time and natural curiosity needed for exploration. Fourth, because children may use computers without a specific goal at times, they may see exploration as valuable. Therefore, computers can provide an opportunity for cognitive growth, positive entertainment, and classroom creativity. However, computers can also be a waste of time in the classroom and at home—at best—or generate negative effects on cognitive and emotional health—at worst. Perhaps the outcome is a simple one, intricate only in its details: Computer technology is only as good in the lives of children as the implementation of that technology—technology is not good or bad, it can simply be used in useful or ultimately frivolous ways. The social implications, then, are not of the technology, but of our *use* of the technology for and by our children.

REFERENCES

America Online/Roper Starch. (1999). *Popular Internet activities.* Retrieved from http://www.media-awareness.ca/english/resources/research_documents/statistics/internet/popular_web_activities.cfm.

Anderson, C. A. (2004). An update on the effects of playing violent videogames. *Journal of Adolescence, 27*(1), 113–122.

Anderson, C. A., & Bushman, B. J. (2001). Effects of violent video games on aggressive behavior, aggressive cognition, aggressive affect, physiological arousal, and prosocial behavior: A meta-analytic review of the scientific literature. *Psychological Science, 12*(5), 353–359.

Anderson, C. A., & Bushman, B. J. (2002). Human aggression. *Annual Review of Psychology, 53,* 27–51.

Anderson, C. A., Carnagey, N. L., Flanagan, M., Benjamin, J., Arlin, J., Eubanks, J., & Valentine, J. C. (2004). Violent video games: Specific effects of violent content on aggressive thoughts and behavior. *Advances in Experimental Social Psychology, 36,* 199–249.

Anderson, C. A., & Dill, K. E. (2000). Video games and aggressive thoughts, feelings, and behavior in the laboratory and in life. *Journal of Personality and Social Psychology, 78*(4), 772–790.

Anderson, C. A., & Murphy, C. R. (2003). Violent video games and aggressive behavior in young women. *Aggressive Behavior, 29,* 423–429.

American Academy of Pediatrics. (2001, November). Media violence. *Pediatrics, 108,* 1222–1226.

Arnaldo, C. A. (2001). *Child abuse on the Internet: Ending the silence.* Paris: Berghahn Books and UNESCO.

Bandura, A. (1994). Social cognitive theory of mass communication. In J. Bryant & D. Zillmann (Eds.), *Media effects: Advances in theory and research* (pp. 61–90). Hillsdale, NJ: Lawrence Erlbaum Asscoaites.

BBC Monitoring International Reports. (2002, April 12). USA: More children choose Internet above other media, p. 3.

Bensley, L., & Van Eenwyk, J. (2001). Video games and real-life aggression: Review of the literature. *Journal of Adolescent Health, 29,* 244–257.

Berkowitz, L. (1993). *Aggression: Its causes, consequences, and control.* New York: McGraw-Hill.

Berkowitz, L., & Heimer, K. (1989). On the construction of the anger experience: Aversive events and negative priming in the formation of feelings. In L. Berkowitz (Ed.), *Advances in experimental social psychology* (Vol. 22, pp. 1–37). San Diego, CA: Academic Press.

Blanton, W. E., Moorman, G. B., & Hayes, B. A. (2000, May 30). *Effects of participation in the Fifth Dimension on far transfer*. Boone, NC: Laboratory on Technology and Learning, Appalachian State University, College of Education.

Bracken, C. C., & Lombard, M. (2004). Social presence and children: Praise, intrinsic motivation, and learning with computers. *Journal of Communication, 54,* 22–37.

Buckingham, D. (2002). The electronic generation? Children and new media. In L. Lievrouw & S. Livingstone (Eds.), *The handbook of new media* (pp. 77–89). London: Sage.

Buckingham, D., Scanlon, M., & Sefton-Green, J. (2001). Selling the digital dream: Marketing educational technology to teachers and parents. In A. Loveless & V. Ellis (Eds.), *ICT, pedagogy and the curriculum: Subject to change*. London: Routledge.

Calvert, S. L., Rideout, V. J., Woolard, J. L., Barr, R.F., & Strouse, G. A. (2004). Age, ethnicity, and socioeconomic patterns in early compute use: A national survey. *American Behavioral Scientist, 48*(5), 590–607.

Cantor, J. (2001). The media and children's fears, anxieties, and perceptions of danger. In D. G. Singer & J. L. Singer (Eds.), *Handbook of children and the media*. Thousand Oaks, CA: Sage.

Cooper, J., & Mackie, D. (1986). Video games and aggression in children. *Journal of Applied Social Psychology,* 16, 726–744.

DFC. (2004). *DFC intelligence forecasts 44% revenue growth for video game industry by 2008*. Retrieved July 25, 2005, from http://www.gameinfowire.com/news.asp?nid=4409

Dill, K. E., & Dill, J. C. (1998). Video game violence: A review of the empirical literature. *Aggression and Violent Behavior, 3*(4), 407–428.

Dominick, J. R. (1984). Video games, television violence and aggression in teenagers. *Journal of Communication, 34,* 136–147.

Drotner, K. (2000). Difference and diversity: Trends in young Danes' media use. *Media, Culture and Society, 22*(2), 149–166.

Facer, K., Sutherland, R., Furlong, R., & Furlong, J. (2001). What's the point of using computers?: The development of young people's computer expertise in the home. *New Media & Society, 3*(2), 199–219.

Fling, S., Smith, I., Rodriguez, T., Thornton, D., Atkins, E., & Nixon, K. (1992). Video games, aggression, and self-esteem: A survey. *Social Behavior and Personality, 20,* 39–46.

Funk, J. B. (1993). Reevaluating the impact of videogames. *Clinical Pediatrics, 32,* 86–90.

Geen, R. G. (1990). *Human aggression.* Pacific Grove, CA: Brooks/Cole.

Greenfield, P. M., Camaioni, L., Ercolani, P., Weiss, L., Lauber, B. A., & Perucchini, P. (1996). Cognitive socialization by computer games in two cultures: Inductive discovery or mastery or an iconic code? In P. M. Greenfield & R. R. Cocking (Eds.), *Interacting with video* (pp. 141–167). Norwood, NJ: Ablex.

Greenfield, P. M., De Winstanley, P., & Kilpatrick, H. (1994). Action videogames and informational education: Effects on strategies for dividing visual attention: Special Issue. Effects of interactive entertainment technologies on development. *Journal of Applied Developmental Psychology, 15,* 105–123.

Huesmann, L. R. (1986). Psychological processes promoting the relation between exposure to media violence and aggressive behavior by the viewer. *Journal of Social Issues, 42,* 125–139.

Interactive Digital Software Association. (2001). *Economic impact of the demand for playing interactive entertainment software.*

Irwin, A. R., & Gross, A. M. (1995). Cognitive tempo, violent videogames, and aggressive behavior in young boys. *Journal of Family Violence,* 10, 337–350.

Kafai, Y. B. (1996). Software by kids for kids. *Communications of the ACM, 39*(4), 38–40.

Kaiser Family Foundation. (2002). *Children and video games.* Washington, DC: Author.

Kaiser Foundation. (2000). U.S. adults and kids on new media technology. In C. Von Feilitzen & U. Carlsson (Eds.), *Children in the new media landscape* (pp. 349–350). Goteborg: UNESCO/Nordicom.

Kellner, D. (2002). New media and new literacies: Reconstructing education for the new millennium. In L. Lievrouw & S. Livingstone (Eds.), *The handbook of new media* (pp. 90–104). London: Sage. Kohlberg, L. (1984). *The psychology of moral development: The nature and validity of moral stages* (Vol. II, pp. 123–139). San Francisco: Harper & Row.

Klopfenstein, B. (2002). The internet and web as communication media. In C.A. Lin & D. Atkin (Eds.), *Communication technology and social change: A adoption and uses* (pp. 353–378). Cresskill, NJ: Hampton.

Kraut, R., Lundmark, V., Patterson, M., Mukopadhyay, T., et al. (1998). Internet paradox: A social technology that reduces social involvement and psychological well-being? *American Psychologist, 53*(9), 1017–1031.

Krcmar, M., & Cooke, M. C. (2001). Moral development and children's interpretations of justified and unjustified television violence. *Journal of Communication, 51*, 300–316.

Krcmar, M., & Valkenburg, P. (1999). A scale to assess children's interpretations of justified and unjustified television violence and its relationship to television viewing. *Communication Research, 26*, 608–634.

Kuchinskas, S. (1999). What's in a game? *Brandweek, 40*(19), 52.

Lievrouw, L., & Livingstone, S. (Eds.). (2002). *The handbook of new media: Social shaping and social consequences.* London: Sage.

Lin, S., & Lepper, M. R. (1987). Correlates of children's usage of videogames and computers. *Journal of Applied Social Psychology, 17*, 72–93.

Livingstone, S. (2001). *Online freedom and safety for children* (Research Report No. 3). London: Institute of Public Policy Research/Citizens Online Research Publication.

Livingstone, S., & Bovill, M. (Eds.). (2001). *Children and their changing media environment: A European comparative study*. Mahwah, NJ: Lawrence Erlbaum Associates.

Media-Awareness. (2000). Canada's children in a wired world: The parents' view. A survey of Internet use in Canadian families. *Media Awareness, 20*(2), 17–18.

Montgomery, K. C. (2000). Children's media culture in the new millenium: Mapping the digital landscape. In *The future of children: Children and computer technology* (Vol. 10, pp. 2, 145–167). Los Altos, CA: David and Lucille Packard Foundation.

National Center for Education Statistics. (2004). *Computer and Internet use. Supplement to 2001 Current Population Survey.* Washington, DC: Author.

NetDay. (2004, March). *Voices and views of today's tech-savvy students: National Report on Netday Speak Up Day for Students 2003*. Retrieved December 29, 2004, from www.netday.org.

Piaget, J. (1970). The stages of intellectual development of the child. In P. H. Mussen (Ed.), *Readings in child development and personality* (pp. 52–71). New York: Harper.

Pulos, S., & Fisher, S. (1987). Adolescents' interests in computers: The role of attitude and socioeconomic status. *Computers in Human Behavior, 3*, 29–36.

Reiber, L. P., Smith, L., & Noah, D. (1998). The value of serious play. *Educational Technology, 38*(6), 29–39.

Rhee, M. C., & Bhavnagri, N. (1991). *Four year-old children's peer interactions when playing with a computer.* Detroit, MI: Wayne State University. (ERIC Document Reproduction Services No. ED342466)

Ribak, R. (2001). "Like immigrants": Negotiating power in the face of the home computer. *New Media & Society, 3*(2), 220–238.

Rideout, V. J., Vandewater, E. A., & Wartella, E. A. (2003). *Zero to six: Electronic media in the lives of infants, toddlers, and preschoolers.* Kaiser Family Foundation Report. Retrieved December 29, 2004, from www.kff.org

Rocheleau, B. (1995). Computer use by school-aged children: Trends, patterns, and predictors. *Journal of Educational Computing Research, 12*(1), 1–17.

RocSearch. (2004–2005). *Video game industry.* London: Author.

Rogers, E. M. (1995). *Diffusion of innovations* (4th ed.). New York: The Free Press.

Salomon, G. (1979). *Interaction of media, cognition, and learning.* San Francisco: Jossey-Bass.

Salomon, G. (1990). Cognitive effects with and of computer technology. *Communication Research, 17*(1), 26–44.

Schmidbauer, M., & Lohr, P. (1999). Young people online. In P. M. Lohr & M. Meyer (Eds.), *Children, television, and the new media* (pp. 63–99). Luton: University of Luton Press.

Schutte, N. S., Malouff, J. M., Post Gorden, J. C., & Rodasta, A. L. (1988). Effects of playing videogames on children's aggressive and other behaviors. *Journal of Applied Social Psychology, 18*, 454–460.

Sherry, J. L. (2001). The effects of violent videogames on aggression: A meta-analysis. *Human Communication Research, 27*, 772–790.

Sherry, J., de Souza, R., Greenberg, B. S., & Lachlan, K. (2003, May). *Why do adolescents play videogames? Developmental stages predicts videogame uses and gratifications, game preference, and amount of time spent in play.* Paper presented at the annual convention of the International Communication Association, San Diego, CA.

Silvern, S. B., & Williamson, P. A. (1987). The effects of videogame play on young children's aggression, fantasy, and prosocial behavior. *Journal of Applied Developmental Psychology, 8*, 453–462.

Subrahmanyam, K., & Greenfield, F. M. (1994). Effect of videogame practice on spatial skills in girls and boys. *Journal of Applied Developmental Psychology, 15*, 13–32.

Subrahmanyam, K., Kraut, R. E., Greenfield, P. M., & Gross, P. M. (2000). The impact of home computer use on children's activities and development. *Children and Computer Technology, 10,* 123–144.

Tamborini, R., Eastin, M., Lachlan, K., Skalski, P., Fediuk, T., & Brady, R. (2001, May). *Hostile thoughts, presence and violent virtual videogames.* Paper presented at the annual conference of the International Communication Association, Washington, DC.

The Home Technology Monitor. (2003). *How children use media technology.* Knowledge Networks Statistical Research. Retrieved December 29, 2004, from www.sri.knowledge networks.com.

U.S. Department of Commerce. (2002). *A nation online: How Americans are expanding their use of the Internet.* Washington, DC: Author.

Von Feilitzen, C., & Carlsson, U. (2000). *Children in the new media landscape: Games, pornography, perceptions.* Goteborg: UNESCO/Nordicom.

Vryzas, K., & Tsitouridou, M. (2002). The home computer in children's everyday life: The case of Greece. *Journal of Educational Media, 27*(1–2), 9–17.

Walsh, D., Gentile, D., Gieske, J., Walsh, M., & Chasco, E. (2004). *Ninth Annual Mediawise Video Game Report Card.* National Institute on Media and the Family. Retrieved December 29, 2004, from www.mediafamily.org

Weiser, E. B. (2000). Gender differences in Internet use patterns and Internet application preferences: A two-sample comparison. *Cyber Psychology and Behavior, 3*(2), 167–178.

Williams, D. (2002). Structure and competition in the US home videogame industry. *The International Journal of Media Management, 4*(1), 41–54.

Wilson, B. J., & Weiss, A. J. (1991). The effects of two reality explanations on children's reactions to a frightening movie scene. *Communication Monographs, 58*, 307–326.

Zillmann, D. (1983). Arousal and aggression. In R. Geen & E. Donnerstein (Eds.), *Aggression: Theoretical and empirical reviews* (pp. 75–101). New York: Academic Press.

III

WORK AND ORGANIZATIONAL SETTING

5

Information Technology and Organizational Telework

David J. Atkin and T. Y. Lau

Our conceptions of work, leisure, and communication are changing dramatically, as industrial-age modes of operation are being transformed by a "communication revolution" fueling the growth of an "information age" (e.g., Stipp, 1998; Williams, 1988). Rogers (2002) and Atkin (2002) are among several chronicling the genesis of this knowledge economy, one in which labor-intensive smokestack industries—and their onsite workforces—are being replaced by a computer-literate workforce equipped with wired and wireless communication channels. Stewart (1994) estimates that this information sector became preeminent in 1991, "the first year that companies spent more on computing and communications—the 'capital goods of the new era'—than on industrial, mining, farming, and construction machines" (p. 70; see also Atkin, 2002). Social scientists (e.g., Dizard, 2000; Porat, 1977) predict that roughly two thirds of the workforce could be subsumed under the information sector, which encompassed nearly half (45.8%) of the workforce as far back as 1981 (Hepworth, 1990). Bell (1989) sees this growth flowing from information, telecommunication, and related technologies.

Given the pervasiveness of this emerging information economy, few can dispute the decisive role that new media technologies—including telematic modalities facilitating telework—can play in our home and workplaces. As one telecommuter recently observed, "(T)hree or even two years ago, you'd have to get in a car or find some way to get into the office. Now you can do anything at home that you can do from your office" (Rhoads & Silver, 2005, p. B4). This chapter assesses the impact of new media at the juncture of those domains and the role of telecommuting in shaping the present and future of the business world. We first address the origins of telecommuting—including the various manifestations and contexts in which it can be found—alongside theories and research addressing the influence of telecommuting on workers, their employers, and the larger society. The chapter focuses, in particular, on communication infrastructure changes in the United States because it stands among the most "wired" computer-literate nations in the world.

BACKGROUND

The Emerging Information Economy

Bates (1988) outlines two basic technological trends underpinning the knowledge economy: (a) the rapid development and diffusion of microelectronics and other information technologies, and (b) that information products are assuming greater importance to Western economies. Strover (1988) suggests that the information economy is predicated on three ideas: (a) The information economy is weighted toward the service sector; (b) productivity gains are realized from research and development in control technologies such as computers and robotics; and (c) people in many jobs will experience greater need for, access to, and use of information. Scholars (e.g., Hudson & Leung, 1988; Jeffres, 1997) used Porat's (1977) definition of the primary information sector to measure the growth of the knowledge or information industry in several subsectors (e.g., "information services"), all of which are amenable to telework.[1]

Defining Telework

Nilles (1975) first coined the term *telecommuting* over 30 years ago when describing workers who work at home for some portion of their work week, using telecommunications as the link to their business office (see Nilles, Munushian, & Carey, 1976). In exploring those origins, Bailey and Kurland (2002) maintain that the term is interchangeable with the notion of *telework*, as both have been lauded as a "cure for a variety of organizational and social ills" (pp. 383–384), including:

1. Helping organizations decrease real-estate costs (e.g., Egan, 1997).
2. Responding to worker needs for a healthy balance between work and family (e.g., Shamir & Solomon, 1985).

[1]According to those authors, the primary information sector includes the following categories: (a) research and development; (b) information services, which include advertising agencies, private information services, search and nonspeculative brokerage industries, nonmarket coordinating institutions, insurance industries, and speculative brokers; (c) media, including mass media as well as theatrical producers; (d) information technology, including "software" such as tapes and CDs as well as hardware; and (e) information trade, which includes retail stores trading in radio and TV sets, cameras, books, papers, and film theaters (see Jeffres, 1997, p. 342).

3. Facilitating compliance with the Americans With Disabilities Act (e.g., Matthes, 1992).
4. Reducing traffic congestion and air pollution (e.g., Handy & Mokhtarian, 1995; Novaco, Kliewer, & Broque, 1991).

According to Sullivan (2003), *telework* refers generically to any form of substitution of information and communication technologies (ICTs) for work-related travel. Other variants like *center-based telecommuting* (Mokhtarian, 2002) refer to the use of office space close to home where the employee works without direct supervision.

By the turn of the millennium, some 11.5 million Americans were teleworking. Pearlson and Saunders (2001) note that, although the notion of telecommuting has been discussed for decades—generating some excitement and experimentation—"the concept has never quite taken off" (p. 117). Even so, the workforce sectors most likely to adopt telework—including "management, business, and financial" (MB&F) as well as "professional" sectors—are slated to grow in the coming decade (e.g., Baltes, 2004). In particular, Bureau of Labor Statistics data suggest that, of the 145.6 million jobs in the economy, 15.5 million are MB&F, while another 26.8 million are in other professions (supported by nearly 24 million administrative assistants; Bailey & Kurland, 2002). At present, a majority of all workers—81.8 million (or 56% of the workforce)—are manipulating some type of information in their work.

Focusing on a more conservative workplace definition—outlining the number of teleworkers who worked at home during business hours at least 1 day per month—the American Interactive Consumer Survey documents the following telework trend frequencies (ITAC, 2003):

1997: 11.6 million employed teleworkers; 18.3 self-employed teleworkers.
1999: 14.4 million employed teleworkers; 19.0 self-employed teleworkers.
2001: 16.8 million employed teleworkers; 19.9 self-employed teleworkers.
2003: 23.5 million employed teleworkers; 23.4 self-employed teleworkers.

As these data indicate, the number of workers telecommuting at least part time doubled between 1997 and 2003, although growth in the self-employed sector was only about one tenth that level. Growth rates were greatest between 2001 and 2003 (40% growth overall and 18% for the self-employed). All told, the data (ITAC, 2003) suggest that 22% of employee teleworkers work from home daily or "nearly every day," whereas 42% of the employees work from home at least 1 day per week.

Around the globe in 2005, some 82.5 million workers worldwide performed their jobs at home 1 day per month, doubling the figure from 2000; 25 million of them were in the United States—comprising 23% of the domestic workforce—up from 12% in 1990 (Rhoads & Silver, 2005). Perhaps the biggest impetus for this growth lies in the dramatic growth in broadband technology since 2000 (e.g., Atkin, Lau, & Lin, 2006). Reinsch (1997) notes, however, that these underlying economic and technological trends do not "foreordain" continued growth for telecommuting, the continued success of which hinges on the extent to which organizations address human factors.

THEORETICAL EXPLANATION

Scholarly discussions of telecommuting are rooted in the oil crisis of the 1970s. Focusing on transportation–technology trade-offs, commentators (e.g., Nilles, Carlson, Gray, & Hanneman, 1976) noted that if one in seven urban commuters dropped out, the United States would have no need to import oil (see Mann & Holdsworth, 2003). This was followed by "blue-sky" visions (e.g., Toffler, 1980) outlining a benevolent "electronic cottage" sector facilitated by emerging communication technologies. Huws, Korte, and Robinson (1990) point to another strain in early telecommuting research, one that outlined new organizational structures and strategies—encompassing mobile work and the virtual organization—facilitated by information communication technologies. As Ellison's (1999) review suggests, this work addressed ways in which such arrangements might help organizations "lower costs, attract or retain employees desirous of geographical or temporal flexibility, and increase productivity" (p. 339).

In their extensive review of the telework literature, Bailey and Kurland (2002) lament that the bulk of research in this area is largely atheoretical, with the proportion of studies testing theory comprising little more than 10% of published studies. Although case studies focused on descriptive findings dominated this early work, the authors conclude that the number of studies that test explicit hypotheses (or build models) has grown steadily since then. The most widely used framework to explore telecommuting adoption is diffusion theory (e.g., Ruppel & Howard, 1998). Four elements in this model—innovation, communication channels, time, and social system—are identified as being critical for technology adoption.

Digital communication networks, delivered over dedicated satellite or fiber networks to provide high-speed broadband connections between offices and homes, constitute the backbone of this telecommuting infrastructure. According to the logic of diffusion theory, enlightened knowledge-intensive companies with highly educated workers are likely to be aware of the relative advantages of an innovation like telework, and those with more resources could of course better afford any costs of adoption.

Scholars (e.g., Kraut, 1989; Kraut, Steinfield, Chan, Butler, & Hoag, 1998) suggest that in order to obtain a more complete model of telework diffusion, characteristics of the specific innovation must be considered. As we explore in our summary of empirical findings in the section to follow, such attributes are important predictors of telecommuting adoption behavior (e.g., Bailey & Kurland, 2002). Even so, the inconsistency of these innovation predictor sets, particularly in organizational contexts, has complicated the task of molding an all-encompassing theory that explains the adoption and diffusion processes in this context (Ruppel & Howard, 1998).

Organizational Innovativeness

Although a worker's decision to adopt telework is ultimately an individual one (Ruppel & Harrington, 1995), organizational factors—such as the decision to allow workers that opportunity—represent an important, system-level facilitator variable (e.g., Rice & Webster, 2002). In particular, past work suggests that the adoption of such innovations of telework is a function of the culture or climate of an organization (Boyton, Zmud, & Jacobs, 1994; Prescott & Conger, 1995). Managerial attitudes toward telework, however, represent a barrier to its adoption (Harris, 2003; Kurland & Bailey, 1999), particularly when fueled by concerns over reduced control and supervision, as we explore later (Fairweather, 1999; Jones, 1996).

Adoption and diffusion dynamics may be related to individual factors as well; Prescott and Conger (1995) underscore the importance of a champion for an innovation. This is particularly true of *administrative innovations*, which can be defined as "those that affect the organization's administrative process, organizational structure or management style" (Ruppel & Howard, 1998, p. 8; citing Damanpour, 1987; Drury & Farhoomand, 1996).

Rogers (2003) suggests that the adoption process in organizations can be identified in a matching stage, which is contingent on knowledge about the specific traits of a given innovation. In cases where an innovation is not initially well suited for an organization, a process of reinvention (i.e., redefinition or restructuring) may occur. Such reinvention is contingent on a rationalization and perhaps modification of the particular traits of an innovation, if not a modification of the organizational structure of the adopting firm. Ruppel and Howard (1998) refer to these specific characteristics of an innovation as "facilitator variables," further suggesting that the factors suggested by this literature as important to the use of telework include:

1. The availability of rich telecommunications media (e.g., video teleconferencing) for telework and the manager (Hotch, 1993).
2. The planning of telework arrangements (Jones 1992).
3. The training of teleworkers (Lavelle, 1993; Szappanos, 1993).
4. The training of managers to remotely manage teleworks (Jones, 1992; Mistuka, 1992; Nilles, 1992).
5. The existence of perceived adequate security measures (Mistuka, 1992; Nilles, 1992).
6. The existence of a career ladder for teleworkers (Knight, 1994; Nilles, 1992; Verespej, 1994).

The adoption and diffusion of telework, then, represent a complex interplay of variables that range in scope from characteristics of the innovation—including telematic technologies that enable telework—and individual as well as organizational adopter traits. As the section to follow details, the inability of predictive models to explain the seemingly slow adoption of telework stands testament to the theoretical complexity of the innovation at hand. The section to follow details empirical findings on the adoption and diffusion of telework, beginning with projections of its aggregate impact on the workplace, through pioneering empirical work based on case studies, to more theoretically driven work focused on individuals and organizations.

Adoption Typology

Reviews of the organization innovation literature (e.g., Damanpour, 1991; Duxbury & Neufeld, 1999; Harris, 2003; Kowalski & Swanson, 2005; Pearlson & Saunders, 2001; Watad, 2003; Wolfe, 1994) classify studies to find similarities and differences across various innovation categories, but find it difficult to account for inconsistent patterns of results. Some differences in adoption can be found within a given category, such as radical (e.g., full migration to a virtual office) versus incremental (e.g., part-time adoption of telecommuting) options. Other innovations may be administrative (e.g., time clocks) or technical (e.g., pagers) in nature. In addition, Ruppel and Howard provide a review of other adoption factors, including the scope of the innovation, the level at which the innovation is adopted and diffused (i.e., individual, organization, workgroup), and the type of innovation under study. Their review suggests that the information technology (IT) innovation literature has employed similar typologies—to similar effect—in emulating those conclusions.

Theory development in this realm has been aided by expansions in the organizational innovation literature that include such concepts as absorptive capacity (Boyton, Zmud, & Jacobs, 1994; Cohen & Levinthal, 1990). This concept states that adoption and diffusion of such an innovation are constrained by a firm's capacity to "recognize the value of new information, assimilate it, and apply it to commercial ends" (Cohen & Levinthal, 1990, p. 128). As Lim and Teo (2000) observe, those who perceive more advantages accruing from teleworking—or adoption of a particular telework modality like the videophone—are more likely to support the concept.

A related concept to this knowledge importance dimension within an organization involves the application of organizational learning to the diffusion process. Huber (1996), for instance, outlined the importance of this concept in the context of technology-critical organizations (e.g., software firms), which require the frequent adoption of new technologies. The concept of organizational learning has also been applied to the adoption and diffusion of computer-aided software engineering (CASE).

Knowledge about teleowork technology capabilities can also help ameliorate organizational hesitancy to adopt the innovation. For instance, Fairweather (1999) outlines vehicles that can facilitate employer surveillance of employee work habits, the knowledge about which can help overcome management concerns about lost productivity in the home environment. Of course, measures taken to enhance management surveillance may undermine employee satisfaction with telework, the implications of which we consider in our final section.

Worker Innovativeness

Today's teleworkers enjoy access to a growing repertoire of communication technologies, extending beyond traditional channels like the telephone and fax to include wire (and wireless) computer/Internet modalities—including wide area networks (WANs) and local area networks (LANs)—as well as cell phones, pagers, handheld devices, and multifunctioning work stations (see Rice & Schneider, chap. 6, this volume). Owing to these versatile technologies, "where a person works these days is not as important as the work performed" (Robertson, Maynard, & McDevitt, 2003, p. 111). Diffusion theory suggests that the telework innovation would need to be compatible with employee and employer values, goals, and lifestyle preferences (LaRose, 1988).

Although adoption of telework is clearly a function of access to telematic technologies, the investigation of such enabling technologies as computer or integrated services digital network (ISDN) access remain ancillary to work-related factors in the literature (Mannering & Mokhatarian, 1995; Mokhtarian & Salomon, 1996, 1997). In addition to the organizational attributes reviewed earlier—including available resources, absorptive capacity, and the like—individual factors also represent important facilitators of telework (e.g., Ruppel & Howard, 1988). Focusing on individual-level variables, Pearlson and Saunders (2001) suggest telecommuting would be compatible with the needs of young, upwardly mobile parents, offering them the chief advantage of working whatever hours they choose. Telework also releases workers and businesses from the assumption that all business work falls within the standard 9 to 5 office day, providing flexibility in terms of the work environment. The authors stress other benefits for telecommuters, including the ability to work in their alternative workspaces in whatever clothes they want, at whatever pace they want, and with whatever environmental factors they want.

Communication Presence

Wellman, Salaff, and Dimitrova (1996) maintain that the computer networks on which telecommuting is predicated are social networks, although ones that are built on media that have limited social presence (e.g., asynchronous modalities like e-mail and computerized conferencing). *Presence* (Marvin, 1980) was first used to describe teleoperation technology that provides "remote presence," and, in the words of Bracken and Lombard (2004), the ability to "see or feel what is happening there" (p. 24), including the following dimensions identified by Lombard and Ditton (1997): (a) presence as social richness (the "warmth" or "intimacy" possible via a medium), (b) realism (perceptual and/or social), (c) transportation (the sensations of "you are there," "it is here," and "we are together"), (d) immersion (in a mediated environment), (e) social actor within medium (parasocial interaction), and (f) medium as social actor. We might expect, then, that positive evaluations of telecommuting would be related to conceptions that enabling technologies have sufficient levels of presence to facilitate office interaction, which brings us to the issue of user satisfaction.

For instance, Kraut, Galegher, Fish, and Chalfonte (1992) maintain that modalities such as face-to-face (Ftf) communication permit rich communication—offering both expressiveness and interactivity—whereas modalities such as text annotation or e-mail limit both. Given that most information exchanged in a conversation is nonverbal, we might expect that employees would rank more highly those technologies that include more visual cues (e.g., videoconferencing) relative to text-only modalities (e.g., e-mail), while real-time audio conferences might fall somewhere in between those extremes. Kraut et al. (1992) suggests that *contingency theory* provides the best fit for this calculus, as it states that "successful task performance depends on matching the design of technologies and of social and organizational structures to the requirements of particular tasks" (p. 378). The authors suggest that, in the realm of media choice (e.g., Trevino, Daft, & Lengel, 1990), workers will prefer interactive media modalities when faced with greater uncertainty (or equivocality) in their work situations. In the context of media presence, workers will prefer interactive channels because they represent richer media that enable more effective communication, relative to one-way modalities (e.g., paper; see Rice & Schneider, chap. 6, this volume).

EMPIRICAL FINDINGS

This section first discusses the background behind telecommuting, including the various definitions and contexts in which it occurs. Second, we examine the effects on the worker, including home life and separation from work life. Third, the effects on companies and society are reviewed, including the benefits and disadvantages. Finally, we discuss the difficulties encountered when studying the diffusion of telework.

Teleworker Profile

Much of the early work on telework focused on demographic profiles of early adopters of the innovation. Consistent with the upscale adopter profile outlined in diffusion theory, the typical American teleworker has a median household income of $45,000 and is 42 years of age (Bailey & Kurland, 2002). Pioneering work on telecommuting adoption in California found that 65% of teleworking participants were men and most were mid-level professionals (Olszewski & Mokhtarian, 1994). Similarly, a Finnish study (Luukinen, 1996) discovered that telecommuters were primarily highly educated, affluent male professionals. More recent work suggests that demographic differences between adopters and nonadopters are leveling (e.g., Ellison, 1999; Francese, 2002), although gains in work task competency attributable to telework were most dramatic for older workers (Sharit et al., 2004).

The gender gap has narrowed in recent years, however, as the International Telework Association and Council (2000) identified a gender distribution that slightly favors men (51%) over women (49%), an even distribution that is echoed in other studies (Tremblay, 2003). Focusing on telework motivations across the genders, Bailey and Kurland (2002) note that women regarded it as something that enabled them to be closer to their children, whereas men chose it because they prefer the home environment, concluding that managers "found it harder to accept that a woman who is teleworking is just as much at work as when they go to the office. Men did not seem to experience these problems, perhaps because most of their male friends were office-based and thus not around to intrude anyway" (p. 388).

Effects on Teleworkers

Although full-time telecommuting is far from reaching the "critical mass" of its diffusion curve, the literature has accreted a sufficient number of mature trials to provide a picture of its long-term impacts. Research suggests that telecommuting can alter relationships with coworkers as well as family members (e.g., Huws, et al. Hill, 1995). Reinsch (1997) found that telecommuters believe that the advantages of telecommuting far outweigh its disadvantages, although the relationship between telecommuter and manager may "deteriorate after an initial honeymoon phase has passed" (p. 343). Consistent with our earlier review, it is useful to divide our discussion between supply-push (i.e., employer-centered organization) and demand-pull (employee user) perspectives, which are explored in turn.

Many of the pioneering telework studies, focused on clerical workers, uncovered a loss in corporate affiliation and benefits that "mirror what legions of contract workers face today" (Bailey & Kurland, 2002, p. 384). These disadvantages in workplace contact are at least partially mitigated by advantages uncovered by the Labor Department, which concluded that teleworkers benefit from easier commutes and greater job satisfaction ("Telecommuting Grows," 2001). A study performed by Illinois Bell Telephone (Teschler, 1991) concluded that telecommuters "feel they are more productive, more relaxed, and think they manage their time better working at home" (p. 4). This observation reinforces the conventional wisdom that telecommuters appreciate the less distracting home work environment, one that is free from the constant interruptions by fellow coworkers as well as the noise of the office.

The opportunity to spend more time with family represents another benefit for home workers. Bailey and Kurland's (2002) review found little support for the proposition, derived from the literature, that telecommuting could enable professionals with families the flexibility to balance the demands of work and home (see Britton, Halfpenny, Devine, & Mellor, 2004). Mann and Holdsworth (2003, p. 196) found a negative emotional impact of telework—driven by such emotions as loneliness, irritability, worry, and guilt—noting that teleworkers experience "significantly more mental health symptoms of stress than office workers and slightly more physical health symptoms."

Other work (e.g., Hill, Ferris, & Martinson, 2003; Nilles, 1996) suggests that the influence of telework is mostly positive on aspects of one's work as well as personal life. In particular, Madsen (2003) found that teleworkers indicated lower levels of several dimensions of work–family conflict. The perceived benefits were based on the logic that, in the words of Mann and Holdsworth (2003), telecommuters would be able to "choose the hours they work, enabling them to take advantage of off-peak supermarket shopping or gym membership, to collect the kids from school or simply of working at times they are more productive" (p. 197). Perhaps for these reasons, the Families and Work Institute found that 70% of employees were willing to change employers—and 81% willing to change employers—in order to realize flexible work arrangements such as this (Rose, 1988).

As far as employees are concerned, some express concern that they may find it difficult to separate work and home, perhaps even venting office frustrations in the family context. As Harris (2003) notes, "Home-based working reverses arrangements that have prevailed since the Industrial Revolution for work and home to develop as distinct domains with different rules, thought patterns and behaviors" (p. 422). Tietz and Musson (2003) conclude that telework led to either (a) an increasing "bureaucratization" of time for women, where boundaries between "work" and "the household" had to be protected; or (b) more task-oriented approaches to the coordination of all activity, which require more "elastic" temporal boundaries.

Of course, social isolation is another commonly cited drawback for telecommuters. As Mann and Holdsworth (2003) suggest, teleworkers may miss important peer feedback in the absence of colleagues, as "(t)he reduction of this barometer or measure of ourselves is significant for teleworkers" (p. 198). A related disadvantage for telecommuters may involve career progression because their physical visibility is reduced in the home work context. As Bailey and Kurland (2002) suggest, a telecommuter's contribution is likely to be overlooked by dint of the person's remote location. Along those lines, Goelman (2004) concludes that, despite benefits in control afforded employees, lack of "face time" may hinder their chances for promotion.

Other policy remedies designed to reduce the disadvantages of telecommuting have been proposed in Europe. The European Commission, for instance, crafted an agreement encouraging member states to formalize work arrangements that impose "equal rights" regarding social and working conditions for teleworkers and nonteleworkers (Peters & den Dulk, 2003). Such measures might include provisions for comparable pay for comparable work, limitations on work hours, and other measures designed to prevent the emergence of an exploitive "electronic cottage"—reminiscent of low-wage work done out of medieval cottages—for home workers (Toffler, 1980).

Effects on Organizations

Duxbury and Neufeld (1999) found that, despite some exceptions, telework has little impact on intraorganizational communication. For instance, contrary to expectations, home access to e-mail was not a requirement for a successful telework experience. Even so, telecommuters report higher satisfaction with office communication than nontelecommuters (Fritz, Narasimhan, & Rhee, 1998). Reinsch (1997) found that the relationship between a manager and worker may "deteriorate" however, over time. Harris (2003) concludes that a telecommuter's circumstances need to be rationalized, to those that clearly recalibrate the boundaries between home and work, if positive employee relations are to emerge.

To the extent that employee satisfaction is enhanced by the telework benefits already mentioned, a Department of Labor report concluded that employers equate it with increased retention, reduced absenteeism, and lower office overhead ("Telecommuting Grows," 2001). One key perceived benefit for telework adoption among employers stems from the fact that, by adopting telework, employers can potentially make considerable savings in the costs of office space. Empirical evidence for this proposition was found in past work (Harris, 2003; Peters & den Dulk, 2003).

Peters, Tijdens, and Wetzels (2004) note that the management literature emphasizes the benefits of telecommuting, which is believed to be "strongly related to changes in organizational structures, in coordination systems and, in task specifications within organizations" (p. 470). They stress how these changes—which include delayering (e.g., integration of units), downsizing, business process reengineering (e.g., streamlining of production), the shift to a core/periphery model, project-based (i.e., task dedicated) structures, and growing interorganizational networks (e.g., wide area networks)—are associated with new styles of control in which direct supervision is replaced by an internally motivated employee. For instance, a work team assembled of workers from several different units—tasked with interacting virtually over distance—would require different managerial models than their "bricks and mortar" counterparts. These new workers enjoy a wider span of control, with Peters et al. finding that most employees preferred telework, although its practice remains a function of limited opportunities for adoption.

Although some productivity benefits have accrued, and employees seem to favor the practice, many managers do not (Pearlson & Sanders, 2001). Perceived disadvantages to telecommuting were initially investigated largely from the perspective of the employer. For instance, Britton et al. (2004) cite organizational concerns that home working would resound negatively for professionals who were unavailable to participate in important, hastily arranged, face-to-face meetings. Perceived gains may be tempered by employer perceptions that telecommuters could abuse their freedom to work at home.

Telecommuting can thus prompt managers to fear a loss of control over subordinates (Fairweather, 1999), particularly their ability to observe, police, and interact with the worker. In addition, telecommuting requires managers to make structural accommodations in the workplace (Fritz, Narasimhan, & Rhee, 1998), maintain closer supervision of work schedules to ensure adequate workload coverage (Powell & Mainiero, 1999), and learn more about the worker's personal and work habits (Pearlson & Saunders, 2001). Peters and den Dulk (2003) conclude that workers' time–spatial flexibility is likely to increase managerial uncertainty about whether telework is being done correctly, given the increased latitude for employees to behave in an untrustworthy way.

A recent survey of executives in major U.S. corporations, for instance, found that 26% thought that telecommuting could compromise job performance, whereas 36% saw no difference in productivity levels between onsite employees and telecommuters ("Out of Sight, Out of Mind," 2001). Researchers (e.g., Fairweather, 1999) thus find that managers placing greater trust in their employees are more sanguine about adopting telework, although fears regarding such use could set potential teleworkers, employers, and trade unions against telework, whether justified or not.

SOCIAL CHANGE IMPLICATIONS

As the knowledge economy unfolds, the pervasiveness of information technologies (ITs) is promoting relationships in the business world to become less personal, more formal, and more task oriented (Zolkiewski & Littler, 2003). Such downsides may help answer an age-old question, posed by Ruppel and Howard (1998), which inquires, "Why does an organization not adopt a potentially beneficial innovation such as telework?" (p. 6). The findings reviewed here should help us obtain clues as to why this diffusion has been so slow, but they also point to positive trends (e.g., enabling technologies) that promise more rapid adoption in the years to come.

The benefits of telecommuting are many, with Fairweather (1999) concluding that workers are separated from the distractions of the physical office and no longer suffer from the fatigue that accompanies physical commuting. To this, we might add the benefits of cost, in fiscal as well as environmental terms. Local and federal government entities have long promoted telework in order to realize the benefits in terms of fuel savings, reduced congestion, and reduced air pollution (e.g., U.S. Department of Labor, 2001).

Organizational Paradoxes

Pearlson and Saunders (2001) suggest that telework has not caught on, in part, because of the paradoxes associated with it: a focus on individuals as well as teamwork, an increase and decrease in control, and an increase in both structure and flexibility for the worker. The authors conclude that paradoxes in control can be ameliorated by focusing on the work to be done, rather than on the worker, and by not mandating structure to address employee flexibility (i.e., agreeing on what would be done when, but not how).

In this emerging era of telecommuting, then, the discipline of assuring work productivity will shift from the manager to the worker. As Tietze and Musson (2003) conclude, "Home based teleworkers need to rely on their self-discipline to devise a temporal structure for their days spent at home. In the absence of any external default signals, they have to act as both control and executing authority of internalized time-disciplines at the same time" (p. 438). Simply put, these workers are both the supervisors and the supervised; to the extent that one might prove to be one's own harshest taskmaster, the process of self-monitoring may result in a work day that does not ever fully end.

One proposed remedy for the prospects of isolation facing teleworkers is discussed by Goelman (2004), who outlines a "co-workplace," or telework center located near a population of workers, providing them the same benefits of working near home while reducing the loneliness and alienation associated with telework. These centers also may help overcome some of the employer concerns about the lack of accountability or potential for surveillance characteristic of home-based teleworkers.

Organizational Implications

If telecommuting became commonplace, how might society be impacted? Using the diffusion of enabling telematic technologies as a guide, we might expect that the impact of telecommuting would be more "evolutionary" rather than "revolutionary." Critical scholars of technology (e.g., Toffler, 1980) prophesize dystopian views where innovations like telework facilitate "electronic cottage" industries, where isolated home workers are denied the benefits of collective bargaining and remain at the mercy of employers in oppressive, low-wage jobs. But much like Orwell's (1948) prophecies of "Big Brother" surveillance by the 1980s, these dark sides of the communication revolution have yet to be realized, and the literature already discussed here suggests that the benefits of technology thus far outweigh the costs.

Read (2003) suggests that call centers engaging in telework are able to "lessen turnover, raise productivity, and cut costs" within a field that has turnover rates as high as 75% to 100%. As a consequence of this emerging telework phenomenon, a nascent home office market is surfacing to supply these at home workers.

Roosevelt (2002) recounts how "Toll Brothers"— the nation's biggest builder of luxury houses—estimates that nearly all of its new units now have home offices or studies; this figure is up from about 75% a decade earlier. The time-shifting benefits accruing to workers seeking a balance in their family lives—outlined at the onset and in Table 5.1—present clear quality-of-life benefits to telecommuters. Even professionals handling security-intensive tasks (e.g., intelligence), heretofore barred from telecommuting on security grounds, may soon enjoy greater mobility owing to state-of-the-art encryption technologies (e.g., Rhoads & Silver, 2005).

In addition to office spaces, the electronics and IT machines that facilitate home work are also generating an economic impact. With the diffusion of personal computers increasing nearly 1000% during the 1980s and the Internet growing at a similar pace in the 1990s—to say nothing of expansion in the broadband and office supply markets—the constraints to the growth of telecommuting are no longer technological (Lin & Atkin, 2002; Teschler 1991). A question remains, however, as to who ultimately will pay for this new office infrastructure. Although the availability of such resources has been a barrier to past adoption of telecommuting, the continuing decline in costs for facilitating technologies (e.g., computers) bodes well for future adoption. By way of summarizing the organizational implications of telecommuting, the "pros and cons" of telecommuting are outlined in Table 5.1.

Social Implications

The innovation of telecommuting will bring with it immutable social change, just as one of its facilitating technologies—the telephone—did a century before. In particular, when exploring the impact of telework on organizational communication patterns, we can expect to see changes reminiscent of the early telephone—the first instance of telecommuting—about which Bates, Washington, and Jones (2002) noted:

> Where cities had once been created to facilitate the process of human communication, the telephone (aided and abetted by the automobile) made physical proximity optional. The city of local neighborhoods—where social, work and commercial needs were all within walking distance—became the city of segregated districts (financial, industrial, commercial, residential). The telephone prompted the population shift out of the inner cities while helping to expand business operations within cities. (p. 70)

Telecommuting may take this process one step further, enabling professionals to live in even the most remote, picturesque areas (e.g., Vail, Colorado), with the aid of telematic technologies via both wired and wireless telecommunication networks. As Britton et al. (2004) suggest, the two chief consequences of the introduction of information technology into the workplace—dispersal of the

TABLE 5.1

Advantages and Disadvantages of Telecommuting

Pros	*Cons*
Reduces traffic congestion and air pollution (Tremblay, 2003; Kurland, 1999; Westfall, 2004)	Inability for managers to supervise program/employees (Fairweather, 1999)
Better balanced work and family (Tremblay, 2003; Taylor 2005; Potter, 2003; Mann, 2003)	Isolation of employees (McCall, 2004; Harpaz, 2002; Kurland, 1999; Mann, 2003)
For companies, reduced expense for work space and other expenses associated with office-bound work. (Fox, 1996; McCall, 2004; Harpaz, 2002)	Employer must find employees who can work productively without supervision. (McCall, 2004; Harpaz, 2002)
Increases employee productivity (Fox, 1996; McCall, 2004; Potter, 2003; Kurland, 1999; DuBrin & Barnard, 1993)	Safeguarding network security, trying to maintain corporate culture (Potter, 2003)
Gives organizations access to national talent (McCall, 2004)	Can create ambiguity in employee performance expectations (Potter, 2003)
Gives flexibility to employees (Fox, 1996; Potter, 2003; Taylor 2005; Harpaz, 2002; Kurland, 1999; Mann, 2003)	Can create uncertainties in employer-employee relationships. (Potter, 2003)
Employees can avoid potential work threats (Potter, 2003)	No separation between work and home (Harpaz, 2002; Mann, 2003; Ward, 2001)
Saves commute time/expenses for employees (Potter, 2003; Kurland, 1999; Mann, 2003)	Possible damage to commitment to organization (Harpaz, 2002)

(Continued)

TABLE 5.1 (Continued)

Advantages and Disadvantages of Telecommuting

Pros	*Cons*
Attracts and retains talented employees (Potter, 2003)	Impeded career advancement (Harpaz, 2002; Mann, 2003)
Increased autonomy and independence for employees (Harpaz, 2002)	Investment needed in training and new supervision methods (Harpaz, 2002)
Decreases in absence levels (Potter, 2003; Harpaz, 2002)	Lack of professional support (McCall, 2004; Harpaz, 2002; Kurland, 1999; Mann, 2003)
Increases motivation/satisfaction of employees (Harpaz, 2002)	Creation of a detached society (Harpaz, 2002)
Creates positive organizational images (Harpaz, 2002)	Interruptions and distractions from family, etc. (DuBrin & Barnard, 1993; Ward, 2001)
Employees more satisfied with general working conditions (DuBrin & Barnard, 1993)	Difficult and unfavorable task demands, unstable work flow. (DuBrin & Barnard, 1993)
Better-rested employers can avoid the fatigue associated with physical commuting and are removed from many office distractions (Fairweather, 1999)	Potential for "electronic cottage" via lengthened work day, lack of employee opportunity to collectively organize (e.g., Schiller, 1981)

workforce and individualization of work—including teleworking—may undermine a key rationale for the city: collectivizing large numbers of people in order to choreograph their work. Even so, their own work indicates minimal influences of telework on corporate decisions to locate in a city center, suggesting that these changes may be evolutionary rather than revolutionary.

On balance, the research reviewed in this chapter profiles a telecommuting innovation that remains relatively foreign to many companies and employees, one whose prospects for adoption are a function of a company's structure and line of work. Aside from situational factors, perhaps the most important determinant of telework's success lies in the presence of commitment to the innovation, on the behalf of the company as well as the employee. The latter must be willing to display an ability to keep work separate from domestic affairs, avoid feelings of isolation, and remain involved in the company from a distance. For its part, the company needs to display a measure of trust in the worker and, in the absence of office supervision, a willingness to support and cooperate with the worker when task equivocality may be high.

As the technology facilitating telework continues to increase, alongside the proportion of the workforce that can benefit from it, prospects for the diffusion of this innovation should improve. However, as Tremblay (2003) notes, emerging media applications like telecommuting may accelerate the use of self-employment and subcontracting outside of the firm. It will be important, then, for social planners to ensure that telecommuting can help supplement—rather than supplant—a skilled, well-paid workforce.

REFERENCES

Anonymous. (2005). Telecommuters numbers growing. *Furniture Today, 0,* 7. Retrieved January 30, 2005, from http://www/il.proquest.com/proquest/

Apgar, M. (1998, May/June). The alternative workplace: Changing where and how people work. *Harvard Business Review, 76*, 3–16.

Atkin, D., Lau, T. Y., & Lin, C. A. (2006). Still on hold: Prospects for competition in the wake of the Telecommunication Act of 1996 on its 10 year anniversary. *Telecommunications Policy*, 30, 80–95.

Bailey, D. E., & Kurland, N. B. (2002). A review of telework research: Findings, new directions, and lessons for the study of modern work. *Journal of Organizational Behavior, 23*(4), 383–400.

Baltes, S. (2004). Teleworking continues to grow, but some employers still leery. *Des Moines Business Record, 22*(47), 1. Retrieved February 1, 2005, from http://proquest.umi.com/

Bates, B. J., Albright, K., & Washington, K. D. (2002). Not your plain old telephone: New services and new impacts. In C. A. Lin & D. J. Atkin (Eds.), *Communication technology and social change: Audience adoption and uses* (pp. 91–124). Cresskill, NJ: Hampton.

Belanger, F. (1999). Communication patterns in distributed work groups: A network analysis. *IEEE Transactions on Professional Communication, 42*(4), 261–275.

Belson Goluboff, N. (1996). Telecommuting and alternative work arrangement that really works. *Complete Lawyer.* Retreived February 3, 2005, from http://www.abanet.org/genpractice/compleat/sp9gol.html

Bently, K., & Yoong, P. (2000). Knowledge work and telework: An exploratory study. *Internet Research, 10*(4), 346.

Bridgmon- Kavila, L., Burke, M., & Fernandez de Jauregui, L. (2005). *Telecom program at SUNYIT.* http://www.tele.sunyit.edu/teleworking/Factors.html

Cascio, W. F. (2003). Managing a virtual workplace. *Academy of Management Executive,* 14(3), 81–91.

Conroy, G. (2005). *The Gazette.* http://user.itl.net/~gazza/telecomm.htm

Cooper, R.C. (1996). Telecommuting: The good, the bad, and the particulars. *Supervision.* 57, 10–19.

Desrochers, S. (2005). *Boston College.* http://www.bc.edu/bc_org/avp/wfnetwork/rft/wfpedia/wfpBBTent.html

Dess, G., Rasheed, A. M. A., McLaughlin, K. J., & Prienm, R. L. (1995). The new corporate architecture. *Academy of Management Executive,* 9, 7–18.

Deutscher, H. (2005). *Prairie Public Television.* http://www.prairiepublic.org/features/ebusiness/future/telecommuting.html

Di Martino, V., & Wirth, L. (1990). Telework: A new way of working and living. *International Labour Review, 129*(5), 529–554.

Dimitrova, D. (2003). Controlling workers: Supervision and flexibility revisited. *New Technology, Work, and Employment, 19*(3), 181–195.

DuBrin, A., & Barnard, J. (1993). What telecommuters like and dislike about their jobs.*Business Forum, 18*(3), 13–17.

Duxbury, L., & Neufeld, D. (1999). An empirical evaluation of the impacts of telecommuting on intra-organizational communication. *Journal of Engineering and Technology Management, 16*(1), 1–28.

Ellison, N. B. (1999). Social impacts: New perspectives on telework. *Social Science Computer Review, 17*, 338–356.

Fairweather, N. B. (1999). Surveillance in employment: The case of teleworking. *Journal of Business Ethics, 22*, 39–49.

Flexible work grows as a work/life solution. (2004, October). *HR Focus, 81*(10), 1–2.

Fox, A. (1996). Home is where the office is (a look at the pros and cons of the growing phenomenon of teleworking, and ideas for making a success of your organizations work-at-home program). *CMA, 70*(1), 19.

Francese, P. (2002, February). The American work force. *American Demographics,* pp. 40–41.

Fritz, M. W., Narasimhan, S., & Rhee, H. (1998). Communication and coordination in the virtual office. *Journal of Management Information Systems, 14*(4), 7–28.

Gainey, T. (1999). Telecommutings impact on corporate culture and individual workers: Examining the effects of employee isolation. *SAM Advanced Management Journal, 64*(4), 4. Retreived January 28, 2005, from http://web6.infotrac.galegroup.com/itw/infomark/660/299/60292799w6/4!help_InfoMark

Gainey, T., Kelley, D., & Hill, J. (1999). Telecommutings impact on corporate culture and individual workers: Examining the effect of employee isolation. *SAM Advanced Management Journal, 64*(4), 4–10.

Gao, G. (2005). *Operations and Information Management, The Wharton School, University of Pennsylvania.* http://opim.wharton.upenn.edu/~lhitt/telecommute.pdf#search=\telecommuting,%20theory

Gurstein, P. (2001). *Wired to the world, chained to the home.* Vancouver: UBC Press.

Handy, S. L., & Mokhtarian, P. L. (1995). Planning for telecommuting: Measurement and policy issues. *Journal of the American Planning Association, 61,* 99–111.

Harpaz, I. (2002). Advantages and disadvantages of telecommuting for the individual, organization, and society. *Work Study,* 51 (2/3), 74–80.

Harris, L. (2003). Home-based teleworking and the employment relationship: Managerial challenges and dilemmas. *Personnel Review. 32* (4), 422–39. Retreived February 01, 2005, from http://proquest.umi.com/

Herschel, R.T., & Andrews, P.H. (1997). Ethical implications of technological advances on business communications. *Journal of Business Communication*, 34(2), 160–170.

Hill, E. J. The perceived influence of mobile telework on aspects of work life and family life: An exploratory study (Ph.D. dissertation,Utah State University, Logan, UT). *Dissertation Abstracts International, 56*(10), 4161A. (University Microfilms No. DA9603489).

Hill, E. J., Ferris, M., & Martinson, V. (2003). Does it matter where you work? A comparison of how three work venues (traditional office, virtual office, and home office) influence aspects of work and personal/family life. *Journal of Vocational Behavior, 63*(2), 220–241.

Hudson, N. (2005). *TELEWORKanalytics international, inc.* http://www.teleworker.com/papers/economics.html

Hylmo, A. (2002). Telecommuting as viewed through cultural lenses: An empirical investigation of the discourses of utopia, identity, and mystery. *Communication Monographs, 69*(4), 329–357.

Illegems, V. (2004). Telework: What does it mean for management? *Long range Planning, 37*(4), 319. Retrieved January 30, 2005, from http://www/sciencedirect.com/science?_ob=HomePageURL&_method=userHomePage&_btn=Y&_acct=C000029718&_version=1&_urlVersion=0&_userid=582538&md5=6fbf3e3e7572ec1a2c5c64e113ccf161

Illegems, V., & Verbeke, A. (2003). *Moving towards the virtual workplace: Managerial and societal perspectives on telework.* Cheltenham, UK: Edward Elgar.ITAC. (2003). Home-based telework by U.S. employees grows nearly 40% since 2001. http://www/telecommutect.com/content/homebased.htm

Jablin, F., & Putnam, L. (1999). *Handbook of organizational communication (2nd ed.).* Thousand Oaks, CA: Sage.

Karnowski, S., & White, B. J. (2002). The role of facility managers in the diffusion of organizational communication. *Environment and Behavior, 34*(3), 322–334.

Khan, M., Tung, L., & Turban, E. (1997). Telecommuting: Comparing Singapore to Southern California. *Human Systems Management, 16*(2), 91–98

Kistner, T. (2004). Mixed messages of teleworks future. *Network World.* Retrieved February 02, 2005, from http://web6.infotrac.galegroup.com/itw/infomark/660/299/60292799w6/8!help_InfoMark

Kowalski, K. B., & Swanson, J. A. (2005). Critical success factors in developing teleworking programs. *Benchmarking: An International Journal, 12*(3), 236–249.

Kraut, R. (1989). Telecommuting: The trade-offs of home work. *Journal of Communication, 39*(3), 19–47.

Kraut, R., Galegher, J., Fish, R., & Chalfonte, B. (1992). Task requirements and media choice in collaborative writing. *Human-Computer Interaction, 7*, 375–407.

Kurland, N., & Bailey, D. (1999). Telework: The advantages and challenges of working here, there, anywhere, and anytime. *Organizational Dynamics, 28*(2), 53–68.

Kurland, N. B., & Egan, T. D. (1999). Telecommuting: Justice and control in the virtual organization. *Organization Science*, *10*, 1–31.

Langhoff, J. (2005). *June Langhoffs Telecommuting Resource Center.* http://www.langhoff.com/faqs.html#growth

Lim, Vivien K. G., & Teo, Thompson S. H. (2000). To work or not to work at home: An empirical investigation of factors affecting attitudes towards teleworking. *Journal of Managerial Psychology, 15*(6), 560–586.

Madsen, S. (2003). The effects of home-based teleworking on work-family conflict. *Human Resource Development Quarterly, 14*(1), 35–38.

Mann, S., & Holdsworth, L. (2003). The psychological impact of teleworking: Stress, emotions, and health. *New Technology, Work, and Employment, 18*(3), 196–211.

McCall, K. (2004). Remote control: Allowing sales reps to work off-site definitely has its perks, but before you try it, be sure telecommuting makes sense for your business. *Entrepreneur, 32*(12), 92–94.

McClean Parks, J., Kidder, D. L., & Gallagher, D. G. (1998). Fitting square pegs into round holes: Mapping the domain of contingent work arrangements onto the psychological contract. *Journal of Organizational Behavior, 19*, 697–730.

Mirchandani, K. (1999). Legitimizing work: Telework and the gendered reification of the Work–nonwork dichotomy. *The Canadian Review of Sociology and Anthropology, 36*, 87–107.

Mokhtarian, P. (2005). *Telecommunications and Travel Behavior Research Program.* http://www/its.ucdavis.edu/telecom/r11/litrev.html

Montero, P. (2003). 5 steps to approaching your current employer about telecommuting. *YouCanWorkFromAnywhere.com* Retrieved January 30, 2005, from http://www.youcanworkfromanywhere.com/articles/pf_approach_employer.htm

Montero, P. (2004). Democratic convention traffic nightmare drives telecommuting solutions. *You Can Work From Anywhere.* Retrieved January 20, 2005, from http://www.youcanworkfromanywhere.com/press/2004/release040621.htm

Moskowitz, R. (2005). *Valley People.* http://www/yoyow.com/rmoskowitz/telecommuting157.html

Neufeld, D. (2005). *IEEE Computer Society.* http://csdl.computer.org/comp/proceedings/hicss/2004/2056/01/205610043c.pdf#search=\telecommuting,%20theory

Nilles, J. M. (1996). What does telework really do to us? *World Transportation Policy & Practice, 2*(1–2), 15–23.

Nilles, J.M., Carlson, F. R., Gray, P., & Hanneman, G. (1976). *The telecommunications transportation tradeoff: Options for tomorrow*. New York: John Wiley.

Nilles, J. M. (1998). *Managing telework: Strategies for managing the virtual workplace.* New York: Wiley.

Olson, M. H. (1987). Telework: Practical experience and future prospects. In R. Kraut (Ed.), *Technology and the transformation of white-collar work* (pp. 135–152). Hillsdale, NJ: Lawrence Erlbaum Associates, Inc.

Out of sight, out of mind? Telecommuting receives mixed reviews from executives, survey Shows. (2001, February 8). Office team press release, www.officeteam.com.

Pearlson, K. E., & Saunders, C.S. (2001). There is no place like home: Managing telecommuting paradoxes. *Academy of Management Executive*, *15*(2), 117– 129.

Pelton, J. N. (1981). The future of telecommunications: A delphi survey. *Journal of Communication, 31*(4), 177–189.

Pelton, J. N. (2004). To telecommute, or not to telecommute. *The Futurist, 38*(6), 56. Retrieved January 28, 2005, from http://web6.infotrac.galegroup.com/itw/infomark/660/299/60292799w6/8!help_InfoMark

Peralta, C. (2002). The shape of digital things to come. *Planning, 68*(9), 24–26. Retrieved February 1, 2005, from http://web6.infotrac.galegroup.com/itw/infomark/660/299/60292799w6/8!help_InfoMark

Peters, P., Tijdens, K. G., & Wetzels, C. (2004). Employees opportunities, preferences, and practices in telecommuting adoption. *Information and Management, 41*, 469–482.

Potter, E. (2003). Telecommuting: The future of work, corporate culture, and American society. *Journal of Labor Research, 24*(1), 73–84.

Raines, J., & Leathers, C. (2001). Telecommuting: The new wave of workplace technology will create a flood of change in social institution. *Journal of Economic Issues, 35*(2), 307–313.

Reymers, K. (2005). *The Infospace of Kurt Reymers.* http://www/acsu.buffalo.edu/~reymers/telecomm.html

Reinsch, Jr., N. L. (1997). Relationships between telecommuting workers and their managers: An exploratory study. *Journal of Business Communication, 34*(4), 343–369.

Reinsch Jr., N. L. (1999).Selected communication variables and telecommuting participating decisions: Data from telecommuting workers. *Journal of Business Communication, 36*(3), 247–260.

Rhoads, C., & Silver, S. (2005, Dec. 29). Working at home gets easier. *Wall Street Journal*, p. B4.

Rose, K. (1998, Autumn). Work/life flexibility: A key to maximizing productivity. *Compensation and Benefits Management*, pp. 27–32.

Ruppel, C. P., & Harrington, S. J. (1995). Telework: An innovation where nobody is getting on the bandwagon? *Data Base Advances, 26*(2 &3), 87–104.

Ruppel, C., & Howard, G. (1998). Facilitating innovation adoption and diffusion: The case of telework. *Information Resource Management Journal, 11*(3), 5–15.

Savage, J. A. (1990). Taking the "place" out of workplace [Electronic version]. *Computerworld, 24*, 67.

Sharit, J., Czaja, S. J., Hernandez, M., Yang, Y., Perdoma, D., Lewis, J., Lee, C. C., & Fair, S. (2004). An evaluation of performance by older persons on a simulated telecommuting task. *Journal of Gerontology, 59*, 305–316.

Stanek, D. M., & Mokhtarian, P. L. (1998). Developing models of preference for home-based telecommuting: Findings and forecasts. *Technology Forecasting and Social Change, 57*, 53–74.

Stipp, H. (1998). Should TV marry PC? *American Demographics, 20*, 16–21

Streeter, L. A., Kraut, R. E., Lucas, Jr., & Caby, L. (1996). How open data networks influence business performance and market structure. *Communication of the ACM, 39*(7), 62–73.

Symons, F. (2000). Telework and bandwidth. *Canadian Journal of Communication, 25*(4), 553.

Taylor, C. (2005). Life in the balance. *Incentive, 179*(1), 16–19.

Telecommuting grows. (2001). *Wall Street Journal*, p. A1.

Teleworking: An assessment of socio-psychological factors. *Facilities, 19*(1/2), 61.

Tietze, S., & Musson, G. (2003). The times and temporalities of home-based telework. *Personnel Review, 34*(4), 438–533.

Toffler, A. (1980). *The third wave*. New York: William Morrow.

Townsend, A. M., DeMarie, S. M., & Hendrickson, A. R. (1998). Virtual teams: Technology and the workplace of the future. *Academy of Management Executive, 12*, 17–29.

Trembley, D. G. (2003). Telework: A new mode of gendered segmentation? Results from a study in Canada. *Canadian Journal of Communication, 28*(4), 461.

Ward, N., & Shahaba, G. (2001). Teleworking: An assessment of socio-psychological factors. *Facilities, 19*, 459–472.

Watad, M., & Will, P. (2003). Telecommuting and organizational change: A middle-managers perspective. *Business Process Management Journal, 9*(4), 459–472.

Wellman, B., Salaff, J., & Dimitrova, D. (1996). Computer networks as social networks: Collaborative work, telework, and virtual community. *Annual Review of Sociology, 22*, 213–218.

Westfall, R. (2004). Does telecommuting really increase productivity? *Communication of the ACM, 47*(8), 93–96.

Westfall, R. (2005, February 1). *Ralph D. Westfall.* http://www/cyberg8t.com/westfalr/dis_abst.html

Westfall, R. (2005, February 2). *European Telework Online.* http://www/eto.org.uk/faq/faq03.htm

Westfall, R. (2005, February 2). *United States of America Department of Transportation, National Transportation Library.* http://ntl.bts.gov/DOCS/telecommute.html

Westfall, R. (2005, February 3). *International Telework Association and Council (ITAC).* http://www/telecommute.org/news/pr090204.htm

Westfall, R. (2005, January 21). *General Services Administration.* http://www/telework.gov/index.asp

Westfall, R. (2005, January 22). *Answers.com.* http://www/answers.com/telecommuting&r=67

Westfall, R. (2005, January 29). *The Transportation Management Association Group.* http://www/tmagroup.org/TelGuide.html

Whincup, J. (1997). Be armed with the right tools. *Computing Canada, 23*(6), 38. Retrieved January 30, 2005, from http://web6.infotrac.galegroup.com/itw/infomark/840/732/60292702w6/5!help_InfoMark

Wicks, D. (2002). Successfully increasing technological control through minimizing workplace resistance: Understanding the willingness to telework. *Management Decision, 40*(7), 672–681.

6

Information Technology: Analyzing Paper and Electronic Desktop Artifacts

Ronald E. Rice and Sara Schneider

A new organizational information/communication technology (ICT) brings changes to an individual's processes, work practices, and organizational processes. Especially challenging is the transition from a set of established and physical work processes to a set of new and cognitive work processes. These changes may create different job demands, and serious conflicts and misalignments for work practices. Such change is often unsettling and requires effort on the part of the individual to cope, interpret, adjust, and resolve. This chapter analyzes the role of desktop artifacts as an organizational communication medium about, and symbolic indicators of, these misalignments and adjustments. It uses examples from the case of an electronic document management system, available through desktop personal computers (PCs), implemented to replace paper-based work flow and a batch-oriented mainframe system. Conceptual analyses of the misalignments and adjustments in this setting identified seven categories of desktop artifacts and four conceptual dimensions of desktop artifacts. By understanding the forms, uses, and significance of desktop artifacts—as well as some of their disadvantages or misuses—better designed information systems could be developed, and researchers could better understand how people adjust to changes in organizational systems by communicating with and through artifacts.

BACKGROUND

Along with oral communication (such as face-to-face meetings), documents are a pervasive communication channel in organizations. Because of computing and networking technology, most documents can now either be initially created in electronic form or converted from paper to electronic form. *Electronic document management* generally refers to the processing of documents as digital data files, in text or image form, or a combination of the two. If documents are initially created electronically on a computer, then that computer file can be indexed, stored, searched for, retrieved, and disseminated across organizational units and locations. Documents may even be used as subsequent transaction triggers, whereby some aspect of the document activates a program to perform some other task. This may occur through a work-flow system, in which when one electronic document or form is completed, the system makes that document available to the next person in the work process flow to use. Or, the document may be designed as a highly structured form in which some areas of the document represent database fields, and the data entered in those fields become input to a subsequent process. For example, when one checks specific boxes on a form and faxes this completed form back to a company, the receiving system can scan it and then perform certain functions depending on which boxes are checked (e.g., sending the originator some specific materials or product). In this sense, documents can become one type of internal organizational communication-facilitation medium, as well as a component of an integrated information system process.

Similarly, if printed documents are scanned in and stored as an electronic image on a computer, the textual portions of the scanned page image can be converted to electronic text through an optical character recognition (OCR) device. The text images that are processed through an OCR device are typically formatted using the code that is commonly used by a computer's default word processor (e.g., Word or Wordperfect). Hence, the OCR converted text document is also available for additional processing as a computer file in subsequent information-processing tasks.

A familiar computer-based example is a document posted on the Internet or sent as an e-mail attachment in ".pdf" format, that is, "portable document format," as an image file. The user can read and print the document, but cannot usually edit or process it, so it is a useful format for distributing final reports or copyrighted documents.

Besides improving storage, retrieval, and accuracy, shifting from paper to a computer system also presents significant benefit of reducing storage and delivery costs of documents (Sellen & Harper, 2003, p. 28). For example, almost 3% of all paper documents are misfiled, 8% are eventually lost, and one third of all forms are obsolete before they are used. Electronic document images can be tagged and processed, viewed simultaneously by multiple users across terminals and communication networks, and distributed much more rapidly and at less cost than paper documents. See Sprague (1995) for a review of electronic document management, and Åborg and Billing (2003) for a discussion of associated physical and psychosocial disorders such as increased workload, decreased autonomy, decreased physical movement, and system use problems.

The shift from paper to paperless offices seems, then, to be an inevitable byproduct of the process of "becoming digital," a hallmark of the emerging knowledge economy (e.g., Negroponte, 1995). Tenner (1998), however, suggests that there are practical reasons for the persistence and growth of paper, including that information growth is exponential; paper—especially acid-free—lasts longer than computer memory and media storage; paper backup is required by many laws and regulated procedures; personal files are made more possible by photocopying, compared to the prior, now obsolete, carbon copies; and digital photocopiers now also function as primary printers.

THEORETICAL EXPLANATIONS

Two concepts seem especially useful in understanding the user and organizational changes that may be associated with the implementation of an electronic document management system. The first is the *transformation of documents from the realm of physical processing of paper to the realm of cognitive processing of symbols*. The second is the role that *desktop artifacts play in communicating about the progress and implications of this transformation*. Brief theoretical explanations of these two concepts follow.

Transforming From the Physical to the Cognitive Realm

Kendall and Kendall (1992) define an *information system* as an entity composed of people, software, and hardware, which together support a broad spectrum of organizational tasks including decision making and analysis. Early on in the diffusion of office information systems (especially desktop work stations and PCs), Weick (1985) argued that "electronic processing has made it harder, not easier, to understand events that are represented on screens" (p. 51). Human understanding of events and information relies not only on an account of the information such as letters or numbers displayed on the computer screen, but on the whole event, including the extra-event information such as procedures and tangible items (Suchman, 1995).

When users must process information by different and new means or procedures (such as in digital form, represented by file names, icons, images, folder/file directories, file types, storage and retrieval protocols, etc., through the display interface of a computer), this creates new demands and often a vague sense of unease (Weick, 1985). The information access, processing, or transfer is no longer primarily physical (involving the paper), but rather is primarily cognitive. Such a change involves not only different media and formats (such as for entering data or reading reports), but also a different set of behavioral skills. For example, the protocol, formats, and styles used in composing, processing, filing, and retrieving a paper memo are different from writing, sending, reading, and filing electronic mail, representing different organizational communication genres (Yates & Orlikowski, 1992).

Moreover, the shift from paper-based work flows to computer-based electronic information and processes provides opportunities to move from simply *automating* a job to *informating* it (Zuboff, 1985). Automation is the process of using technology to replace functions formerly done by humans. It does not use any processing potential to gather information about the function, the participants, or how well the work process is performed. Informating involves collecting information about the processes (meta-information) and making that available to other work processes, to the users, and to other organizational members. Informating increases the control over, and efficiency and effectiveness of, work practices, such as by providing the user with information about the job, one's performance, and the client (Davidow & Malone, 1993). Without gathering and analyzing this information, the complexities and interdependencies of a new system may make a job less comprehensible and create conflicts between preceding and subsequent work processes.

The Desktop and Its Artifacts—A Context for Understanding Information System Use

The major component of a person's work environment is the physical work space, the area that houses a worker's furniture, supplies, equipment, decorative items, and any other items in the physical work space (Sundstrom & Sundstrom, 1986). The ability for people to personalize the spaces they use (Sommer, 1969)—by changing and having some control over their immediate physical environments (Archea, 1977; Lucas, 1991)—is important because the physical and social work environment affects an individual's attitudes, behavior, and perceptions (Barker & Associates, 1978; Zalesny & Farace, 1987). The desktop represents an employee's personal task and processing space (for better or worse, ranging from a dynamic and virtual organizational network, to a monitored cell-like cubicle; Malone, 1983).

Artifacts (such as furniture, paper, office accessories, and clothes) can "communicate information about the organization and the people who work there" (Davis, 1984, p. 277). In this chapter, the term *artifact* refers both to physical objects and to the symbolic byproduct of other processes and phenomena (Rice, 1999). That is, artifacts are both the material medium (such as a Post-It note) and the social constructions of the medium—and it exists largely because of some other social circumstance or phenomenon (such as a Post-It note representing a warning about a specific undocumented problem with an ICT). Another way to think of this is that artifacts—and media technologies generally—represent both channels for exchanging messages, and a symbolic message themselves (Rice, 1987). That is, they serve both as signal (the denotative content) and symbol (the connotative cues; Feldman & March, 1981; Sitkin, Sutcliffe, & Barrios-Choplin, 1992).

Furthermore, artifacts are often crucial for the conduct of work—especially when tasks are interdependent (Suchman, 1995)—and are used to facilitate what

Gasser (1986) calls "augmentation" or "workarounds" (e.g., phenomena that can creatively achieve task efficiency or avoid potential pitfalls in the task processes). Indeed, some have argued that work practices should take advantage of media technologies to provide greater and more explicit support for sharing and communicating through and about visual artifacts (and for studying those processes; Suchman & Trigg, 1991).

Gibson (1979) introduced the concept of *affordance*—a possibility for action available through characteristics and uses of objects (in particular, technologies and media). Hartson (2003) extended the concept of affordances by distinguishing cognitive, physical, sensory, and functional affordances. However, regardless of what types of functional technical or technological "affordances" may be available, a particular user in a given context may not perceive, use, or value that affordance. Instead, a user of a new information-processing system may continue to use familiar technologies to accomplish their tasks, by adjusting, bypassing, changing, or reinventing aspects of a new system that do not meet their familiar affordances, even if those familiar technologies create significant costs, errors, and interdependencies (Goodman, Griffith, & Fenner, 1990; Johnson & Rice, 1987; Majchrzak, Rice, Malhotra, King, & Ba, 2000; Rice & Gattiker, 2000).

For example, paper use persists for rational, practical, emotional, and symbolic reasons. Paper has many different affordances, especially in combination with other technologies (such as pens or thumbtacks), supporting a wide variety of human actions (Sellen & Harper, 2003). Paper documents allow users to make notes, mark on them, and navigate or lay out the paper for different purposes flexibly, in addition to facilitating the coordination of action among organizational members. Paper, in the form of binders, reports, stacks on the desktop, and so on, can also serve highly important symbolic purposes (Feldman & March, 1981), to indicate, for instance, that the person sitting behind the desk is well prepared, organized, or has access to valuable information. Conversely, the absence of paper could convey that the person is of sufficiently high status that he or she does not need to manage paper, or is technologically savvy, because the desktop computer has replaced the paper.

So although a new technology such as electronic document management represents possibilities for positive organizational social change, current paper-based practices represent not only costs and obstacles to such change, but also highly significant, valid, and symbolic reasons for not changing. Attitudes of users toward a new information system play an important role in how well the users adjust to the system (Nelson, 1990; Rice & Aydin, 1991). These attitudes are formed by the individuals' experience with prior changes, their expectations about a new system, and their experience with the implementation process. Employees who feel that a new technology reduces their control or de-skills their jobs are more resistant prior to, and less satisfied after, the introduction of new technology (Capaldo, Raffa, & Zollo, 1995; Clement, Parsons, & Zelechow, 1991; Goodman, Griffith, & Fenner, 1990; Kraut, Dumais, & Koch, 1989;

Patrickson, 1986; Pava, 1983). Thus, we argue here that desktop artifacts communicate both signals and symbols about these forces for and against change, and about the transition between old and new ways of conducting organizational activities.

EMPIRICAL FINDINGS

To gain a better understanding about how workers assimilate these new paperless technologies in an organization, we examined individual adjustments associated with the implementation of a document imaging service. Information gathered from a specific organizational example such as this, although not generalizable across organizations, should help develop a richer understanding of the social change implications accompanying the diffusion of ICTs into the workplace.

Methods

Participants in this study work for a large company we'll call "Syndicate" in a division we'll call "AKA," an outsourcing service that provides customer support for business users of calling cards. AKA had conducted business in the past by receiving faxes—nearly 10,000 each week—into a central fax room and then, once an hour, distributing via cart the faxes to the appropriate customer service representatives ("reps") by depositing them in a "hot bin" on each representative's desk. As the reps processed these faxes, they documented their work, made a copy for the file room, and placed the original work in an "out bin." The cart person would then pick up the paperwork and take it to the next person, who would perform his or her service on the work, make a copy for the file room, and again place it in an out bin. This would continue until work was completed on the original fax. The result would be copied and filed. Subsequently, there would then be multiple occasions to retrieve the files, as both specific processes and the AKA–customer relationship continued across time.

When analyzing productivity and work procedures, the company determined that it had a problem with distribution and storage of paper and tracking errors. It then decided to adopt fax/document imaging technology. Incoming faxes would be scanned directly into a digital imaging system, and images of the faxes would be electronically indexed and stamped, and made available through a network to the appropriate customer representatives or other personnel. Moreover, the company also upgraded its customer service database processing system and hardware. It replaced remote work terminals on employees' desks—connected to a mainframe computer system (i.e., a very large-scale central processing system hereafter called OCS, for "old customer system")—with personal computers (PCs). These were linked, in turn, to a company-wide local area network (LAN) and a client-server system (hereafter called NCS for "new customer system"). All information about each customer account had to be transferred or converted from the old system to the new one.

NCS would work differently than OCS in many ways. OCS was a text-based system that required the user to type in specific commands at a system prompt on the terminal. All the data input from all users would be kept until the time came to process it in a batch processing fashion during the early morning hours by the mainframe computer system. NCS provided a graphical windows interface, with icons for applications and files, and cursor-controlled dialog boxes. Hence, OCS was used to enter data, whereas NCS was designed to almost instantaneously process and update records. Under the NCS system, card holders/customers are able to get usable cards in 2 to 24 hours, as opposed to a couple of days under the OCS system. AKA employees would have new PC-based graphical work stations, some with the OCS interface, some with the imaging interface, and some with both interfaces on two screens simultaneously.

The study design was based primarily on the work of Kendall and Kendall (1992), Malone (1983), and Weick (1985). Kendall and Kendall's structured observation of the environment (STROBE) approach provides a standard methodology and classification so that analysts may evaluate organizational elements and their influence on decision making. Deciding what desktop items to focus on was adopted from Malone's (1983) and Weick's work (1985).

The study sample was designed to represent different job functions from the five primary processes identified as areas most likely to be affected by NCS and Imaging. The participants were one volunteer from each of the five processes, with a gender ratio that was proportional to company personnel. No two participants had the same job function because the nature of their customer service accounts differed significantly and each required separate information handling procedures.

Four field visits to the AKA company were made to develop relationships with the study participants, gain a better understanding of their organization and tasks, and assess the three phases of the system implementation (two preimplementation visits, during implementation, and postimplementation). During each of these four visits, we interviewed the study participants for 1 hour and took pictures of their desktops. Participants were asked about the significance of the placement of certain items on the desktop, the perceived importance of those items, and any problems relating to the new information system. They were encouraged to discuss issues that were of concern to them, allowing insights into issues not initially identified on the structured interview guide.

Example Implications of System Change to Desktops and Work Processes

For each implementation phase, full analyses were made to identify and describe the desktop contexts and problems noted by the participants or identified through our

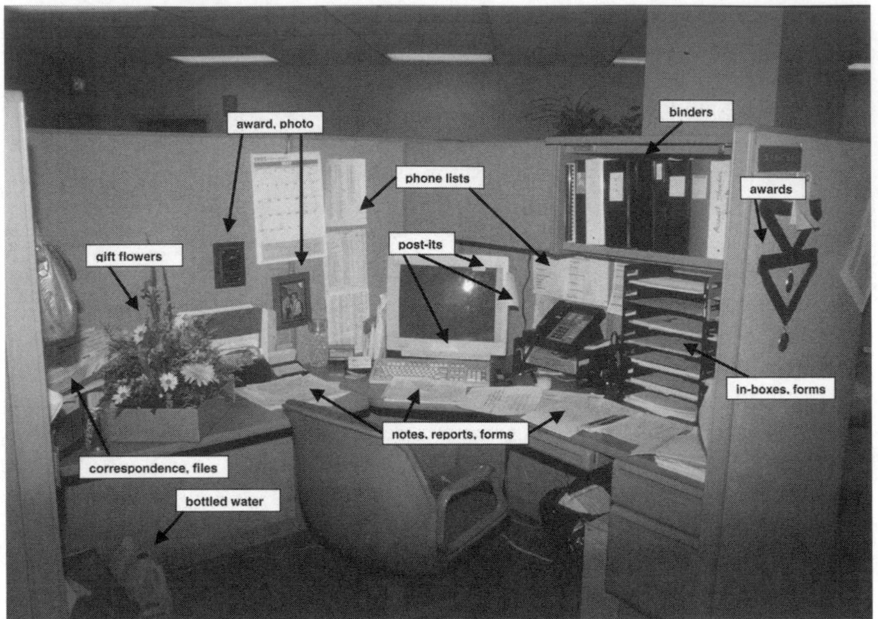

FIGURE 6.1 Example desktop

observations. The following provides a few illustrative examples, describing desktop artifacts, and underlying problems relating to the technical and conceptual switch from prior systems (both paper and computer) to a new (all computer) system.

Desktops. Each person's desktop included a variety of tools and artifacts, such as computer, telephone, Post-Its, binders, files, files, forms, lists, awards, personal photographs, materials stacked on desktop and floor, and so on. Figure 6.1 shows one person's desktop.

One addition to the system was the peer review. AKA has employees use peer reviews to evaluate the performance of their team members and leader. The old procedure involved downloading and printing the forms, filling them out, and then handing them in to the team leader. The new procedure allowed employees to fill out the forms online, and the team leader could later merge them for a final report (cutting and pasting were not necessary).

After several months, the implementation of imaging eliminated faxes arriving to study participants in the form of paper. One participant's old "hot bin" (which previously received new and important faxes delivered from the cart) was now used to hold that day's work to be accomplished. Her old "in bin" became storage for an important account. The other bins became holders for new and old forms. A new Post-It note was on her PC as a customer-specific reminder.

Another study participant had developed and stored a pile of paper—from scrap size to letter size, with all kinds of information that she may need—slightly in front and to the right of her desktop phone. Another change was that she put a Post-It note on her PC as a reminder for an experiment, which she and a client were performing to see how long the system took to respond when information was sent by electronic feed. After the experiment was complete, it stayed on the PC as a reminder to check the system for a different reason, but for the same client contact.

A third study participant had removed all but three of her Post-It notes from the prior implementation stages. She used her hot bin to temporarily store completed forms—a practice discouraged by the organization—which she used when entering data and managing accounts to reduce the uncertainty of changing to the new system. She also kept a Post-It note pad by her PC to write case numbers of faxes, whose images become obscured when the NCS window is opened simultaneously with the imaging system. That is, a flaw in system design generated the need for temporary paper artifacts.

Work Processes. When it introduced the new systems, the organization announced that it hoped to eliminate large amounts of paper as well as the errors and retrieval problems associated with it. However, all participants reported concerns that the amount of paper they had seemed to be increasing, with new binders and multipage handouts of system update information. For example, some Post-It notes were one visible indication that the new imaging system in fact offered no quick or easy reminders as to how to perform certain work functions.

Although the trial system was easier to view (because of its graphical interface) than the OCS, it required double and triple entry of the same information (such as customer contact name), whereas the old system automatically filled in the repeated fields. Finding filed faxes through the imaging system was difficult, although one of the motivations for the imaging system was to overcome difficulties in finding paper faxes. This is because when the fax was scanned into the imaging system, it was electronically "stamped" with a sequential case number that was different from the confirmation numbers possessed and used by the external clients. As both the study participants and their clients were accustomed to using the same confirmation numbers, the new system-assigned sequential case numbers created new retrieval problems instead of solving the old ones.

It often took longer to find information on NCS because the customer service representatives were less familiar with where the information was stored and the system prompts would not allow users to exit without completing every field on the screen. The new system's extensive reliance on a graphical user interface and the mousing function, in fact, slowed down the whole process, when compared to the quickness of executing the keystrokes for data entry used by the OCS system before the conversion.

Although the organization expected that processing time would be reduced significantly, the time required for NCS to interface with another processing system—the one that controlled the actual ordering of new calling cards by processing these orders in batch jobs each night—increased. As a result, the high-speed NCS actually created a bottleneck on this other processing system, which operates at a considerably slower speed. This particular system interdependency had not been thought of in advance, although it was eventually fixed. There was also an unexpected problem under NCS, one that involved a missing feature, which was needed to segment the main file into subfiles that could be easily retrieved, processed, and stored. This problem was caused by the interfacing system sending the company one large file that contained all the necessary information (such as updated information on the client companies, business calling card fees, etc.) each night. While under OCS, the system would receive the file and then break up the file into usable self-contained subfiles. Once discovered, the problem was corrected by giving NCS the capability to break down the file.

Several study participants reported that although the system seemed to run more rapidly, it would sometimes lock up, leaving the user unable to do anything. This was due to system problems associated with having more users on the system than it was designed to handle (which again was paradoxical given that another goal of the system was to eventually also allow all customers to access their own accounts online at any given time). The design team was trying to solve this problem.

There was yet another problem involving the system's inability to properly input or encode low-resolution or blurry information that came from the electronic feed of scanned paper images, among others. This problem created a backlog of paper for re-input and filing, resulting in the need to have the study participants (or customer service representatives) redo every transaction over the phone and on paper. Hence, the new system helped generate additional media-based transactions, as well as media transformations that had not existed before.

One study participant also reported that not all client contact names for each account properly transferred in conversion between the OCS and NCS. Consequently, if a client request was not a preestablished choice on the screen, the customer service representative would have to choose a different (incorrect) option to designate how the call came in and from whom, while adding an extra notation in the account to explain how the transaction was completed and by whom. Thus, system and design flaws created certain situations that generated not only more information and informal paper work than the traditional process, but also more opportunities for error and delay.

Conceptual Analyses

The rich over-time detail from the interviews, observations, and photographs provided the basis for two kinds of conceptual analyses, derived from the theoretical explanations described earlier. These included the transformation of paper documents from the physical to the cognitive, desktop artifacts (as signal and symbol, as medium and content), affordances, and adaptation/reinvention.

Typology of Seven Desktop Artifacts. By inspecting the participants' comments and our photographs of their desktops, we identified seven types of desktop artifacts. (a) *System and task-related information, such as procedures for processing work,* was usually provided by company management to the employees in the static form of paper (with unchanging information) in binders. However, the system and task-related procedures not only changed, but one of the goals of the new system was to make changes in procedures available more quickly, eventually online. (b) *Personal* items ranged from the static awards and family photographs to frequently changing notes about errands and social events (including coworker birthdays). (c) *Temporary* information included items generated by the management personnel (e.g., memos or one-page notices) and by the study participants (e.g., Post-It notes). This type of information could be generated while someone was waiting for something to be adapted/updated, a note about a file folder that was moving between offices, or an unscheduled system problem. (d) *Process-related* items were usually in the form of Post-It notes, generated because of a temporary change in system protocol (hence not included in the binders), or because the individual wanted a quick reference for how to execute a particular process. (e) *Frequency/importance of use* of a set of information was indicated in many cases by how close or far the paper information was relative to their central focus point—the PC (similar to what Malone found in his 1983 analysis of how people organized their desktops). So the PC becomes sort of a window on information, literally and figuratively, representing not just a focal point for information access and processing, but also a symbolic landmark for indicators of the importance of other paper information. (f) *Unsupported by the system* describes those items (usually Post-It notes) that individuals used to keep track of information that was unavailable on the system, deemed "difficult" to access (i.e., representing an error or a temporary solution), or unable to be seen or stored in the system.

The last of the seven types of desktop artifacts was (g) *reminders.* Reminders created by the study participants typically referenced a particular event or idea. Some information needed to be constantly available, both for immediate use as well as for potential future use. Important phone lists were created by all participants (in one case, made accessible to all employees through the LAN) and were shorter versions of the ones issued to all employees in a binder by the company. Even when information about a process or the system would come to participants on letter-sized paper, depending on the perceived importance of the information, some participants made themselves another reminder on a smaller piece of paper and posted that on their bulletin boards or elsewhere on their desktop.

When participants were asked whether they used the notepad feature on their PCs—which allowed users to make notes and store them on the system—each answered that it was inconvenient to use (noting that a reminder stored within the system's notepad feature was not really a reminder). All, even the most technical participant, said that it was more convenient to have a piece of paper that was visible by just turning one's head or moving one's eyes instead of opening another application on the computer.

Table 6.1 provides some examples of new desktop information artifacts that are appearing or old artifacts that are disappearing through the different implementation phases of the study period.

Four Conceptual Dimensions of Desktop Information Artifacts

1. Paper vs. Electronic Information. As discussed earlier, paper and computer files have different "affordances," which may be more or less useful to different people and different jobs. Despite the move to the imaging system and the graphical customer service system, paper still exists on the desktops of the AKA information workers. This is because the complex system commands made it difficult to locate and retrieve the computer versions and users could not compare documents side-by-side on the screen. Further, in some instances, new paper records and forms were created as a by-product of the implementation, to make up for limitations of the system or to provide an interface between the old and new systems. Hence, excess space (to store even worthless paper) is a cheaper resource than scarce time (to filter and evaluate and toss old paper). Groleau (1995) also found that even after computerization, office workers continued to use paper files that the new system was supposed to replace.

2. Materiality and Complexity of Information. As Weick (1985) and others noted, electronic information artifacts are no less pervasive or important than paper-based artifacts, but are less visible and tangible. Both electronic and paper artifacts manifest two conceptually different aspects of information: amount/presence (materiality), and complexity/uncertainty. With respect to *materiality*, too great an amount of material (such as paper) or electronic information (such as individual, compressed, or concatenated files) leads to problems of storage (desktop space, file cabinets, network memory capacity, or disk storage), retrieval (finding and indexing), delay (time to find, convert to a useful format, delivery), and understanding (both cognitive and organizational overload). Note that similar problems of retrieval seem to take on different forms in paper versus electronic environments: Not being able to find a paper fax corresponds, in some ways, to when the system goes down or locks up or response time is poor; in neither case can the customer representative retrieve the customer's records.

The main advantage of accessing paper-based information is that the delivery system (the paper) generally needs no additional processing in order for a person to use the information on it; the obvious disadvantage here is that the information takes up physical space. That is, the artifact of its material delivery system generates problems and costs that are not inherent in the contained information (Rice, 1999). By comparison, electronic information separates the substantive information from the material artifact. A notable disadvantage here is that information

TABLE 6.1

Example Disappearing and New Desktop Artifacts, by Category of Desktop Information and Implementation Phase

Categories of desktop information	*Implementation phase*			
	Initial	*Pre-Implementation*	*Implementation*	*Post-Implementation*
System (NCS/Imaging) & M&Ps information	Imaging binders in cabinets		NCS binders in cabinets	*All old binders (by person who was leaving)* *Peer review forms replaced by an online process*
Personal	Post-It notes			
Temporary			Post-It notes with passwords, backup logins, processes for NCS transition Post-It note for an account which grew in importance A Post-It note that replaced an OCS form that was about to be phased out	Post-It notes for fax case numbers Post-It notes for packages sent that day or expected to be received *Post it notes with implementation-transition information*
Process-related	OCS process post-It note		Processes for NCS-transition OCS process Post-It notes	Paper "to dos" and forms stored in the old "hot bin"

(Continued)

TABLE 6.1 (Continued)
Example Disappearing and New Desktop Artifacts, by Category of Desktop Information and Implementation Phase

Categories of desktop information	*Implementation phase*			
	Initial	*Pre-Implementation*	*Implementation*	*Post-Implementation*
Frequency of use			Paper copy of list of necessary information (also more difficult to view in NCS compared to OCS) Binder of M&Ps moved closer to PC	
Unsupported by the system — NCS or Imaging		Post-it notes with fax number matched to case numbers	M&Ps not online, have to use the binder Paper forms required for NCS but not residing in NCS or on the LAN	Pieces of paper and Post-It notes with information for rep who has no good way to store them
Reminders	Fax numbers on Post-It notes to remind rep to check status of fax	New Post-It notes *Post-It note of a completed task*	Post-it note of the path (filenames) to access LAN phone list Taped down Post-It note	Post-it notes with procedures and passwords Post-it notes for customer-specific issues *Wall lists of accounts with input restrictions*

Notes. Italics disappearing items; **bold** new items.

requires sometimes considerable additional processing before it can be accessed, display, and interpreted.

With respect to *complexity* (e.g., what Rogers, 2003, defines as the user's perception of difficulty, uncertainty, or learning costs of an innovation), an information artifact may represent a simplification of an inherently complex situation (such as information about a system error or an unfamiliar procedure) that may require further system encoding and decoding before its meanings could be clearly understood. This artifact may also indicate the unpredictability yet immediacy of some information (such as a reminder to call an important customer when a certain transaction occurs) that may demand an instant system response to accommodate its temporal nature. Information artifacts may also represent a temporary, necessary but costly media transformation (such as writing down a system-generated document number from a computer database field on to a piece of paper, a number that will then be further communicated through, for instance, a phone call to a customer, who must then again write it on a piece of paper or type it into a computer; Rice & Bair, 1984).

Moreover, information artifacts may even serve as an indication of poor synchronization across processes, actors, or information sources that should have been coordinated or integrated in advance to ensure an unequivocal information flow. For example, although the new system was intended to provide near-immediate updating, the upstream input system still operated on an overnight batch update process, causing delays of up to a day, prompting the reps to generate various notes and backup paper to keep track of what was supposed to be processed and when.

3. Paper and Electronic Forms as Communication Media. Information artifacts are, in many ways, a communication medium—between information elements, people, task processes, work units, interfaces, systems and organizations. This is especially true of forms (Sless, 1988), which are transformation and transaction interfaces between different communication systems, processes, or entities. Indeed, the costs of printing forms are usually less than 5% of all costs associated with using those forms (Barnett, 1988). They can shape the nature and content of organizational information, as they require, constrain, and filter inputs. They "present a picture of the organization's activities" (Barnett, 1988, p. 12; see also his intriguing short history of forms, chap. 1).

One advantage of paper forms is that they are physically decoupled from other system processes, so that incomplete paper forms can be temporarily set aside, while a user performs other tasks. Some AKA workers continued to use old paper forms during the implementation process, even when new versions were available through the system. Here, the old forms were desktop artifacts serving as symbolic interfaces for the individual between two system regimes, providing a sense of continuity and security during a time of uncertainty. From the perspective of diffusion theory, this usage of these old paper forms would be considered highly *compatible* with user needs and values (what Rogers, 2003, refers to as the extent to which potential adopters think an innovation fits in with prior norms, values, preferences, ways of doing things, and integration with other systems).

By comparison, on-screen forms, especially in real-time processing systems, may fail or create problems. When a system presents such forms on the entire screen within a sequential process, the system becomes unavailable for other activities (including searching for the information) while the user is waiting for the necessary information. However, computer forms have a wide variety of benefits, including making the entered information available throughout the organization, providing on-screen error detection and help, preventing the use of out-of-date forms, and so on.

Thus, both paper and online forms are part of the desktop, often serving multiple purposes, as well as providing areas of concern, limits, and inflexibility for system designers, implementers, users, and customers. Indeed, informating (measures collected from an information system about a work process; Zuboff, 1985) may be used to provide users control over their processes, or to monitor, control, and routinize work (Lehr & Rice, 2005). Similar, more general points about the dual role of information technologies in both integrating as well as controlling work and information flow have been made by many others (see e.g., Clement et al., 1991; Orlikowski, 1991).

4. Artifacts as Meta-Information. The content, or even the mere presence, of desktop artifacts may be serving as an indicator of other issues, problems or implications of system change—that is, meta-information about information (Rice, 1999)—in much less intentional or intended ways than "informating" (Zuboff, 1985). Changes in the shape, size, color, and location of artifacts in relation to desktop surface may be meta-information about the shifting relevance of certain kinds of information—such as a Post-It directly on the PC—or they may appear or disappear as system problems arise and are resolved. Artifacts may be meta-information about "augmentation" or "work-arounds" (Gasser, 1986), reflecting how people work around inadequate computing systems by adjusting data or procedures or using backup (manual or computer) systems. An old paper form, for example, can serve as a backup system to a new on-screen electronic document that a user doesn't completely trust.

Other researchers have noted that paper artifacts (such as Post-Its) are frequently used to draw one's attention to a problem, explain cryptic system information, or explain otherwise tacit system knowledge (such as what an error message on a photocopier means; Sellen & Harper, 2003). Indeed, an internal economic analysis may well find that both desktop and electronic landscapes are strewn with negative externalities or shadow costs (costs or negative consequences that users and customers pay for but are never accounted for through traditional system costs; Rice & Bair, 1984; Ryan & Harrison, 2000). Artifacts can be some of the indicators of shadow costs—time and money involved in a process that do not directly contribute to the value of the process, and are not included in any accounting record (Rice & Bair, 1984).

Finally, some artifacts endure because they continue to refer to some underlying common issue, account, or problem, even though the surface information is interpreted differently or has changed (Sellen & Harper, 2003; referring to Kidd, 1994). For example, a Post-It note that used to serve a reminder about an updated customer number remains because it now serves as a reminder of a different problem relating to that same contact's company. Conversely, an artifact may continue in its material form, but be used to represent or store different information, because its old uses have been transformed by a new information system.

SOCIAL CHANGE IMPLICATIONS

The transformation of documents from the physical paper realm to the cognitive symbolic realm generated reasons for users to create desktop artifacts (mostly paper, but also some using the system itself) to help track no longer physically visible work practices and system misalignments. People created and used these desktop artifacts, as both signal and symbol, medium and content, to help them adapt to the new electronic document management system. The affordances of both paper and electronic documents are not mutually exclusive, and are highly contextual, regardless of the obvious benefits of a new ICT. And organizational members adapted, reinvented, and personalized both preexisting artifacts, and new ones, to communicate to themselves and with others about the new cognitive demands as well as system adjustments and misalignments. These included system and task-related information, personal items, temporary information, process-related items, frequency of use indicators, items unsupported by the system, and reminders.

Major conceptual dimensions of such artifacts that contribute to understanding changes associated with a new ICT include paper versus electronic information, materiality and complexity of physical and electronic information, paper and electronic organizational forms (literally, the forms used within organizational settings) as communication media, and artifacts as meta-information. The items placed or posted there communicated what was not working well, what information was missing, and what processes were causing the individuals difficulty. Indeed, system implementation strategies should consider evaluating and tracking the presence of and changes in desktop artifacts, as part of initial user information needs assessment and systems analysis, as part of system evaluation and ongoing adaptation, and as evidence about possible organizational and social changes associated with new ICTs.

What are some wider potential implications of electronic document management systems and desktop artifacts? Certainly this specific ICT is not as inherently significant as the development of carbon paper, filing systems and cabinets, standardized paper form, the typewriter, electricity, the telephone, the elevator, and other innovations that transformed the nature of office work in the late 1800s (Johnson & Rice, 1987); the development of scientific management and organizational communication genres such as the memo, paper and information filing systems, and mechanical adding machines (Yates, 1989, 2005); and the rise of electronic and computer media such as the radio, television, facsimile, calculators, magnetic storage typewriters, word processors, and electronic data processing that transformed the nature of office work in the 1970s (see Chandler & Cortada, 2000; Johnson & Rice, 1987; Yates & van Maanen, 2001). Yet it incorporates many aspects of each of these transformational technologies, processing and facilitating both information processing and communication interaction. Thus, many of the same work, organizational and social changes may be associated with the transformation of paper documents to computer icons, and the transient desktop artifacts that accompany the implementation of such systems.

Such changes may range from the personal electronic and physical landscape of individuals' desktops, to the flow and processing requirements of communication and information within and among organizational units, to how external customers (both other organizations and individual customers) interact with and conceptualize their relationship with organizations (including government agencies, etc.). For example, the transformation to work-flow computer symbols from physical paper may increase organization-wide access to customer information and responses to customer requests, but may also make the representative–customer relationship more abstract and ephemeral. Because desktop artifacts are indicators of possible system problems and difficulties in individual cognitive processing of ICT routines, individuals may have more difficulty transferring to new positions, and their replacements may have more difficulty figuring out how to actually accomplish the work.

The transfer of work flow and relationship indicators from physical paper to internal symbols may reduce the number of and accessibility to cues about dysfunctional system processes, making the work experience and organizational functioning more cryptic and difficult to diagnose and repair. This transformation may also reduce many sources of error and delay associated with unnecessary media transformations, and constraints of time and space associated with physical materials. And the shift from paper to online forms may both avoid errors and confusion from using out-of-date forms, while also quickening the pace at which forms change because there is no need to reprint and redistribute paper forms throughout an organization. These are just a few examples of the wider implications of the implementation of, and adaptation to, electronic document management systems in particular, and organizational ICTs in general.

REFERENCES

Åborg, C., & Billing, A. (2003). Health effects of "the paperless office": Evaluations of the introduction of electronic document handling systems. *Behaviour & Information Technology, 22*(6), 389–396.

Archea, J. (1977). The place of architectural factors in behavioral theories of privacy. *Journal of Social Issues, 33*(3), 116–137.

Barker, R. G., & Associates (1978). *Habitats, environments, and human behavior*. San Francisco: Jossey-Bass.

Barnett, R. (1988). *Managing business forms*. Canberra, Australia: Communication Research Institute of Australia.

Beniger, J. R. (1986). *The control revolution: Technological and economic origins of the information society*. Cambridge, MA: Harvard University Press.

Capaldo, G., Raffa M., & Zollo, G. (1995). Factors influencing successful implementation of computer based technologies in knowledge-intensive activities. *Information Resources Management Journal, 8*(4), 29–36.

Chandler, A., Jr., & Cortada, J. (Eds.). (2000). *A nation transformed by information: How information has shaped the United States from Colonial times to the present*. New York: Oxford University Press.

Clement, A., Parsons, D., & Zelechow, A. (1991). Toward worker-centred support for desktop computing. In P. Van Den Besselaar, A. Clement, & P. Jarvinen (Eds.), Information system, work and organizational design (pp. 295–305). Amsterdam, The Netherlands: Elsevier Science.

Davidow, W., & Malone, M. (1993). *The virtual corporation*. New York: Harper Collins.

Davis, T. R. V. (1984). The influence of the physical environment in offices. *Academy of Management Review, 9*(2), 271–283.

Feldman, M. S., & March, J. G. (1981). Information in organizations as signal and symbol. *Administrative Science Quarterly, 26*, 171–186.

Gasser, L. (1986). The integration of computing and routine work. *ACM Transactions on Office Information Systems, 4*(3), 205–225.

Gibson, J. (1979). *The ecological approach to visual perception*. Boston: Houghton-Mifflin.

Goodman, P., Griffith, T., & Fenner, D. (1990). Understanding technology and the individual in an organizational context. In P. Goodman, L. Sproull, & Associates (Eds.), *Technology and organizations* (pp. 45–86). San Francisco: Jossey-Bass.

Groleau, C. (1995). *An examination of the computerized information flow contributing to the mobility of tasks in three newly computerized firms*. Unpublished doctoral dissertation, Concordia University, Montreal, Canada.

Hartson, R. (2003). Cognitive, physical, sensory, and functional affordances in interaction design. *Behaviour & Information Technology, 22*(5), 315–338.

Johnson, B. M., & Rice, R. E. (1987). *Managing organizational innovation*. New York: Columbia University Press.

Kendall, K. E., & Kendall, J. (1992). *Systems analysis and design* (2nd ed., pp. 167–173). Englewood Cliffs, NJ: Prentice-Hall.

Kidd, A. (1994). The marks are on the knowledge worker. In *Proceedings of CHI 94: ACM conference on human factors in computing systems* (pp. 186–191). New York: Association for Computing Machinery.

Kraut, R., Dumais, S., & Koch, S. (1989). Computerization, productivity, and quality of work-life. *Communications of the ACM, 32*(2), 220–238.

Lehr, J., & Rice, R. E. (2005). How are organizational measures really used? *Quality Management Journal, 12*(3), 39–60.

Lucas, A. G. (1991). *Effects of desk and chair arrangements on the social climate of business and industry training rooms*. Unpublished doctoral dissertation, Rutgers University, New Brunswick, NJ.

Majchrzak, A., Rice, R. E., Malhotra, A., King, N., & Ba, S. (2000). Technology adaptation: The case of a computer-supported inter-organizational virtual team. *MIS Quarterly, 24*(4), 569–600.

Malone, T. (1983). How do people organize their desks? Implications for the design of office information systems. *ACM Transactions on Office Information Systems, 1*(1), 99–112.

Negroponte, N. (1995). *Being digital*. New York: Knopf.

Nelson, D. (1990, March). Individual adjustment to information-driven technologies: A critical review. *MIS Quarterly*, pp. 79–98.

Orlikowski, W. (1991). Integrated information environment or matrix of control? The contradictory implications of information technology. *Accounting, Management and Information Technologies, 1*(1), 9–42.

Patrickson, M. (1986). Adaptation by employees to new technology. *Journal of Occupational Psychology, 59,* 1–11.

Pava, C.H.P. (1983). *Managing new office technology*. New York: The Free Press.

Rice, R. E. (1987). Computer-mediated communication and organizational innovation. *Journal of Communication, 37*(4), 65–94.

Rice, R. E. (1999). What's new about new media? Artifacts and paradoxes. *New Media and Society, 1*(1), 24–32.

Rice, R. E., & Aydin, C. (1991, June). Attitudes toward new organizational technology: Network proximity as a mechanism for social information processing. *Administrative Science Quarterly,* pp. 219–244.

Rice, R. E., & Bair, J. (1984). New organizational media and productivity. In R. E. Rice (Ed.), *The new media: Communication, research and technology* (pp. 185–215). Beverly Hills, CA: Sage.

Rice, R. E., & Gattiker, U. (2000). New media and organizational structuring. In F. Jablin & L. Putnam (Eds.), *New handbook of organizational communication* (pp. 544–581). Newbury Park, CA: Sage.

Rogers, E. (2003). *The diffusion of innovation* (4th ed.). New York: The Free Press.

Sellen, A., & Harper, R. (2003). *The myth of the paperless office*. Cambridge, MA: MIT Press.

Ryan, S. D., & Harrison, D. A. (2000). Considering social subsystem costs and benefits in information technology investment decisions: A view from the field on anticipated payoffs. *Journal of Management Information Systems, 16*, 11–41.

Sitkin, S., Sutcliffe, K., & Barrios-Choplin, J. (1992). A dual-capacity model of communication media choice in organizations. *Human Communication Research, 18*(4), 563–598.

Sless, D. (1988). Forms of control. *Australian Journal of Communication, 14*, 57–69.

Sommer, R. (1969). *Personal space*. Englewood Cliffs, NJ: Prentice-Hall.

Sprague, R. (1995, March). Electronic document management: Challenges and opportunities for information systems managers. *MIS Quarterly*, pp. 29–49.

Star, S. L. (1993). Cooperation without consensus in scientific problem solving: Dynamics of closure in open systems. In S. Easterbrook (Ed.), *CSCW: Cooperation or conflict* (pp. 93–106.) London: Springer-Verlag.

Suchman, L. (1995). Making work visible. *Communications of the ACM, 39*(9), 56–68.

Suchman, L., & Trigg, R. (1991). Understanding practice: Video as a medium for reflection and design. In J. Greenbaum & M. Kyng (Eds.), *Design at work* (pp. 65–90). Hillsdale, NJ: Lawrence Erlbaum Associates.

Sundstrom, E., & Sundstrom, M.G. (1986). *Work places.* Cambridge, MA: Cambridge University Press.

Tenner, E. (1998, March-April). The paradoxical proliferation of paper. *Harvard Magazine*, pp. 23–26.

Weick, K. (1985, Autumn). Cosmos vs. chaos: Sense and nonsense in electronic contexts. *Organizational Dynamics*, pp. 50–64.

Yates, J. (2005). *Structuring the information age: Life insurance and technology in the twentieth century*. Baltimore, MD: Johns Hopkins University Press.

Yates, J. (1989). *Control through communication: The rise of system American in management*. Baltimore, MD: Johns Hopkins University Press.

Yates, J., & Orlikowski, W. (1992). Genres of organizational communication: An approach to studying communication and media. *Academy of Management Review, 17*, 299–326.

Yates, J., & Van Maanen, (Eds.). (2001). *Information technology and organizational transformation: History, rhetoric, and practice*. Thousand Oaks, CA: Sage.

Zalesny, M. D., & Farace, R. V. (1987). Traditional versus open offices: A comparison of sociotechnical, social relations, and symbolic meaning perspectives. *Academy of Management Journal, 30*(2), 240–259.

Zuboff, S. (1985, Autumn). Automate/informate: The two faces of intelligent technology. *Organizational Dynamics,* pp. 4–18.

IV

SURVEILLANCE SETTING

7

Media Technology and Civic Life

Leo W. Jeffres

Communities have always been served by a variety of modes of communication, from bulletin boards to community newspapers and the interpersonal grapevine. But it is the newer communication technologies that concern us here, particularly the Internet, but also cable TV. Although the original term is forgotten today, cable initially was called CATV, or community antenna TV, because it was viewed as a community medium. Cable TV's promise for communities has largely been limited to community access channels and dictated through contractual agreements and government pressure on cable companies.

Following a different course of development, only in the early 1990s did the Internet became widely accessible to the public at large and reach a critical mass that allowed it to start playing a role serving communities and neighborhoods. Although there is no comparable pressure to serve communities or neighborhoods, the new medium has attracted many stakeholders in communities—from the media to community organizations, governmental bodies, and commercial interests.

Doheny-Farina (1996) argues that electronic communication pushes people toward a "virtual community" while they ignore communities tied to geographic space. He believes that people will construct communities of interest, education, tastes, beliefs, and skills—what Bellah, Madsen, Sullivan, Swidler, and Tipson (1985) call "lifestyle enclaves"—while their neighbors become strangers.

As each medium of communication has debuted, it has received similar scrutiny over its potential positive and negative effects on individuals (Wartella & Reeves, 1985). However, the potential impact of the Internet on neighborhoods and communities differs from earlier media. Although the newspaper (Edelstein & Larsen, 1960; Hindman, 1998; Janowitz, 1952/1967; Stamm, 1985) and other media—such as radio (Schulman, 1985), cable TV (Higgins, 1999a), and CATV channels (Atkin, Neuendorf, & Jeffres, 1998; Jacobs & Yousman, 1999)—have generally been seen as strengthening community ties, the Internet has been viewed as a more national or international medium, with a weaker community base.

This chapter examines two forms of communication technologies—cable TV and the Internet. Each has served local communities (e.g., municipalities, neighborhoods, grassroots organizations, etc.) in unique ways. We review the theories and empirical research findings that address the impact of these two media on social change.

BACKGROUND

Cable arrived first in Astoria, Oregon, and in the mountainous regions of Pennsylvania that had problems with receiving over-the-air broadcast TV signal (Phillips, 1972). In the 1950s, this cable TV service typically involved setting up an antenna to receive on average three to five over-the-air local broadcast signals and a head end to retransmit these signals, from the mountain top, to homes in these remote communities. In 1976, when HBO and WTBS (owned by Ted Turner) started utilizing satellites to transmit their signals to many of these community cable TV systems, the cable TV industry started to assume the operational structure that exists today. Over the years, local ownership of cable TV systems diminished and was replaced by large media conglomerates, due to public policies favoring deregulation of ownership restrictions and loosening of corporate merger restrictions.

As noted earlier, a modern-day cable TV system is usually operated and owned by a multiple-system owner (e.g., Comcast, Cox, Cablevision, etc.) that receives a franchise agreement from a municipality to operate its service in a community. The system acquires its programs from various program content providers (e.g., CNN, ESPN, HBO, etc.) and is required by law to retransmit local over-the-air broadcast signals (e.g., TV-3, TV-5, etc.) and government access channels (e.g., C-Span). Depending on the franchise agreement with the municipality, each cable system usually agrees to allot channel space for community-based programs to serve the public interest, which could include public access, leased channels, police stations, fire stations, middle schools, high schools, community colleges, universities, community centers, senior community centers, museums, city hall, and so on. Today, digital cable TV, with the use of digital signal compression technology, delivers several hundred channels of programs instead of just a few. Cable TV is now in 65.8% of American TV homes, as satellite TV systems have gradually siphoned away cable subscribers to reach a 19% penetration rate (Bachman, 2005).

The Internet became widely accessible to the American public in 1992, when then Senator Al Gore's bill "liberated" and opened up the Internet from a medium primarily utilized for scientific research purposes by universities, technology laboratories, and the military ("Gore Stumps for Info Highway Bill," 1994). Subsequently, AOL helped popularize the World Wide Web (the graphic-capable hypertext markup language [html] system) in the United States by turning it into a consumer service, along with such portal services (or search engines) as AltaVista and Yahoo, among others. Today, terms like *google me* or *e-mail me* have become part of our cultural lingo, reflecting the Internet's role as an information source and a communication medium in our society. Since then, the Internet has expanded from merely providing information as a mass medium to a platform for communication that crosses interpersonal, organizational, and mass communication boundaries (Jeffres & Atkin, 1996).

With broadband technology stimulating access interest, Internet penetration in the United States has reached nearly three out of four households (Rainie, 2005). However, there are considerable regional variations and differences in access and use by social categories, as documented by surveys conducted as part of the Pew Internet & American Life project (Spooner, 2003). Usage from "third places" such as schools and libraries is highest among the lower education groups, with 45% of those who did not graduate from high school using "third places" in contrast to 17% of those with college degrees (Harwood, 2004). Internet usage in poorer communities is likely to benefit from access via these "third places," including community centers and libraries.

THEORETICAL EXPLANATIONS

Concerns about the "digital divide" and its impact on how technologies affect audiences and communities are reminiscent of earlier work in development and modernization, where communication scholars working in a variety of traditions sought to use the tools and theories of communication to reduce "effects gaps" based on social status.

The bulk of this work occurred in the Third World, where scholars such as Schramm (1964; Schramm & Lerner, 1976), Rogers (1976, 1978, 2003), Hamelink (1985), and Stevenson (1991) applied a variety of concepts and theories to improve people's lives through development programs that employed diffusion, grassroots media, soap operas, and investment in telecommunication. Some focused on issues of inequality (e.g., Hamelink, 1985) between developing and developed nations.

Focusing on mass media, a Minnesota research team (Tichenor, Donohue, & Olien, 1980) posited gaps in knowledge between high- and low-status groups, which widened with increases in the amount of public affairs information that diffused in a community. Borrowing from this "knowledge gap" concept, researchers noted an "effects gap" between groups of different social status (Galloway, 1977), and the solution was sought in popular participation in both planning and execution of programs, with influence flowing from the "bottom up" rather than downward from influentials. The notion of an "effects gap" is that more educated people have more information-processing skills, more background knowledge for understanding incoming messages, more interest in the pertinent content, more social skills for making the content relevant, and stronger print media use patterns than do less educated people; the result is that they learn faster, creating a gap in knowledge or other effects based on such knowledge acquisition.

The Internet and newer mediated communication channels offer a potentially strong vehicle for such "bottom up" influence in neighborhoods and communities in the United States and around the world as they seek to cope with a changing and competitive environment. Of course, in the 1990s, the Internet was in a nascent phase of development and had thus received relatively little attention in the new media literature to that point. The question is whether governmental and other institutions will help support the growth of the Internet use, which is by nature a freewheeling network not easily controlled by any one government, institution, or individual.

Community and Cable TV

Consequences of the digital divide may be more significant in the political than in the economic arena. The new technologies represented by cable and digital transmission have the potential for providing a public space where a democratic forum can emerge (Rennie, 2001). Stein (2002) describes a variety of radical access TV projects that suggest cable can host a range of political communication and act as a forum for alternative views. This hope for an enhanced democracy at the community level is consistent with the traditional normative theory that media fulfill key functions for citizens, including surveillance of the environment and coordination of different sectors of the community, socialization, and entertainment (Lasswell, 1948; Wright, 1960). The first function, often labeled the *watch-dog function,* is associated with journalism and newspapers, but cable newscasts and public access forums can fulfill the same function. Empirical theory also addresses the relationship between community and cable TV, in particular, the work of Tichenor and his colleagues (1980) at Minnesota. Although phrased at the societal level, they tested their theory at the community level, arguing that a town's media reflect the size and diversity of its population, operating to serve the community social system. It is this social structural explanation that underlies much of the concern over the "digital divide."

Community and the Internet

From the very beginning when the Internet entered into our daily lives, it has been the subject of competing visions. The medium's promoters have viewed it as an opportunity for connecting people and bringing them together in communities. Critics of the Web view it as a vehicle that isolates people from their physical communities, as heavy users become isolated while surfing the net. As history has taught us about earlier communication technologies and mass media, the truth is likely to fall somewhere in between. A host of more limited theoretical visions stress the potential of the World Wide Web for offering community services by fulfilling political surveillance functions, helping not-for-profit organizations (Kubicek & Wagner, 2002), enhancing the delivery of governmental services, supporting educational programs, and acting as a community economic engine. But it is the potential for strengthening civic involvement and cohesion or weakening community ties that directs much of the research (Baker & Ward, 2002; Dutta-Bergman, 2005; Pigg, 2001; Shah et al., 2000). Some of this work operates under the concept of *social capital*, a term most recently popularized by Putnam (1995a, 1995b, 1996), who saw a decline in the image of America as a nation of joiners and a consequent decline in community activity. Many of the same theoretical underpinnings of studies on cable TV also apply to research examining the relationship between community and the Internet, including structural explanations underlying the "digital gap" and media effects theories concerned with how the Internet affects community ties (Matei & Ball-Rokeach, 2003).

EMPIRICAL FINDINGS

Although volumes have been written about communication technologies and communities, empirical evidence is fairly limited for both cable TV and the more recent Internet, as this section indicates.

Community and Cable TV

Early evidence on cable TV showed a growth of grassroots programming. Wurtzel (1975) found that New York City's community access channel was locally oriented or directed toward specialized audiences that did not receive such programming via conventional broadcasting. Atkin and LaRose (1991) found that only one of six cable viewers reported watching a community channel (e.g., public access) in the past week, making them perform as well or better than general offerings such as C-SPAN. However, efforts to provide public information about community business persist (e.g., the community of Yorktown, New York, broadcasts live meetings of the town planning and zoning appeals board on cable; in addition, they use e-mail to deliver agendas of board meetings and alert cable viewers if they're interested; Town Board Meetings, Live on Cable TV, 2004). Such local programming can improve residents' knowledge about their communities, and many local access channels sponsor high school newscasts and cover community events, including high school sports. Dordick and Lyle (1971) found increased voter awareness resulted when unlimited time was given to local political candidates on a cablecast channel in Hawaii. Thus, cable TV has fulfilled the surveillance function of the media in a variety of ways.

Increased interest and participation in local community activities also have been linked to cable TV coverage (Atkin et al., 1998; Smith, 1972). Early interactive projects also offered promise for education (Baldwin, Greenberg, Block, & Stoyanoff, 1978; Clarke, Kline, Schumacher, & Evans, 1978) and increasing community involvement (Brownstein, 1978; Burns & Elton, 1978; Lucas, 1978; Moss, 1978). Others argue that cable TV has failed to serve the public interest (Aufderheide, 1992)—for example, failing to overcome public access barriers to the media—or have worried about how the audience fragmentation associated with cable, and more recently with satellites and the Internet, affect public broadcasting and its promise to help communities (Agostino, 1980a, 1980b; Jacobs, 1995).

A study in the 1980s found that merely one showing of a 20-minute cable TV program dramatizing conservation strategies resulted in significant savings on electricity in the community (Winett, Leckliter, Chinn, & Stahl, 1984). Brody (1984) found that cable TV subscription was linked positively to library book loans in one city and negatively in another city. Viswanath, Finnegan, Rooney, and Potter (1990) found civic involvement—encompassing local political and other functions—linked to newspaper subscription, but not to cable TV subscription. A cultivation study found viewing specialized cable channels not related to feelings of fear and mistrust (e.g., concerns about crime; Perse, Ferguson, & McLeod, 1994). A study in Israel found, more recently, that families use cable TV in different ways to socialize children into their culture and community (Tidhar & Nossek, 2002). In more recent years, researchers have shifted their attention to the Internet and its effects on the human landscape. Wartella, Heintz, Aidman, and Mazzerella (1990) found that children's access to a variety of media sources—particularly broadcast, cable, and VCRs—has grown over the years, and cable TV has enhanced their programming choices and provided more diversity.

Community and the Internet

The Internet also can function to provide positive benefits to neighborhoods and cities around the world. A program of public "telecenters" (i.e., free Internet access sites) in São Paulo, Brazil, is fomenting community life in poor neighborhoods by providing both digital and social inclusion. An average of 180 people visit the telecenter daily in Cidade Tiradentes to take courses or use broadband-connected computers; that is 1 of 106 telecenters in poor neighborhoods. In New York City, a man who spent time on a kibbutz in Israel and enjoyed the sense of community created a web site that lets people register and meet their neighbors (Cohen, 2004). Both successes and failures have been noted with similar digital community centers in Mexico City and Colombia, where telecourses enjoy limited appeal (Osava, 2004).

In Florida, residents of Jacksonville neighborhoods can get their own neighborhood-level crime information with a few clicks of the mouse. The crime-mapping site helps provide accountability for local law enforcement (Mitchell, 2004). At a library in Suffolk County, New York, many Hispanics turn to the library's computers to help learn English and find a local job; the library offers classes on web design on computers whose home page has special links for Spanish-language search engines (Dalton, 2004). In the South Bronx (of New York City), a local area network installed by the community development organization, Mount Hope Housing, serves hundreds of poor residents with technology that combines both wired and wireless service. This system is wired with cable inside buildings to move data to and from apartments and transmits the data signals through an antenna on the roof. Tenants are introduced by small classes that focus on how to seek help from Internet resources for employment, paying bills, keeping in touch with relatives, and homework (Fountain, 2004).

Similar wireless networks are sprouting up across the country, from San Francisco to New York City, in some cases attracting residents to move to the area (McCall, 2004; Sidener, 2004). Some communities are installing fiber-optic networks that can connect residential homes. For example, five neighborhoods in York and Lancaster counties in South Carolina joined almost 100 communities in the United States to have high-speed cable hooked up to their homes, providing services such as Internet, phone, TV, and home security systems (Fossi, 2004). Local residents also are using the web to generate income. One woman in a poor Philadelphia neighborhood now uses the Internet to help her sell a line of candles to other people in the neighborhood (Krim, 2004).

A recent innovative use of the Internet that could serve neighborhoods is the weblog: personalized accounts, diaries, essays, or newsletters that have wide distributions, but also allow residents to post their views for neighbors. In New York City, for example, one site designer (*www.nycbloggers.com*) posted a version of the subway map and invited people to link their online diaries to the station nearest to their neighborhood. Thus, subway stations have become terminals for neighborhood residents on the virtual highway (Fallows, 2004; Knight, 2004; Zimmerman, 2004). Another example is found with a web site that allows amateur and professional writers in New York to tell their stories, mostly narratives from personal lives and their neighborhoods (*www.mrbellersneighborhood.com*). The home page is a satellite view of lower Manhattan and Brooklyn, where visitors can navigate their mouse across neighborhoods that turn red to indicate which neighborhood one wishes to visit. Clicking the mouse then takes you to another map of the neighborhood, with red and green dots that announce where a story is situated (Grady, 2000).

More recently, commercial enterprise has developed tools that serve neighborhoods. In August 2004, Yahoo introduced a service—Yahoo! Local—to help people search for local information online. Residents can pinpoint businesses and services around an address or zip code. They also can use a dynamic map to see where businesses and points of interest are located in relation to each other. Residents can rate and review businesses as well. According to the Kelsey Group, as much as 25% of all Internet searches have a local intent. Clearly, neighborhoods and community are important, and people are using the Internet a fourth of the time to connect with their local neighborhood or community in some manner

("Yahoo! Launches Local Search Beta," 2004). Thus, the Internet can connect people with local businesses and institutions, fulfilling a coordinating function.

Evidence also suggests that the Internet can enhance civic involvement and strengthen community ties. Shah et al. (2000) found evidence that the Internet may stimulate civic involvement in a secondary analysis of the 1999 DDB Needham Life Style Survey. They found that time spent on the Internet was a positive predictor of such civic participation as volunteering, attending club meetings, and working on community projects. It also correlated with informal socializing and taking advantage of community leisure opportunities. These findings are consistent with those reported by Jeffres, Lee, Neuendorf, and Atkin (2003), which reveal that there is potential for both Internet use and traditional media to strengthen community identity (e.g., awareness of local school events via web postings or educational access programming).

Pinkett and O'Bryant (2003) describe a university–tenants association partnership that developed a community network of desktop computers and high-speed Internet connectivity to support interests and needs of local community residents. Reports of a community-building effort in Scotland showed that hundreds of individuals, groups, and businesses were connected through a community system, with new relationships established, and discussion groups formed around topics such as employment, health, community initiatives, civic issues, and legal problems (Malina & Jankowski, 2002).

Analyzing data collected by the Pew Research Center for the People and the Press (2000) project, Dutta-Bergman (2005) compared individuals living in communities with Internet access with those in communities absent such access, finding that people were more involved with community activities when they lived in Internet communities. Although the differences are not startling, they are statistically significant. For example, almost 23% of residents in Internet communities were involved in community sports leagues, compared with only 18% of those in non-Internet communities. Even involvement in religious organizations was greater (45.5% vs. 41.9%), as was involvement in community charities (31.6% vs. 25.4%). Thus, there is indirect evidence that merely living in a community with Internet access enhances the chances that one will get more involved in civic activities.

The potential for using public places, often called "third spaces," for enhancing community engagement is found in a Finnish study of Internet cafés called Netcafé, a space that provides opportunity but not the necessity for spontaneous involvement and participation in the Internet society (Uotinen, 2003). The café was set up with the cooperation of 100 different organizations and community groups to encourage dialogue and cooperation among different parties. Visitors included all types of people from a wide range of age groups. Qualitative assessments suggest the NetCafé attracted visitors for a host of personal and work-related reasons.

The Metamorphosis project in Los Angeles uses a variety of methods to examine neighborhood communication, with results showing that the Internet increasingly is incorporated into daily life (Matei & Ball-Rokeach, 2003). The southern

California neighborhoods being examined are largely ethnic communities as well, providing an opportunity to examine those differences too. Thus, for example, the authors find that in predominantly Asian and Latino areas, the Internet is associated with mainstream media and probably contributes to the process of ethnic assimilation as people disengage from their communities while remaining informed about their ethnic communities. As a case in point, the project focuses on "storytelling" and community connections. Results show that people who talk with others about their neighborhoods are also more likely to belong to community organizations.

Among "old-immigrant groups" (White and African American), Internet connectedness is positively associated with membership in community organizations—those connected to the Internet are 1.4 times more likely to belong to community organizations. Among newer immigrant communities, Asians and Hispanics, Internet-connected people are neither more nor less likely to participate in communication environments leading to a sense of belonging (Matei & Ball-Rokeach, 2003). Asian respondents rely on community media, which include the ethnic Chinese and Korean press. Thus, the role of the Internet in community life depends, in part, on the ethnic nature of the community and the significance of older media as well. Other data from the same project show that Internet connectedness varies considerably by ethnic community, with mean time online in seven areas ranging from 7.71 hours per week for the Central American Pico Union neighborhood, and 7.74 for the Mexican community in East Los Angeles, to 8.47 in the largely Caucasian community of South Pasadena (Jung, Qiu, & Kim, 2001).

In Sweden, Ranerup (2001) reports on recent local-government initiatives to provide a "virtual public sphere," finding that some 15% of the 289 Swedish municipalities have provided citizens with an online discussion forum, compared with 43 municipalities in 1999. With few exceptions, the agenda for discussions has been completely open, providing support for the Internet as a tool for people-centered governance.

Summary

These examples of how the Internet and related cable TV technologies are being used by and in their neighborhood communities fail to capture the full range of possibilities. Table 7.1 provides a sample inventory of communication technologies and their potential community functions that can benefit society.

SOCIAL CHANGE IMPLICATIONS

Access to cable TV and the Internet by poor neighborhoods, ethnic communities, and the elderly is generally limited (Jeffres, 2002), creating the potential for gaps in information, services, and opportunities between the advantaged and

TABLE 7.1
Communication Technologies & Their Community Functions

Communication Technology Functions	*Community Functions*
Cable television	
Public access channels Local government channels, e.g. state, city, fire department, police department, libraries, senior centers and other civic agencies	Connecting community constituents—linking churches and parishioners, businesses and customers, employers and employees, officials and the citizenry
Local school and college channels, e.g., elementary schools, secondary schools and colleges	Increase "community capital" by strengthening local organizations, and strengthening ties among residents
Local public libraries Local community organization channels, e.g., private or nonprofit community groups, service groups, recreation centers; leased access channel use possible	Strengthening civic engagement, involvement and participation by providing a forum for discussing community conflict; mobilizing residents to carry out agreed—on social goals; and developing community consensus
Internet	
Local Internet portals via neighborhood or community websites	Surveillance of community leaders and stakeholders and their activities to alert residents to potential public problems and allow for exercising individual influence
Public telecenters for web access and social/economic development	Socialization—pass on traditions, customs of community to newcomers
Listservs and bulletin boards paralleling residential networks	
Community chat rooms for public discussions	Proclaiming and reclaiming community memory—documenting and reporting neighborhood and community history
Community e-mail alerts linking residents to each other and to the web	Reinforce community/neighborhood identity
Neighborhood/community Wi-Fi coverage for residents, visitors	Provide a platform for personal expression

disadvantaged. For example, those citizens who do not receive cable TV service and community-oriented cable channels miss out on a variety of community services and activities featured therein. Viewers could miss out on free flu shot announcements for seniors or coverage of community festivals and celebrations unless they have other means of receiving information about these types of happenings. More important is the potential for cable TV to increase neighborhood cohesion through news and other programs that introduce residents to each other and reinforce ties to the community. Other means of becoming informed involve access to community newspapers or exposure to flyers at community centers and government offices in neighborhoods.

Similarly, unequal access to the Internet by communities and neighborhoods could affect not only knowledge levels of residents, but also the development of job skills so important for the new economy. Various government and nonprofit groups/institutions have been endeavoring to bridge the digital divide. Their efforts often involve improving Internet access in poorer neighborhoods and communities through schools, libraries, community centers, and other community gathering places (e.g., Rogers, 2002).

For instance, in Hartford, Connecticut, Trinity College's Smart Neighborhood Initiative is designed to help neighbors cross the digital divide by offering all nonprofit organizations and small businesses—within a 1-mile radius of the campus—free or at-cost wire connections to the Internet. Small labs with community access have been established, and every resident in a 15-square-block area adjacent to the campus is eligible for a free Internet connection (Goldscheider, 2000).

Another example involved making private cybercafés have the neighborhood access points. Cyber Casa, the Bronx's first and only Internet café, opened in January 2001 amid hopes it would spur high-tech savvy and development in a neighborhood in the nation's poorest congressional district (Crow, 2001). Another cyber café offered Internet access to a largely Polish neighborhood, with a computer and web site offering links to Polish sites both in the United States and abroad. Visitors can check the weather in Gdansk, follow their favorite Polish sports teams, or check out a directory of Polish-American companies, most in the neighborhood being served in New York (Korengel, 2000).

Considering how access to both cable TV and the Internet has contributed to a better informed citizenry when it comes to community matters, it is logical to ask to what extent communication via the Internet and cable TV is used to strengthen identity with community, enhance relationships with others in the community, and reinforce community values. Clearly, this an empirical question that researchers will have to answer, not the pundits or hopeful observers.

Public concerns with community media, particularly the Internet, are growing as towns and cities across the United States and around the world try to find the best way to guarantee universal, affordable Internet access to their residents. New networks are taking root across the country, and most recently they have been used to connect families and provide relief to thousands of community residents displaced by hurricanes (see e.g., www.freepress.net/communityinternet). High-speed Internet access, whether via cable, satellite, Wi-Fi, or other broadband options, is fast becoming a basic public necessity, much like water, gas, and electricity. New community Internet networks cropping up across the country are likely to improve access to information, enhance public safety, speed technological innovation, and bolster local economic development.

With the passage of the Satellite Home Viewer Improvement Act in 1999, satellite companies began installing digital technology to offer more channels and services. To provide local channels to specific markets, the companies use spot-beam satellites that can focus a signal to cover relatively small areas. Thus, satellite technology, like the Internet and cable, is honing in on communities (Taub, 2004).

In summary, if community-oriented cable TV and Internet channels achieve a more pervasive presence, then communities can realize greater social capital, as facilitated by enhanced public integration and civic development. As commentators (e.g., Lin & Atkin, 2002) have noted of communication technology generally, the barriers to advancement are no longer a function of technology scarcity, so much as of political, economic, or social constraints. For instance, with a better informed citizenry—facilitated by cable or Internet-based communication information—a number of community activities can flourish in support of community development projects.

These projects can include any number of activities, including the encouragement of seniors to participate in various daily activities coordinated by area senior centers, community centers, civic clubs, church groups, hospitals, clinics, and the like. Specific presentations could encompass film screenings, history lectures, workshops, exercise classes, health seminars, and financial seminars, as well as volunteer work and travel. Similar activities can be found on various community-based cable TV and Internet sites that invite citizens, families, and children alike to participate in cultural, recreational, learning (e.g., English classes), sports, civic activities (e.g., town hall meetings), volunteer programs, and so on.

All of these media-publicized activities create opportunities to bring people together, perhaps to plant a tree (on Arbor Day), participate in a 5K distance run, join a library ice cream social, volunteer to clean up a park after July 4th fireworks festivities, donate books for a local library fundraiser, visit seniors at retirement homes, deliver meals for shut-ins, attend small business management seminars, and the like.

In essence, communication technologies that are developed help render our communication process more efficient while, over time, allowing more people to communicate with others in their communities and in the larger society. If these technologies can be adequately diffused across various segments of society and into different communities, the end result will be the enhancement of civic development, integration, and democratization.

REFERENCES

Agostino, D. (1980a, summer). New technologies: Problem or solution? *Journal of Communication*, *30*, 198–206.

Agostino, D. (1980b). Cable television's impact on the audience of public television. *Journal of Broadcasting*, *24*, 347–365.

Atkin, D., & LaRose, R. (1991). Cable access: Market concerns amidst the marketplace of ideas. *Journalism Quarterly*, *68*, 354–362.

Atkin, D. J., Neuendorf, K., & Jeffres, L. W. (1998). Reassessing public support for public access cablevision: A faded passion? *Telematics and Informatics*, *15*, 67–84.

Aufderheide, P. (1992). Cable television and the public interest. *Journal of Communication*, *42*(1), 52–65.

Bachman, K. (2005, August 3). *TVB: Satellite TV continues gains against cable*. Media Week. Retreived August 4, 2005, from http://www.mediaweek.com/mw/news/cabletv/article_display.jsp?vnu_content_id=1001008526

Baker, P. M. A., & Ward, A. C. (2002). Bridging temporal and spatial "gaps": The role of information and communication technologies in defining communities. *Information, Communication & Society*, *5*, 207–224.

Baldwin, T. F., Greenberg, B. S., Block, M. P., & Stoyanoff, N. (1978). Rockford, IL: Cognitive and affective outcomes. *Journal of Communication*, *28*(2), 180–194.

Bellah, R. N., Madsen, R., Sullivan, W. M., Swidler, A., & Tipson, S. M. (1985). *Habits of the heart: Individualism and commitment in American life*. New York: Harper & Row.

Brody, E. W. (1984). Impact of cable television on library borrowing. *Journalism Quarterly*, *61*, 687–689.

Brownstein, C. N. (1978). Interactive cable television and social services. *Journal of Communication*, *28*(2), 142–147.

Burns, R., & Elton, L. (1978). Reading, Pa.: Programming for the future. *Journal of Communication*, *28*(2), 148–152.

Clarke, P., Kline, F. G., Schumacher, H., & Evans, S. (1978). *Journal of Communication*, *28*(2), 195–201.

Cohen, J. (2004, November 14). Letting the Internet knock on the door. *New York Times*, late edition, p. 11.2

Crow, K. (2001, February 11). Neighborhood report: Soundview-Buzz. *New York Times*, p. 14.

Cunningham, W. H., Cunningham, I. C. M., & English, W. D. (1974). Sociopsychological characteristics of undergraduate marijuana users. *Journal of Genetic Psychology*, *125*, 3–12.

Dalton, R. J., Jr. (2004, July 4). Libraries offer help: Trying to close digital divide. *Newsday*, p. 8A.

Dordick, H., & Lyle, J. (1971). *Access by local political candidates to cable television: A report of an experiment*. Santa Monica, CA: Rand.

Doheny-Farina, S. (1996). *The wired neighborhood*. New Haven, CT: Yale University Press.

Dutta-Bergman, M. J. (2005). Access to the Internet in the context of community participation and community satisfaction. *New Media & Society*, *7*, 89–109.

Edelstein, A. S., & Larsen, O. N. (1960). The weekly press' contribution to a sense of urban community. *Journalism Quarterly*, *37*, 489–498.

Fallows, J. (2004, May 16). The twilight of the information middlemen. *New York Times*, p. 3.

Fossi, C. (2004, July 20). Comporium to install fiber-optic cables in some neighborhoods. *The (Rock Hill, S.C.) Herald*, p. 4B.

Fountain, K. (2004, June 10). Antennas sprout, and a Bronx neighborhood goes online. *New York Times*, p. 8G.

Galloway, J. J. (1977). The analysis and significance of communication effects gaps. *Communication Research, 4*, 363–386.

Goldscheider, E. (2000, November 1). College initiates program to give back to its neighbors, *New York Times*, p. 15B.

Gore stumps for info highway bill. (1994, June 14). *Broadcasting & Cable*, p. 38.

Hamelink, C. J. (1985). High-tech transfer: Selling the canoe without the paddle. *Development*, *1*, 28–37.

Harwood, P. (2005). *People who use the Internet away from home and work*. Accessed February 1, 2005, from Pew Internet & American Life Project, www.pewinternet.org

Higgins, J. W. (1999). Sense-making and empowerment: A study of the "vision" of community television. *Electronic Journal of Communication*. *9*(2).

Hindman, E. B. (1998). Community, democracy and neighborhood news. *Journal of Communication*, *48*, 27–39.

Jacobs, R. (1995). Exploring the determinants of cable television subscriber satisfaction. *Journal of Broadcasting & Electronic Media*, *39*, 262–274.

Jacobs, R., & Yousman, W. (1999). Understanding cable television community access viewership. *Communication Research Reports*, *16*, 305–316.

Janowitz, M. (1967). *The community press in an urban setting: The social elements of urbanism* (2nd ed.). Chicago: University of Chicago Press. (Original work published 1952)

Jeffres, L. W. (2002). *Urban communication systems: Neighborhoods and the search for community*. Cresskill, NJ: Hampton.

Jeffres, L. W., & Atkin, D. (1996). Predicting use of technologies for communication and consumer needs. *Journal of Broadcasting & Electronic Media*, *40*, 318–330.

Jeffres, L. W., Lee, J.-W., Neuendorf, K., & Atkin, D. (2003, August). *Community and civic values, Communication and social capital: "Bowling alone" as a product of values and communication*. Paper presented to the Communication Theory and Methodology Division at the annual conference of the Association for Education in Journalism and Mass Communication, Kansas City, MO.

Jeffres, L. W., Neuendorf, K. A., Bracken, C., & Atkin, D. (2002, November). *Cosmopoliteness in the Internet age*. Paper presented at the annual conference of the Midwest Association for Public Opinion Research, Chicago, IL.

Jung, J.-Y., Qiu, J. L., & Kim, Y.-C. (2001). Internet connectedness and inequality: Beyond the divide. *Communication Research*, *28*, 507–535.

Korengel, K. (2000, January 9). Neighborhood report: Greenpoint. *New York Times*, p. 14.

Knight, S. (2004, September 12). Neighborhood Report: New York Online. *New York Times*, p. 14.

Krim, J. (2004, August 9). Program aids urban poor in accessing the Internet. *Washington Post*, p. 1A.

Kubicek, H., & Wagner, R. M. (2002). Community networks in a generational perspective: The change of an electronic medium within three decades. *Information, Communication & Society*, *5*, 291–319.

Lasswell, H. D. (1948). The structure and function of communication in society. In L. Bryson (Ed.), *The communication of ideas* (pp. 32–51). New York: Harper.

Lin, C., & Atkin, D. (2002). *Communication technology and society: Audience adoption and uses*. Cresskill, NJ: Hampton.

Lucas, W. A. (1978). Spartanburg, S.C.: Testing the effectiveness of video, voice and data feedback. *Journal of Communication*, *28*(2), 168–179.

Malina, A., & Jankowski, N. (2002). Community-building in cyberspace. In N. W. Jankowski, with O. Prehn (Eds.), *Community media in the information age: Perspectives and prospects* (pp. 271–292). Cresskill, NJ: Hampton.

Matei, S., & Ball-Rokeach, S. (2003). The Internet in the communication infrastructure of urban residential communities: Macro- or mesolinkage? *Journal of Communication*, *53*, 642–657.

McCall, W. (2004, October 24). Network expert wires up free wi fi access. *Associated Press State & Local Wire*, BC cycle, accessed via LexisNexis.

Mitchell, T. (2004, October 15). New map pinpoints crime in your area; anyone with Internet access can find police data about lawlessness on a neighborhood level. *Florida Times-Union*, p. 1A.

Moss, M. L. (1978). Reading, Pa.: Research on community uses. *Journal of Communication*, *28*(2), 160–167.

O'Grady, J. (2000, December 3). A web site reverberates with the din of urban life. *New York Times*, p. 14.

Osava, M. (2004, May 31). *Latin America: Telecentres provide jobs and culture for the poor*. IPS-Inter Press Service, accessed through LexisNexis.

Perse, E. M., Ferguson, D. A., & McLeod, D. M. (1994). Cultivation in the newer media environment. *Communication Research*, *21*, 79–104.

Pew Research Center for the People and the Press. (2000). *Internet sapping broadcast news audience*. Washington, DC: Author.

Phillips, M. A. (1972). *CATELEVISION: A history of community antenna television*. Evanston, IL: Northwestern University Press.

Pigg, K. E. (2001). Applications of community informants for building community and enhancing civic society. *Information, Communication & Society*, *4*, 507–527.

Pinkett, R., & O'Bryant, R. (2003). Building community, empowerment and self-sufficiency. *Information, Communication & Society*, *6*, 187–210.

Putnam, R. D. (1995a). Bowling alone: America's declining social capital. *Journal of Democracy*, *6*, 34–48.

Putnam, R. D. (1995b). Tuning in, tuning out: The strange disappearances of social capital in America. *PS: Political Science and Politics*, *284*, 664–683.

Putnam, R.D. (1996). The strange disappearance of civic America. *The American Prospect*, *24*, 65–78.

Rainie, L. (2005). *The mainstreaming of online life*. Washington, DC: Pew Internet & American Life Project. Retrieved August 18, 2005, from http://207.21.232.103/pdfs/Internet_Status_2005.pdf

Ranerup, A. (2001). Online forums as a tool for people-centred governance: Experiences from local government in Sweden. In L. Keeble & B. D. Loader (Eds.), *Community informatics: Shaping computer-mediated social relations* (pp. 205–221). New York: Routledge.

Rennie, E. (2001). Community television and the transition to digital broadcasting. *Australian Journal of Communication*, *28*, 57–68.

Rogers, E. M. (1976). New perspectives on communication and development. *Communication Research*, 3, 99–106.

Rogers, E. M. (1978). The rise and fall of the dominant paradigm: New approaches to development. *Journal of Communication*, *28*, 64–69.

Rogers, E. M. (2002). The information society in the new millennium: Captain's log, 2001. In C. A. Lin & D. J. Atkin (Eds.), *Communication technology and society: Audience adoption and uses* (pp. 43–64). Cresskill, NJ: Hampton.

Rogers, E. M. (2003). *Diffusion of innovations* (5th ed.). New York: The Free Press.

Schramm, W. (1964). *Mass media and national development.* Stanford, CA: Stanford University Press.

Schramm, W., & Lerner, D. (Eds.). (1976). *Communication and change: The last ten years—and the next.* Honolulu: University Press of Hawaii, East-West Center.

Schulman, M. (1985). Neighborhood radio as community communication (Harlem, New York, City College of New York, Local Media, Communication Theory). *Dissertation Abstracts International, 47* (04), 1102A.

Shah, D., Schmierbach, M., Hawkins, J., Espino, R., Ericson, M., Donavan, J., & Chung, S. (2000, November). *Untangling the ties that bind: The relationship between Internet use and engagement in public life.* Paper presented at the annual conference of the Midwest Association for Public Opinion Research, Chicago.

Sidener, J. (2004, September 20). Neighborhood network; Free, wireless web comes to Golden Hill. *San Diego Union-Tribune*, p. 1E.

Smith, R. L. (1972). *The wired nation. Cable television: The electronic communications highway.* New York: Harper & Row.

Spooner, T. (2003). *Internet use by region in the United States.* Accessed February 1, 2005, from Pew Internet & American Life Project, www.pewinternet.org.

Stamm, K.R. (1985). *Newspaper use and community ties: Toward a dynamic theory.* Norwood, NJ: Ablex.

Stein, L. (2002). Democratic "talk": Access television and participatory political communication. In N. W. Jankowski, with O. Prehn (Eds.), *Community media in the information age: Perspectives and prospects* (pp. 123–140) Cresskill, NJ: Hampton.

Stevenson, H. (1991). A new entrepreneurial paradigm. In A. Etzioni & P. R. Lawrence (Eds.), *Socio-economics: Toward a new synthesis* (pp. 39–359). Armonk, NY: M.E. Sharpe.

Taub, E. A. (2004, October 14). Satellite TV spreads its signals across the landscape. *New York Times*, late edition, p. G6.

Tichenor, P., Donohue, G. A., & Olien, C. N. (1980). *Community conflict and the press.* Beverly Hills, CA: Sage.

Tidhar, C. E., & Nossek, H. (2002). All in the family: The integration of a new media technology in the family. *Communications: The European Journal of Communication Research*, *27*, 15–34.

Town Board Meetings, Live on Cable TV [letter]. (2004, December 12). *New York Times*, p. 14WC.

Uotinen, J. (2003). Involvement in (the information) society—The Joensuu Community Resource Centre Netcafé. *New Media & Society*, *15*, 335–356.

Viswanath, K., Finnegan, J. R., Rooney, B., & Potter, J. (1990). Community ties in a rural Midwest community and use of newspapers and cable television. *Journalism Quarterly*, *67*, 899–911.

Wartella, E., Heintz, K. E., Aidman, A. J., & Mazzarella, S. R. (1990). Television and beyond: Children's video media in one community. *Communication Research*, *17*, 45–64.

Wartella, E., & Reeves, B. (1985). Historical trends in research on children and the media: 1900–1960. *Journal of Communication*, *35*(2), 118–133.

Winett, R. A., Leckliter, I. N., Chinn, D. E., & Stahl, B. (1984). Reducing energy consumption: The long-term effects of a single television program. *Journal of Communication*, *34*, 37–51.

Wright, C. R. (1960). Functional analysis and mass communication. *Public Opinion Quarterly*, *24*, 605–620.

Wurtzel, A. (1975). Public-access cable television: Programming. *Journal of Communication*, *25*(3), 15–41.

Yahoo! Launches local search beta: An innovative way for people to search for local information online. (2004, August 3). Business Wire, Inc., accessed via LexisNexis Academic.

Zimmerman, E. (2004, October 3). Before applying, check out the blogs. *New York Times*, p. 10.

8

Media Technology and the 24-Hour News Cycle

Erik P. Bucy, Walter Gantz, and Zheng Wang

In the 1970s, American households received their televised local news from their local market stations and their national news from the three TV networks: ABC, CBS, and NBC. Today, the choices among national news networks have grown to include the 24-hour cable news channels, from the pioneering Cable News Network (CNN), established in 1979, to CNBC, MSNBC, and, most recently, FOX News. Through satellite services such as DirecTV and the Dish Network, the number of international news channels available to American viewers, including also BBC America and International Television News (ITN), among others, has also grown. Satellite radio, including Sirius and XM Radio, delivers 24-hour radio news to interested subscribers regardless of geographic location.

Driven by economic, technological, and audience-related pressures, the news industry continues to undergo significant transformation, following the buyout of the networks by major conglomerates and heightened expectations about news profitability. The death of ABC's Peter Jennings and retirements in 2005 of long-time network news anchors Tom Brokaw, of NBC, and Dan Rather, of CBS, also symbolized the passing of an era during which the three broadcast networks dominated the nightly news viewing audience. Against the backdrop of the flight of young viewers away from traditional news formats, the explosion of news outlets both offline and online, and the 24-hour news cycle, the industry has also witnessed a rise of tabloid news and opinion journalism. The news industry has been heavily criticized by media watchdog groups as well as academics for focusing on the sensational and trivial, for attracting eyeballs but diminishing civic discourse, and for putting ratings and profits ahead of fostering an informed citizenry.

Offsetting some of these concerns, every network and almost all local TV stations now have companion web sites—as do most newspapers and radio stations—with content that can be downloaded on demand to be accessed via a PC or wirelessly by cell phone, laptop, or personal digital assistant (PDA), or even "podcasted" into a portable media player. Web portals and most major search engines (e.g., Google, Yahoo!, AOL, or MSN) also facilitate custom news retrieval and delivery.

BACKGROUND

The 24-Hour News Cycle

The around-the-clock availability of cable news, combined with the ability to access online content for free (at least the same day) on most news sites—and at a time convenient for the consumer rather than the network or newspaper—has contributed to the flight from traditional media formats especially among younger viewers. However, the networks' web sites attracts millions of online users, so the loss of viewers may be at least indirectly compensated for by the gain in online traffic. Indeed, news organizations often break original stories first on their web sites before publishing or broadcasting them later that evening or the next day (South, 1999).

A central technology in the digital media environment, the Internet has begun to foster an interactive relationship between news producers and audiences (Bucy, 2004). CBS operates a 24-hour online-news on-demand network and offers a blog, or web-based diary, called the "Public Eye," enabling visitors to post public comments (Zurawik, 2005). CNN's *Inside Politics* runs "Inside the Blog," the first regular TV coverage of reporting and opinions about web blogs (Johnson, 2005). Most major newspaper and magazine sites have also established blogging sections for reporters to post their reflections about issues in the news and for visitors to interact with editorial staff (Smolkin, 2004). This adds to the trend of individually tailored news delivery possible on most news sites and portal pages.

News Media Ownership

In the early 1980s, some 50 corporations owned most of the newspapers, magazines, book publishers, movie studios, and radio and TV stations in the United States. That number has shrunk to five or six multibillion-dollar corporations today, with Time Warner occupying the top position as the largest media conglomerate, followed by Disney, News Corporation, Viacom, Bertelsmann, and Vivendi Universal (Bagdikian, 2004). One consequence of this ownership concentration is reduced coverage of international news and hard news, except for breaking news events (e.g., 9/11, the Iraq war), and increased coverage of sensational human-interest news and soft news. These trends have raised questions about news quality.

Patterson (2001) tracked the dramatic increase of soft news (e.g., *The Today Show*, *Dateline NBC,* or 20/20) or news stories lacking public policy content since 1980. His findings show that soft news content jumped from less than 35%

of all stories in 1980 to roughly 50% of stories by 2000; stories with a moderate to high level of sensationalism rose from roughly 25% in the early 1980s to 40% by century's end. Alternative forms of news also emerged to compete with mainstream news coverage. For instance, blogs or Web diaries (Blood, 2003; Singer, 2005) have become "quick-moving, multilinked, interactive venues of choice for millions of people wanting to share information and opinions, commentary and news" (*Nieman Reports*, 2003, p. 59). Mock news is most famously represented by *The Daily Show* on the Comedy Central cable network, an irreverent and satirical twist on public affairs, which has proven highly popular among young audiences—unlike the network newscasts (Pew Research Center, 2004a).

All of these variations on the news pose challenges to creating an informed citizenry with a shared foundation of common knowledge, as opposed to one that is merely self-interested or entertained (see Postman, 1986). Another challenge to information quality involves efforts by the administration to manage the flow of political information and orchestrate the production of video news reports for release by governmental agencies, which stations then pass along as legitimate news (Alterman, 2005).

THEORETICAL EXPLANATIONS

Agenda Setting

Research on agenda setting has shown that the press is influential in shaping the public's view of how important a given issue is by regulating the amount of attention and quality of coverage that issue receives (McCombs & Shaw, 1972). In Cohen's (1963) classic formulation, the press "may not be successful much of the time in telling people what to think, but it is stunningly successful in telling its readers what to think about" (p. 13). In the contemporary information environment, characterized by real-time news gathering and distribution technologies—including the Internet and communication satellites—the agenda-setting function of the news is potentially strengthened and accelerated. As salient news topics are reinforced and repeated across a wide variety of delivery platforms and channel outlets, the scope of issue importance, and hence influence, is arguably expanded.

However, at an individual level, even deepening knowledge of certain topics and growing reliance on online news sources may contribute to an overall narrowing of common knowledge. As Althaus and Tewksbury (2002) observed in a study comparing the print and online editions of the *New York Times*, "the parallel presentation mode of online news outlets might encourage greater depth of exposure to a particular topic even as it inclines people to be more selective in the topics they read about" (p. 198). Online news outlets, they argue, might therefore promote the development of *issue publics*—small groups that acquire expertise in particular subjects.

Internationally, mass media play a significant role in driving public attention to foreign affairs (Soroka, 2003). Studies have documented that the more media coverage a foreign nation receives, such as Iraq, the more likely individuals are to think the nation is vitally important to U.S. interests (see Althaus & Tewksbury, 2002; Soroka, 2003). Similarly, and consistent with research on second-level agenda setting, which examines the influence of media coverage beyond issue prioritizing to focus on the persuasive impact of issue characteristics and attributes (McCombs & Estrada, 1997), the more *negative* coverage a nation receives, the more likely respondents are to think negatively about the country (Wanta, Golan, & Lee, 2004).

In addition to encouraging individual depth in specialized knowledge, the new technology environment has led to an unexpected institutional outcome—the reduced Western control of international news flow. Until recently, the dominant flow of news influence was from American media to the rest of the world (see Hallin, 1994). Now, with satellite and Internet broadcasting, foreign news media are able to transmit and distribute alternative views and information into America to influence the news agenda in this country.

For instance, the Arab-language satellite network Al Jazeera (or "The Island") has a potential agenda-setting influence far outside of its "island" (peninsula, actually) in the Persian Gulf. This small Arabic TV channel, based in Qatar, gained worldwide attention following the September 11, 2001, terrorist attacks, when it broadcast videos showing Osama bin Laden and Suliman Abu Ghaith defending and justifying the attacks.

News Diffusion

The integration of networked information and communication technologies (ICTs) into the news gathering and dissemination process is likely to affect how fast news travels, how audiences first hear about particular news items, how many people ultimately find out about news events, and how individuals follow up on news items of personal interest. The study of these processes falls within news diffusion research.

Greenberg's (1964a, 1964b) study of the assassination of U.S. President John Kennedy discovered that the tragic event triggered an avalanche of continuous coverage. As such, the diffusion of the assassination news was virtually universal, and nearly everyone was aware of the tragedy within 90 minutes. Moreover, personal communication was an integral part of the initial diffusion process, as about half of the population first heard about that assassination from personal sources.

Likewise, Gantz's (1983) study of the attempted assassination of U.S. President Ronald Reagan also found that nearly 50% of the public learned the news through personal communication channels and that a similar percentage of people tried to pass the news along to others. Investigation of news diffusion patterns during the

1991 Persian Gulf war further supported the importance of personal communication for events of great magnitude (Gantz & Greenberg, 1993).

Beyond the importance of an event, a number of structural, temporal, and individual level factors affect the diffusion process. Political and media systems can either facilitate or impede diffusion, as well as influence the interplay between mediated and personal channels. When the press has limited access to news events or the public has limited access to the media, the diffusion process is slowed while personal channels increase in importance. Relatively unrestrained press access to news events and an abundance of readily accessible media channels speed the process and, in general, make personal communication channels less significant (Allen & Colfax, 1968; Fathi, 1973; Fine, 1975; Gantz, 1983; Rosengren, 1987).

Convenience, habit, and trust play a role in the diffusion process, too, particularly with everyday news events—the sort of important but routine stories that don't warrant blanket coverage or pique a nation's attention (Gantz & Greenberg, 1993; Gantz, Krendl, & Robertson, 1986). Media that are convenient to access and can be easily integrated into people's busy schedules also can become part of their daily routines. TV news can be turned on in the background at mealtimes or as people get ready for bed—or for work the next morning; radio news functions as an audio accompaniment to and from work. Internet news services offer similar and additional convenience features, which are easy to check while online for other purposes (Johnson & Kaye, 2002; Lin, Salwen, Garrison, & Driscoll, 2005).

With sites that can be bookmarked or tabbed with a browser, Net news is easier to scan and search than TV or radio news. Online news can also be accessed via an on-demand model (Albarran, Chan Olmsted, & Wirth, 2006). For example, Internet users can have online news sites (e.g., NYT.com) send daily headline or news alerts to their e-mail. Internet portals such as Yahoo! allow their users to automatically receive news from a number of news sources on various news topics (e.g., international news, national news, health news, entertainment news) based on user preferences. Because the Net can be accessed with an Internet-connected computer or a personal digital assistant (PDA) on a continuous basis, it is easier to check Net news than offline newspapers or other print and electronic news outlets.

Beyond this, the Internet is, at least to date, regarded as a credible source of news and information (Flanagin & Metzger, 2000). In addition, Net news may serve to enhance the credibility of traditional media when the two are used conjointly: Compared with control group ratings, cross-platform media use has been associated with more favorable perceptions of TV and Net news credibility than the use of either TV or the Web in isolation (Bucy, 2003; see also Johnson & Kaye, 1998, 2002). An important question that arises next, then, has to do with whether and how dependency on 24-hour *online* news sources may displace, supplement, or complement use of 24-hour *offline* news sources.

Media Displacement

Discussions about media displacement date back to classic communication studies conducted in the 1940s. Lazarsfeld (1940) examined whether the emergence of radio announced the end of print media; Lasswell (1948) examined how the emergence of TV affected radio. Media displacement research grew rapidly with the rise and widespread diffusion of TV in the late 1940s and 1950s (e.g., Belson, 1959, 1961; Bogart, 1956; Coffin, 1948; Himmelweit, Oppenheim, & Vince, 1958; Maccoby, 1951). Subsequently, there have been waves of inquiry on media displacement effects whenever a new medium or technology gained popularity. This can be seen with cable TV from the late 1970s (e.g., Jeffres, 1978; Kaplan, 1978; Lin, 1993; Reagan, 1984; Sparkes, 1983), VCRs in the mid- and late 1980s (e.g., Harvey & Rothe, 1985; Henke & Donohue, 1989; Scherer, 1989), computers from the mid-1980s (e.g., Finholt & Sproull, 1990; Hu, 1998; Perse & Dunn, 1998; Robinson, Barth, & Kohut, 1997; Robinson & Kestnbaum, 1999; Schweitzer, 1991; Vitalari, Venkatesh, & Gronhaug, 1985), and the Internet and Web technology from the mid-1990s (Kayany & Yelsma, 2000; Lin, 1999, 2001, 2002; Nielson Media Research, 1999; Stempel & Hargrove, 1996; Stempel, Hargrove, & Bernt, 2000).

The underlying logic for these studies is straightforward. Media have to compete for limited resources, including advertising dollars and the public's expenditure of time and money (McCombs & Shaw, 1972). Hence, the introduction of a new medium or technology should result in a corresponding reduction in the resources allocated to existing media (Robinson, 1969; Robinson & Godbey, 1999). Whether the pie represents time or money, it is not likely to increase. Instead, with each new medium or technology, the pie is likely to be cut into more and smaller slices. With that in mind, displacement effects generally have been operationalized in two ways: the amount of time spent with an existing medium that has been displaced by a new medium (time displacement) and/or the perceived functional equivalence between the emerging and existing media (Kayany & Yelsma, 2000).

Displacement effects have been observed for more than a half century, although effects are not uniform across existing media. For example, computers displaced more TV viewing time than time devoted to reading the newspaper. Himmelweit et al. (1958) introduced three principles to account for this: (a) the displacement of marginal activities principle—new media are more likely to displace marginal rather than central activities, (b) the principle of functional equivalence—new media perceived as more effective than existing media in satisfying the same needs will displace existing media, and (c) the transformed activities principle—traditional media that change in order to satisfy needs not met by a new medium will not be significantly displaced. Furu (1971) added the principle of physical proximity: Activities that share the same physical space will interfere with each other; the one that better satisfies an individual's needs will triumph.

Three of those principles can be applied to the large-scale emergence of 24/7 news. For the principles of functional equivalence, studies suggest that providing information is an important function of online media. Online news can satisfy user information needs in ways similar to offline news (Kayany & Yelsma, 2000; Kraut et al., 1998). Kayany and Yelsma (2000) found that, although traditional media such as TV and newspapers remained important for information needs, those who rated online sources as important for news tended to downgrade the importance of TV. Compared to offline news, speed and convenience are credited as the essential advantage of online news in terms of satisfying the information needs of users (Lin et al., 2005). Among those who say they read the online version of the newspaper, 73% cited its convenience (Pew Research Center, 2005a). To survive the competition brought about by 24/7 cable channels and strictly online news outlets, TV stations and networks, newspapers, and news magazines have set up their own web sites. Opening up such outlets on the Web points to the transformed activities principle. Computers and TV sets may share the same living space in a household, so conflicts associated with competing activities in the same physical space are likely to exist. If we assume that news consumption is not a marginal, unimportant activity, the remaining principle (marginal activities) is not likely to apply.

Uses and Gratifications

The conceptual model of the uses and gratifications tradition assumes (a) that the audience for news and other genres of media content is active and goal oriented, (b) that media are an important source of need gratification whose fulfillment lies with audience choices, and (c) that media compete with other sources of need satisfaction (Katz, Blumler, & Gurevitch, 1974). Although elaborated during the TV era, the assumptions of this research tradition are particularly applicable in the Internet age, when the role of electronic media audiences has evolved from passive "viewers" or "listeners" of media content into active "users" of information and communication technologies (see Newhagen & Rafaeli, 1996).

Research on uses and gratifications has identified five general categories of needs fulfilled by media use, including cognitive (acquiring information, knowledge, and understanding), affective (emotional, pleasurable, or aesthetic experience), personal integrative (strengthening credibility, confidence, stability, and status), social integrative (strengthening contacts with family, friends, etc.), and tension release (escape and diversion; Katz, Gurevitch, & Haas, 1973). The question of just what constitutes a basic psychological need that might be satisfied by different forms of communication has never been completely resolved, although Perse and Courtright (1993) identified a list of 11 specific needs potentially fulfilled by mass, interpersonal, or computer-mediated communication, including to relax, be entertained, forget about work or other things, have something to do with friends, learn things about self and others, pass the time, feel excited, and feel less lonely.

A key distinction made in the uses and gratifications literature is the difference between instrumental and ritualized media use, which was first proposed in relation to TV viewing. Rubin (1984) contended that ritualized viewing involved more habitual use of television for purposes of diversion and greater attachment with the medium. Instrumental viewing, on the other hand, constituted more goal-oriented use of TV content to gratify information needs or motives. Ritualized and instrumental media use is not neatly dichotomous, but may overlap at times depending on situational demands (Rubin, 1984). Studies of Internet use conducted from a uses and gratifications perspective (e.g., Papacharissi & Rubin, 2000) have found similar distinctions between diversionary and more directed use of the medium, suggesting that networked communication may serve as a functional alternative to face-to-face interaction.

Over the years, the uses and gratifications approach has come under fire for being too descriptive and insufficiently theoretical, and for relying too heavily on audiences for reporting their true motivations for media use (see Ruggiero, 2000). Further development in this area will require less reliance on individual interpretations and more direct observation of actual audience behavior—combined with the use of indirect measures to accurately assess user motivations.

EMPIRICAL FINDINGS

Agenda Setting

In a digital communication era, the direction of information flow is no longer one way. Increasingly, news buffs and online enthusiasts—notably, the blogging community—are calling attention to news items that would otherwise go unnoticed or underreported and provide a service by relentlessly fact-checking and stoking interest in stories that would otherwise fade (Singer, 2005). Indeed, there are fewer than 1 million professional journalists in United States, but an estimated 8 million active bloggers—and room for millions more (Welch, 2005).

In contrast to traditional journalism, which is supposed to be "disinterested" and report the "facts fairly" (Andrews, 2003), blogs are "unedited, unabashedly opinionated, sporadic and personal" (Palser, 2002). Importantly, blogs frequently work in conjunction with mainstream news by linking to articles on online news sites that might have gone unnoticed. So-called *filter blogs* comment on and connect to content found elsewhere on the Web (Zeller, 2005). In this manner they often amplify, illuminate, or interconnect the news rather than create it.

A recent study by the Pew Research Center (2005b) found that blogs do indeed have substantial agenda-setting influence, or the capacity to generate "buzz," but only under certain conditions. Bloggers are more likely to exert influence when they analyze, circulate, and call attention to stories that are initially reported, but might have been underplayed by mainstream news, than when they take the lead themselves on initially reporting stories. In essence, bloggers are most effective as amateur journalists when they put an intense spotlight on an already dimly lit news item and turn it into a newsworthy story (Zeller, 2005).

Blogs are read by millions but are under no professional or ethical obligation to be fair and accurate. One of the first blogs to have influence on the mainstream news agenda was the *Drudge Report* (www.drudgereport.com), published by conservative gossip news purveyor Matt Drudge. Drudge is credited for revealing salacious aspects of the Monica Lewinsky sex scandal that rocked the White House during the late 1990s (and for getting many other facts wrong), aspects that helped keep the Lewinsky story on the top of the national news agenda for months.

In 2002, when mainstream news discounted former Senate Majority Leader Trent Lott's praise of Strom Thurmond's 1948 segregationist presidential campaign (Kurtz, 2005), bloggers kept the story alive. By compiling a "rap sheet" of Lott's other comments over the years, these bloggers helped contribute to a climate that eventually forced Lott from his leadership position. During the 2004 presidential campaign, bloggers helped to cast doubt about the authenticity of a series of memos on which CBS based a report that was critical of President Bush's alleged failure in reporting for duty to his Texas Air National Guard unit; CBS was eventually forced to admit its reliance on an uncorroborated source (Hedges, 2005).

These examples clearly illustrate how the news agenda, particularly the online news agenda, is beginning to be dominated by the interests and efforts of these journalism amateurs. Bloggers, not having to abide by any specific journalistic ethic or reporting standard, in turn act as a sort of external check on the establishment press—or as an untamed rumor mill.

News Diffusion

In the case of the first Gulf War in 1991, news that the war had started broke on the East Coast in the early evening (when many were home watching TV) and on the West Coast in late afternoon (while many were still at work). According to Greenberg, Cohen, and Li (1993), East Coast respondents who participated in a survey were more likely to say they first heard the invasion from TV. By comparison, a larger proportion of West Coast survey respondents said they first heard the war news from personal channels than did those out east. Research on the influence of cable news and its 24-hour news cycle on network television viewing uncovered evidence of a displacement effect (Bae, 2000; Baldwin, Barrett, & Bates, 1992).

As use of new technology for news increases, the Internet will naturally play a greater role in the initial news diffusion process. Similar to 24/7 cable news outlets, the Internet should speed up the diffusion process. It is also likely to reduce TV's role as a first source of awareness of breaking events. A recent national study determined that 29% of the public turned to the Internet at least 3 days per week for news—up from 13% in 1998 and 2% in 1995 (Pew Research Center, 2004b). Hence, by the time of the 9/11 terrorist attacks, research attention had turned to the Internet.

News of the 9/11 attacks spread quite rapidly, and its diffusion was a function of four factors: the event's news value, the attacks' relevance to all, the time of day when they occurred, and the blanket coverage they triggered. According to a survey study (Greenberg, Hofshire, & Lachlan, 2002), initial diffusion of the 9/11 attacks was just about complete within 2 hours of the attacks. Half of those surveyed said they first heard the shocking news from someone else, one in three said they first heard from TV, and another third turned to the Internet for information about the attacks. Yet only 2% of respondents described the Internet as their most valuable information source.

With a national sample, Jones and Rainie (2002) found that nearly 50% said they first heard about the attacks from TV. Only 1% said they first learned about the attacks from the Internet. Although many people stayed off the Internet to watch events initially unfold on TV, about half of Internet users in Jones and Rainie's study turned to the Internet for news about the attacks within the first three weeks following 9/11. For many, then, the Internet served as a supplement to TV as well as to personal communication channels such as the telephone.

Functional Displacement

Studies examining potential displacement effects have documented displacement as well as supplemental effects. Some studies have found no relationship between use of existing and new media (see Reagan & Lee, chap. 10, this volume; Grant, chap. 9, this volume).

A good number of studies support the displacement effect (e.g., Hu, 1998; Kayany & Yelsma, 2000; Kaye, 1998; Lee & Kuo, 2002; Lin et al., 2005; Nie, 2001; Nie & Erbring, 2000; Perse & Dunn, 1998). For instance, a 1998 survey conducted by the Pew Research Center found that heavy Internet news consumers watched relatively less TV news than nonusers, even when factoring in media habits, interest in politics, and demographics (Pew Research Center, 1999). Strong evidence for displacement effects on existing news outlets comes from an investigation that relied on surveys conducted over time of TV, radio, daily and weekly newspaper, Internet, and cable use (Waldfogel, 2002). This evidence suggests that the Internet substituted for broadcast TV and daily newspapers; cable also served as a substitute for daily newspapers. Based on a nationwide survey of 792 respondents, Lin and colleagues (2005) found displacement effects as well. Newspapers were the most displaced by online news access. Cable news networks were the least displaced, with displacement effects on local TV news, radio news, and broadcast network evening news falling in between.

A number of studies suggest that online news and offline news can coexist; that online news is a supplement for offline news (e.g., Althaus & Tewksbury,

2000; Lin, 1999, 2001; Pew Research Center, 2005a; Robinson et al., 1997; Robinson & Kestnbaum, 1999; Stempel et al., 2000). Relying on national surveys of adults conducted in 1995 and 1999, Stempel et al. (2000) examined the consumption patterns of 10 media: local TV news, network TV news, daily newspapers, radio news, radio talk shows, news magazines, political magazines, grocery store tabloids, the Internet, and online services. They concluded that the decline in the use of network and local TV news, newspapers, and news magazines could not be attributed to the Internet. They found no significant difference between Internet users and nonusers on both network and local TV news consumption. Moreover, and certainly counter to the displacement perspective, they found that Internet users were more likely to read newspapers and listen to radio news than were those who did not use the Internet. A 2005 Pew poll similarly found that most Americans said they currently read the print version as much as or more often than they did before they began reading the newspaper's online version (Pew Research Center, 2005a).

Several early studies suggested that use of traditional news media was not related to the use of new media (Bromley & Bowles, 1995; Jeffres & Atkin, 1996). A recent experimental examination of the time displacement effects (Cai, 2005) found no evidence for the claim that computer use displaces time spent with traditional media.

It may be a matter of time before significant displacement effects occur. This has happened in the past. For example, according to Roper's public opinion polls, half (51%) of those studied in 1937 reported that newspapers were their primary news source. By 1945, 62% said their primary news source was radio. In 1972, the primary news source for most adults (60%) was TV (Basil, 1990). That still appears to be the case, although change may be afoot. According to a 2005 Pew poll (Pew Research Center, 2005a), 24% cited the Internet as their main source of news; 74% of respondents selected TV. (In that survey, respondents were able to list up to two choices.) This echoes findings by Lin et al. (2005), where only 1.4% reported online outlets as their primary source of news.

Although the vast majority of respondents (95%) to recent Pew Research Center surveys (2005b) classify themselves as regular consumers of some news, audiences are multitasking or telewebbing (watching video content streamed through their computer) more than ever before—they are becoming news samplers with little loyalty to any single medium. Indeed, some 36% of the American public are regular consumers of *four or more* different news media, including network and local TV, newspapers, cable, radio, the Internet, and magazines (Project for Excellence in Journalism, 2005).

As the audience for print and broadcast news continues to decline, the audience for online news has grown. In 2004, some 92 million Americans, or 42% of adults, went online for news—and about two thirds of those visited news sites three times a week or more. Getting news is now one of the most popular online activities, next only to using e-mail (Pew Research Center, 2005a). The flight of young adults away from the news may not just be content related—flagging interest in public affairs—but based on medium preferences as well. Brown (2005) notes that "young people are moving away not just from television news to the Internet, but also away from television in general."

A survey of 18- to 34-year-olds commissioned by the Carnegie Corporation in May 2004 showed that, with the exception of Web portals, the vast majority of young adults do not turn to the news on a daily basis in *any* medium (Brown, 2005). Although 44% check online portals at least once a day for news, just 37% watch local TV news on a daily basis; 19% check network, newspaper, or cable TV web sites; 18% tune into cable news networks; and just 16% watch the national broadcast networks (Brown, 2005).

Uses and Gratifications

The uses and gratifications approach assumes that audiences actively use media to satisfy a wide variety of needs (Katz et al., 1974). With interactive features and an actively engaged audience, the Internet seems like a natural fit for the uses and gratifications paradigm (Morris & Ogan, 1996). Indeed, quite a few scholars have employed the approach when examining Internet use (Charney & Greenberg, 2002; Eighmey, Ferguson, & Perse, 2000; Flanagin & Metzger, 2001; Korgaonkar & Wolin, 1999; Lin, 1999; McCord, 1998; Papacharissi & Rubin, 2000; Parker & Plank, 2000; Perse & Courtright, 1993; Perse & Dunn, 1998; Stafford, Stafford, & Schkade, 2004).

To date, few studies have focused on uses and gratifications directly associated with 24/7 *news* on the Internet (e.g., Lin, Salwen, & Abdulla, 2005). Yet uses and gratifications studies about the Internet provide some clues of its functionality here. Flanagin and Metzger (2001) found that conversation capabilities of the Internet were similar to mediated interpersonal technologies such as the telephone, and, more to the point, the Internet's information-retrieval and information-giving functions were similar to those of newspapers and TV. Such similarities run counter to earlier studies which suggested that computer-mediated communication was mostly unrelated to gratifications associated with traditional media (Perse & Courtright, 1993). Similarly, Lin (1999) found that three motivations common for traditional media—surveillance; escape, companionship, and identity; and entertainment—predicted 47% of the variance in the likelihood to adopt online services. Knowing this, online news consumption should be related to the surveillance function of traditional media.

Charney and Greenberg (2002) identified eight gratification dimensions associated with the Internet: keeping informed, diversion and entertainment, peer identity, good feelings, communication, sights and sounds, career, and "coolness." Keeping informed and communication explained 36% of the variance in the amount of time spent using the Internet. Among the eight dimensions, at least half (keeping informed, diversion and entertainment, communication, and career) should be closely related with online news consumption. Kaye and Johnson (2002) explored political information needs during the 1996 presidential election. Respondents were active Internet users who were also interested in politics. Four primary motivations were identified: guidance, information seeking, entertainment, and social utility. These four needs can be extended to consumption of general online news. Based on a survey of customers for a prominent Internet service provider, Stafford et al. (2004) derived three dimensions of uses and gratifications, one of which may be uniquely suited to the Internet: use of the medium as a social environment. This gratification dimension should also apply to the use of interactive news channels such as news discussion forums and newsgroups. Indeed, such outlets may provide details (and speculation) about an event not found elsewhere.

Uses and gratifications associated with online and other 24/7 news services will vary on the basis of the widespread availability and technological capacity of these outlets. Uses and gratifications associated with those outlets are also likely to vary on the basis of user demographics. For example, it has been consistently found that different age cohorts use the Internet in different ways—and for different purposes. Generally, older users tend to go online for news and information, whereas younger users access the Web for a wider variety of purposes, including information, but also entertainment and socializing (Severin & Tankard, 2001). A study exploring the similarities between TV and Web use among college students, specifically assessing whether Internet surfing is a functional alternative to TV viewing, found that, after school-required activities, the most frequent use of the Web was for entertainment; additional motivations included passing the time, relaxation, socializing, and information purposes (Ferguson & Perse, 2000). National surveys of Internet users in the United States have found that young people are more likely than older users to have a positive attitude about the Internet, especially in thinking of the online world as a desirable place to go for entertainment (Pew Research Center, 2004b).

The commonality of functions between the Internet and traditional news media, including 24/7 cable and satellite TV outlets, coupled with its convenience and value for social and other information functions, suggests that the Internet will—over time—displace some use of traditional news media, including those that offer 24/7 news on demand.

SOCIAL CHANGE IMPLICATIONS

Even as satellite networks, mobile communication devices, and the Internet have facilitated instantaneous, real-time, and 24-hour on-demand news diffusion, the American public remains markedly uninformed about current events. According to a recent poll, more than half of adults polled did not know how many U.S. senators serve in Congress; among American teenagers, more knew the stars of the motion picture "Titanic" than the name of our vice president; more teenagers named all three of the Three Stooges than the three branches of government (National Constitution Center, 1997).

With bookmarked sites and easy-to-digest news content available to those with the interest and access, the Internet theoretically has the potential to level the current events playing field and serve as the knowledge equalizer that other news media never became. A host of economic, technological, social, and individual level factors make this unlikely to happen, however (see Bucy & Newhagen, 2004). Others have argued that the Internet continues to fragment the audience and reduce the amount of shared knowledge (Bonfadelli, 2002; Graber, 2001).

Because many adults are not particularly interested in hard news, they are not likely to bookmark hard news sites on the Internet. Instead, as the country shifts to a celebrity culture (witness the explosion and popularity of celebrity-based magazines such as *People*, *Us*, and *Entertainment Weekly* and programs such as "E" and "Entertainment Tonight"), many Internet users are likely to flag these entertainment-oriented companion web sites that cover and celebrate the day's hot stars (Jeffres, Atkin, Bracken, & Neuendorf, 2004).

A likely consequence of this phenomenon then is a widening *knowledge gap* that reflects a digital news divide between well-informed and poorly informed information-age news consumers. The knowledge gap concept was initially coined to describe an information phenomenon that manifests over time: As information about an issue gets introduced and disseminated over time, the gap in knowledge between higher and lower socioeconomic groups tends to grow rather than diminish (Tichenor, Donohue, & Olien, 1970).

Customization features of Web portals, search engines, and news sites enable users to control what they see on their monitors and shape the news they choose to retrieve online but also create the potential for another "digital news divide." Net news sites routinely offer specialized packages of information tailored to individual interests, allowing users to provide the parameters for specific automated news searches. With that algorithm in hand, these sections deliver only those stories that are likely to interest the user.

There is, of course, something to be said for this, including arguments about efficiency, precision, and user satisfaction. At the same time, exposure to these *Daily Me* forms of news may be all that users turn to for news. If adopted by a majority of users, this would create a cascading number of small "fenced in" communities in cyberspace, a further disaggregation and fragmentation of the audience (Gitlin, 1998; Turow, 1997). This phenomenon, Pavlik and McIntosh (2004) note, "could fragment audiences into small groups of like-minded individuals who do not interact with other groups or with society as a whole and choose to receive only the news and information that reinforces their beliefs and values" (p. 24).

If a critical mass of users creates and relies on their own *Daily Me* for news, the general surveillance and civic information function of the Internet will be minimized (Lasica, 2002). This then creates another type of digital knowledge gap—one that exists between those who are immersed in "specialized" news and those who are schooled in "mainstream" news. New technology may thus have the unintended consequence of contributing to the fragmentation of society and polarization of social discussion, rather than strengthening and reinforcing a shared sense of community.

A contrasting view of personalized news holds that, by allowing the user to decide what is important and relevant, personalization reduces clutter and promises a shift in the balance of power between news providers and news consumers (Lasica, 2001). Custom information and modifiable interfaces may also generate renewed interest in the news. As part of a broader effort to harness interactive technology to bring audiences "closer to the news," personalization represents "an opportunity to engage the next generation of news consumers in the worlds of news and public affairs" (Brown, 2000, p. 26; see also Bucy, 2004).

In all, the Internet and other information technologies that deliver 24-hour news services are not likely to make news fans out of nonfans or turn soft-news seekers into hard-news enthusiasts. The trends in news consumption indicate that audiences are multitasking more than ever before, with over a third of the American public regularly sampling four or more different types of news media. Much as is the case with other media, some incidental exposure to news, whether on portal pages or search engine sites, may be difficult to bypass without at least a fleeting glance. Accidental exposure such as this has been linked with current events knowledge (Tewksbury, Weaver, & Maddex, 2001).

As more people engage with ICTs on a frequent basis—and as those connections become more mobile, less dependent on wired connections, and remain accessible 24 hours a day—the Internet will assume an increasingly important role in the news diffusion process. The challenge to informed citizenship remains whether the sampling implied by such news grazing has the capacity to actually inform, as opposed to merely cultivating a superficial sense of knowing about important developments in the world.

REFERENCES

Albarran, A., Chan-Olmsted, S., & Wirth, M. (2006). *Handbook of media management and economics*. Mahwah, NJ: Lawrence Erlbaum Associates.

Allen, I. R., & Colfax, J. D. (1968). The diffusion of news of LBJ's March 31 decision. *Journalism Quarterly*, *45*, 321–324.

Alterman, E. (2005, May 9). Bush's war on the press. *The Nation*, pp. 11–20.

Althaus, S. L., & Tewksbury, D. (2000). Patterns of internet and traditional news media use in a networked community. *Political Communication*, *17*, 21–45.

Althaus, S. L., & Tewksbury, D. (2002). Agenda setting and the "new" news: Patterns of issue importance among readers of the paper and online versions of the *New York Times*. *Communication Research, 29*, 180–207.

Andrews, P. (2003). Is blogging journalism? *Nieman Reports, 57*(3), 63–64.

Bae, H. S. (2000). Product differentiation in national TV newscasts: A comparison of the cable all-news networks and the broadcast networks. *Journal of Broadcasting & Electronic Media, 44,* 62–77.

Bagdikian, B. H. (2004). *The new media monopoly*. Boston: Beacon.

Baldwin, T. F., Barrett, M., & Bates, B. (1992). Influence of cable on television news audiences. *Journalism Quarterly, 69*, 651–658.

Basil, M. D. (1990). Primary news source changes: Question wording, availability, and cohort effects. *Journalism Quarterly, 67*, 708–722.

Belson, W. A. (1959). Effects of television on the interests and initiative of adults viewers in greater London. *British Journal of Psychology, 50*, 145–158.

Belson, W. A. (1961). The effects of television on the reading and buying of newspapers and magazines. *Public Opinion Quarterly, 25*, 366–381.

Blood, R. (2003). Weblogs and journalism: Do they connect? *Nieman Reports, 57*(3), 61–63.

Bogart, L. (1956). *The age of television: A study of viewing habits and the impact of television on American life*. New York: Frederick Ungar.

Bonfadelli, H. (2002). The Internet and knowledge gaps. *European Journal of Communication, 17*, 65–84.

Bromley, R. V., & Bowles, D. (1995). Impact of Internet on use of traditional news media. *Newspaper Research Journal, 16*(2), 14–27.

Brown, M. (2000). Bring people closer to the news. *Brandweek, 26*.

Brown, M. (2005). Abandoning the news. *Carnegie Reporter, 3*(2). Retrieved July 29, 2005, from http://www.carnegie.org/reporter/10/news/index.html.

Bucy, E. P. (2003). Media credibility reconsidered: Synergy effects between on-air and online news. *Journalism & Mass Communication Quarterly, 80*, 247–264.

Bucy, E. P. (2004). Second generation Net news: Interactivity and information accessibility in the online environment. *International Journal on Media Management, 6*, 102–113.

Bucy, E. P., & Newhagen, J. E. (Eds.). (2004). *Media access: Social and psychological dimensions of new technology use*. Mahwah, NJ: Lawrence Erlbaum Associates.

Cai, X. (2005). An experimental examination of the computer's time displacement effects. *New Media and Society, 7*, 8–21.

Charney, T., & Greenberg, B. (2002). Uses and gratifications of the Internet. In C. Lin & D. Atkin (Eds.), *Communication, technology and society: New media adoption and uses* (pp. 383–406). Cresskill, NJ: Hampton.

Coffin, T. E. (1948). Television's effects on leisure-time activities. *Journal of Applied Psychology, 32*, 550–558.

Cohen, B. C. (1963). *The press and foreign policy*. Princeton, NJ: Princeton University Press.

Eighmey, J., & McCord, L. (1998). Adding value in the information age: Uses and gratifications of sites on the World Wide Web. *Journal of Business Research, 41*, 187–194.

Fathi, A. (1973). Diffusion of a "happy" news event. *Journalism Quarterly, 50*, 271–277.

Ferguson, D. A., & Perse, E. M. (2000). The World Wide Web as a functional alternative to television. *Journal of Broadcasting & Electronic Media, 44*, 155–174.

Fine, G. A. (1975). Recall of information about diffusion of a major news event. *Journalism Quarterly, 52*, 751–755.

Finholt, T., & Sproull, L. (1990). Electronic groups at work. *Organization Science, 1*, 41–64.

Flanagin, A. J., & Metzger, M. J. (2000). Perceptions of Internet information credibility. *Journalism and Mass Communication Quarterly, 77*, 515–540.

Flanagin, A. J., & Metzger, M. J. (2001). Internet use in the contemporary media environment. *Human Communication Research, 27*, 153–181.

Furu, T. (1971). *The function of television for children and adolescents*. Tokyo: Sophia University Press.

Gantz, W. (1983). The diffusion of news about the attempted Reagan assassination. *Journal of Communication, 33*, 56 66.

Gantz, W., & Greenberg, B. S. (1993). Patterns of diffusion and information seeking. In B. S. Greenberg & W. Gantz (Eds.), *Desert Storm and the mass media* (pp. 166–181). Cresskill, NJ: Hampton.

Gantz, W., Krendl, K. A., & Robertson, S. R. (1986). Diffusion of a proximate news event. *Journalism Quarterly, 63*, 282–287.

Gitlin, T. (1998). Public sphere or public sphericles? In T. Liebes & J. Curran (Eds.), *Media, ritual and identity* (pp. 168–174). London: Routledge.

Graber, D. (2001). *Processing politics: Learning from television in the Internet age*. Chicago: University of Chicago Press.

Greenberg, B. S. (1964a). Diffusion of the news about the Kennedy assassination. *Public Opinion Quarterly, 28*, 225–232.

Greenberg, B. S. (1964b). Person-to-person communication in the diffusion of news events. *Journalism Quarterly, 41*, 489–494.

Greenberg, B. S., Cohen, E., & Li, H. (1993). How the U.S. found out about the war. In B. S. Greenberg & W. Gantz (Eds.), *Desert Storm and the mass media* (pp. 145–152). Cresskill, NJ: Hampton.

Greenberg, B. S., Hofshire, L., & Lachlan, K. (2002). Diffusion, media use and interpersonal communication behaviors. In B. S. Greenberg (Ed.), *Communication and terrorism* (pp. 3–16). Creskill, NJ: Hampton.

Hallin, D. C. (1994). *We keep America on top of the world: Television journalism and the public sphere*. New York: Routledge.

Harvey, M., & Rothe, J. (1985). Video cassette recorders: Their impact on viewers and advertisers. *Journal of Advertising Research, 25*, 19–27.

Hedges, M. (2005, January 11). Panel raps CBS News for story on Bush; Network axes four involved with the show. *Houston Chronicle*, p. A-1.

Henke, L. L., & Donohue, T. R. (1989). Functional displacement of traditional television viewing by VDR owners. *Journal of Advertising Research, 29*, 18–23.

Himmelweit, H. T., Oppenheim, A. N., & Vince, P. (1958). *Television and the child: An empirical study of the effects of television on the young*. London: Oxford University Press.

Hu, J. (1998). Net users eclipsing TV. *CNET News*. Retrieved July 30, 2005, from http://news.com.com/Study+Net+use+eclipsing+TV/2100-1023_3-209661.html.

Jeffres, L. W. (1978). Cable TV viewer selectivity. *Journal of Broadcasting, 22*, 167–177.

Jeffres, L. W., & Atkin, D. (1996). Predicting use of technologies for communication and consumer needs. *Journal of Broadcasting & Electronic Media, 40*, 318–330.

Jeffres, L. W., Atkin, D., Bracken, C. C., & Neuendorf, K. (2004). Cosmopoliteness in the Internet age. *Journal of Computer Mediated Communication, 10*(1), Article 2.

Johnson, P. (2005, March 21). It's prime time for blogs on CNN's "Inside Politics." *USA Today*, p. D-4.

Johnson, T. J., & Kaye, B. K. (1998). Cruising is believing? Comparing Internet and traditional sources on media credibility measures. *Journalism & Mass Communication Quarterly, 75*, 325–340.

Johnson, T. J., & Kaye, B. K. (2002). Webelievability: A path model examining how convenience and reliance predict online credibility. *Journalism & Mass Communication Quarterly, 79*, 619–642.

Jones, S., & Rainie, L. (2002). Internet use and the terror attacks. In B. S. Greenberg (Ed.), *Communication and terrorism* (pp. 3–16). Cresskill, NJ: Hampton.

Kaplan, S. J. (1978). The impact of cable television services on the use of competing media. *Journal of Broadcasting*, *22*, 155–165.

Katz, E., Blumler, J. G., & Gurevitch, M. (1974). Ulilization of mass communication by the individual. In J. G. Blumler & E. Katz (Eds.), *The uses of mass communications: Current perspectives on gratifications research* (pp. 19–32). Beverly Hills, CA: Sage.

Katz, E., Gurevitch, M., & Haas, H. (1973). On the use of the mass media for important things. *American Sociological Review, 38*, 164–181.

Kayany, J. M., & Yelsma, P. (2000). Displacement effects of online media in the sociotechnical contexts of households. *Journal of Broadcasting & Electronic Media, 44*, 215–229.

Kaye, B. K. (1998). Uses and gratifications of the World Wide Web: From couch potato to web potato. *New Jersey Journal of Communication*, *6*, 21–40.

Kaye, B. K., & Johnson, T. J. (2002). Online and in the know: Uses and gratifications of the Web for political information. *Journal of Broadcasting & Electronic Media*, *46*, 54–71.

Korgaonkar, P., & Wolin, L. (1999). A multivariate analysis of Web usage. *Journal of Advertising Research*, *39*, 53–68.

Kraut, R., Patterson, M., Lundmark, V., Kiesler, S., Mukopadhyay, T., & Scherlis, W. (1998). Internet paradox: A social technology that reduces social involvement and psychological well-being? *American Psychologist*, *53*, 1017–1031.

Kurtz, H. (2005, June 27). Blogged into submission; the political aggrieved let a thousand forums blame. *Washington Post*, p. C-1.

Lasica, J. D. (2001) Weblogs: A new source of news. *Online Journalism Review*. Retrieved August 11, 2005, from http://ojr.usc.edu/content/story.cfm?request=588.

Lasica, J. D. (2002). The promise of the Daily Me. *Online Journalism Review*. Retrieved July 30, 2005, from http://www.ojr.org/ojr/technology/1017778824.php.

Lasswell, H. D. (1948). The structure and function of communication in society. In Bryson (Ed.), *The communication of ideas* (pp. 37–51). New York: Harper.

Lazarsfeld, P. (1940). *Radio and the printed page*. New York: Duell, Sloan & Pearce.

Lee, W., & Kuo, E. C. Y. (2002). *Internet and displacement effect: Children's media use and activities in Singapore*. Retrieved July 30, 2005, from http://unpan1.un.org/intradoc/groups/public/documents/APCITY/UNPAN013634.pdf.

Lin, C. (1993). Modeling the gratification-seeking process of television viewing. *Human Communication Research*, *20*, 224–244.

Lin, C. (1999). Online-service adoption likelihood. *Journal of Advertising Research*, *39*, 79–89.

Lin, C. (2001). Audience attributes, media supplementation, and likely online service adoption. *Mass Communication & Society*, *4*, 19–38.

Lin, C. (2002). A paradigm for communication and information technology adoption research. In C. A. Lin & D. J. Atkin (Eds.), *Communication technology and society: Audience adoption and uses* (pp. 447–475). Cresskill, NJ: Hampton.

Lin, C. A., Salwen, M. G., & Abdulla, R. A. (2005). Uses and gratifications of online and offline news: New wine in an old bottle? In M. B. Salwen, B. Garrison, & P. D. Driscoll (Eds.), *Online news and the public* (pp. 221–236). Mahwah, NJ: Lawrence Erlbaum Associates.

Lin, C., Salwen, M. B., Garrison, B., & Driscoll, P. D. (2005). Online news as a functional substitute for offline news. In M. B. Salwen, B. Garrison, & P. D. Driscoll (Eds.), *Online news and the public* (pp. 237–256). Mahwah, NJ: Lawrence Erlbaum Associates.

Maccoby, E. (1951). Television: Its impact on school children. *Public Opinion Quarterly, 15*, 421–444.

McCombs, M. E., & Estrada, G. (1997). The news media and the pictures in our heads. In S. Iyengar & R. Reeves (Eds.), *Do the media govern? Politicians, voters, and reporters in America* (pp. 237–247). Thousand Oaks, CA: Sage.

McCombs, M. E., & Shaw, D. L. (1972). The agenda-setting function of the press. *Public Opinion Quarterly, 36*, 176–187.

Morris, M., & Ogan, C. (1996). The Internet as mass medium. *Journal of Communication, 46*(1), 39–50.

National Constitution Center. (1997). *Startling lack of constitutional knowledge revealed in first-ever national poll.* Retrieved July 30, 2005, from http://www.constitutioncenter.org/CitizenAction/CivicResearchResults/NCCNationalPoll/index.shtml.

Newhagen, J. E., & Rafaeli, S. (1996). Why communication researchers should study the Internet: A dialogue. *Journal of Communication, 46*(1), 4–13.

Nie, N. H. (2001). Sociability, interpersonal relations, and the Internet: Reconciling conflicting findings. *American Behavioral Scientist, 45*, 420–435.

Nie, N. H., & Erbring, L. (2000). *Internet and society: A preliminary report.* Retrieved August 11, 2005, from http://www.stanford.edu/group/siqss/Press_Release/Preliminary_Report 4 21.pdf.

Nielsen Media Research. (1999). *TV viewing in Internet households.* Retrieved August 11, 2005, from http://www.nielsen-netratings.com/.

Nieman Reports. (2003). Weblogs and journalism. *Nieman Reports,* 57(3), 61–98. Retrieved July 30, 2005, from http://www.nieman.harvard.edu/reports/contents.html.

Palser, B. (2002, July/August). Journalistic blogging. *American Journalism Review.* Retrieved on July 14, 2005, from http://ajr.org/article.asp?id=2571.

Papacharissi, Z., & Rubin, A. M. (2000). Predictors of Internet usage. *Journal of Broadcasting & Electronic Media, 44*, 175–196.

Parker, B. J., & Plank, R. E. (2000). A uses and gratifications perspective on the Internet as a new information source. *American Business Review, 18*, 43–49.

Patterson, T. E. (2001). *Doing well and doing good.* Working paper, RWP01-001. Kennedy School of Government, Harvard University. Retrieved July 31, 2005, from http://ksgnotes1.harvard.edu/Research/wpaper.nsf.

Pavlik, J. V., & McIntosh, S. (2004). *Converging media: An introduction to mass communication.* Boston: Allyn & Bacon.

Perse, E. M., & Courtright, J. A. (1993). Normative images of communication media: Mass and interpersonal channels in the new media environment. *Human Communication Research, 19*, 485–503.

Perse, E. M., & Dunn, D. (1998). The utility of home computers and media use: Implications of multimedia and connectivity. *Journal of Broadcasting & Electronic Media*, *42*, 435–456.

Pew Research Center. (1999). *The Internet news audience goes ordinary*. Retrieved July 30, 2005, from http://www.people-press.org/tech98mor.htm.

Pew Research Center (2004a). *Cable and Internet loom large in fragmented political news universe*. Retrieved July 30, 2005, from http://www.people-press.org/reports.

Pew Research Center (2004b). *The Internet and daily life*. Retrieved July 30, 2005, from http://www.pewinternet.org/pdfs/PIP_Internet_and_Daily_Life.pdf.

Pew Research Center (2005a). *Trends 2005*. Retrieved July 30, 2005, from http://pewresearch.org/trends.

Pew Research Center (2005b). *Buzz, blogs, and beyond*. Retrieved July 30, 2005, from http://www.pewinternet.org/ppt/BUZZ_BLOGS__BEYOND_Final05-16-05.pdf.

Postman, N. (1986). *Amusing ourselves to death: Public discourse in the age of show business*. New York: Penguin Books.

Project for Excellence in Journalism. (2005). *State of the news media*. Retrieved July 30, 2005, from http://stateofthemedia.org/2005/index.asp.

Reagan, J. (1984). Effects of cable television on news use. *Journalism Quarterly*, *61*, 317–324.

Robinson, J. P. (1969). Television and leisure time: Yesterday, today and (maybe) tomorrow. *Public Opinion Quarterly*, *33*, 210–222.

Robinson, J. P., Barth, K., & Kohut, A. (1997). Social impact research: Personal computers, mass media, and use of time. *Social Science Computer Review*, *15*, 56–82.

Robinson, J. P., & Godbey, G. (1999). *Time for life: The surprising ways Americans use their time* (2nd ed.). University Park: Pennsylvania State University Press.

Robinson, J. P., & Kestnbaum, M. (1999). The personal computer, culture and other uses of free time. *Social Science Computer Review*, *17*, 209–216.

Rosengren, K.E. (1987). The comparative study of news diffusion. *European Journal of Communication*, *2*, 227–255.

Rubin, A. (1984). Ritualized and instrumental television viewing. *Journal of Communication, 34*(3), 67–77.

Ruggiero, T. E. (2000). Uses and gratifications theory in the 21st century. *Mass Communication and Society*, *3*, 3–37.

Scherer, C. W. (1989). The videocassette recorder and information inequity. *Journal of Communication*, *39*, 94–103.

Schweitzer, J. C. (1991). Personal computers and media use. *Journalism Quarterly*, *68*, 689–697.

Severin, W. J., & Tankard, J. W. (2001). *Communication theories: Origins, methods, and uses in the mass media* (5th ed.). New York: Addison Wesley Longman.

Singer, J. B. (2005). The political j-blogger: "Normalizing" a new media form to fit old norms and practices. *Journalism*, *6*, 173–198.

Smolkin, R. (2004, June/July). The expanding blogosphere. *American Journalism Review*. Retrieved July 30, 2005, from http://www.ajr.org/article.asp?id=3682.

Soroka, S. N. (2003). Media, public opinion, and foreign policy. *Harvard International Journal of Press/Politics*, *8*(1), 27–48.

South, J. (1999, October 14). Extra! Makes a comeback: Breaking local news online. *Online Journalism Review*. Retrieved July 30, 2005, from http://www.ojr.org/ojr/business/1017967313.php.

Sparkes, V. M. (1983). Public perception of and reaction to multi-channel cable television service. *Journal of Broadcasting*, *27*, 163–175.

Stafford, T. F., Stafford, M. R., & Schkade, L. L. (2004). Determining uses and gratifications of the Internet. *Decision Sciences*, *35*, 259–288.

Stempel, G., & Hargrove, T. (1996). Mass media audiences in a changing media environment. *Journalism and Mass Communication Quarterly*, *73*, 549–558.

Stempel, G., Hargrove, T., & Bernt, J. (2000). Relation of growth of use of the Internet to changes in media use from 1995–1999. *Journalism and Mass Communication Quarterly*, *77*, 71–79.

Tewksbury, D., Weaver, A. J., & Maddex, B. D. (2001). Accidentally informed: Incidental news exposure on the World Wide Web. *Journalism and Mass Communication Quarterly, 78*, 533–555.

Tichenor, P. J., Donohue, G. A., & Olien, C. N. (1970). Mass media flow and differential growth in knowledge. *Public Opinion Quarterly*, *34*, 159–170.

Turow, J. (1997). *Breaking up America: Advertisers and the new media world.* Chicago: University of Chicago Press.

Vitalari, N. P., Venkatesh, A., & Gronhaug, K. (1985). Computing in the home: shifts in the time allocation patterns of households. *Communications of the ACM*, *28*, 512–522.

Waldfogel, J. (2002). *Consumer substitution among media.* Philadelphia: Federal Communications Commission Media Ownership Working Group.

Wanta, W., Golan, G., & Lee, C. (2004). Agenda setting and international news: Media influence on public perceptions of foreign nations. *Journalism & Mass Communication Quarterly*, *81*, 364–377.

Welch, M. (2005). Who gets to play journalist? *Reason Online*. Retrieved July 14, 2005, from http://www.reason.com/0506/co.mw.who.shtml.

Zeller, T., Jr. (2005, May 23). Are bloggers setting the agenda? It depends on the scandal. *New York Times*, p. C-5.

Zurawik, D. (2005, July 13). CBS announces move toward 24-hour online video news; division aims to take on rivals ABC, NBC. *Baltimore Sun*, p. C-3.

V

ENTERTAINMENT SETTING

9

Video Technology and Home Entertainment

August E. Grant

In 1999, the Internet boom prompted great interest in finding ways to distribute entertainment content over the Web. In a meeting with the president of an Internet-related startup company, an entrepreneur shared his vision of distributing his broadband hardware to consumers at no cost, paying for the hardware with revenues from providing pay-per-view content to consumers over the Internet. Fueling his plan was a set of economic analyses based on the earnings revenues from a substantial portion of the 7 to 8 hours of TV consumed in the average household each day. The plan fell apart as soon as he learned that the average household rented an average of less than four movies a month, less than one movie per week. After paying for the rights to distribute movies, hardware and network costs, and marketing costs, he realized that less than a dollar per month would be left to pay for his device. He dropped the idea.

The primary lesson in this story is that, in order to understand the consumption of home video technology and programming, a wide range of organizational, economic, and technical factors must be examined. This chapter attempts to provide such an understanding, contextualizing home video technology through the use of economics, media consumption theories, and other related factors.

BACKGROUND

One way to help understand a phenomenon as complex as TV and home video is to study the set of components that make up that phenomenon. The "Umbrella Perspective on Communication Technology" (Grant, 2004) specifies that, in order to understand any communication technology, you must separately analyze five different components of that technology: hardware, software, organizational infrastructure, social system, and individual users. Each of the elements of perspective illustrates a different dimension of the technology, and relationships among them are implicit in the discussions that follow.

The first step in understanding home video is detailing the specific technologies that are discussed in this chapter. This section defines the technologies in terms of the elements just described, exploring equipment (hardware), programming (software), the organizations that provide both of these (organizational infrastructure), and how they are used by individual audience members. The discussion begins with the technology that is most fundamental to home video—the TVdisplay. It then expands to include peripheral devices that connect to the TV display, moving from there to explore programming and the complex issues surrounding the interplay of organizational behavior and consumption of programming by individual users.

Home Video Entertainment Technologies

The range of new TV display technologies has put a great deal of attention on TV hardware in recent years. Although consumers have long demonstrated that programming and economics are much more important than picture quality in choosing entertainment programming, the dramatic differences in TV display technologies have drawn a great deal of attention.

For the first 30 years of consumer availability, TV receivers were differentiated on only two dimensions: how large the picture tube was and whether the picture tube displayed color or monochrome images. As the demand for larger TV sets increased and the limits of tube technology intervened, TV manufacturers developed various projection systems that were quite different form the traditional TV picture tube, allowing much larger pictures. In the process of enlarging these pictures, however, the limitations in the quality of TV transmission systems became apparent. As researchers (e.g., Dupagne, 2002) recount, this desire for larger and clearer pictures led to the development of high-definition TV, which, in turn, led to the development of digital TV.

Digital Broadcasting and High Definition TV. For purposes of this discussion, it is important to consider the implications of the impending abandonment of analog TV in favor of digital TV. Based on the most recent congressional action, the U.S. Senate has set February 18, 2009, as the deadline for full implementation of converting all terrestrial TV broadcasting services from an analog to a digital standard (Dupagne & Seel, 2006). This deadline signifies the time when all analog TV sets will become obsolete and will only be able to receive the digital broadcast signals with the aid of a stand-alone digital-to-analog converter. However, given that digital television (DTV) receivers are several times more expensive than the analog receivers that they will replace, questions concerning the viewers' marginal utility for the new technology remain unanswered (e.g., Atkin, Neuendorf, Jeffres, & Skalski, 2003).

The most obvious implication of this transition is that traditional analog TV sets—which use cathode-ray tubes—are slowly being replaced by digital TV displays that usually provide a much larger picture, a thin display, or both. The key question to consider, however, is the level of motivation for consumers to upgrade to digital technology. The engineers who design these high-definition displays might point to the superior resolution, the improved color, or the ability to better display motion. But, as suggested in the discussion of "social information processing" discussed elsewhere in this book, consumers are much more likely to point to the thin nature of flat panel displays or the wider aspect ratio possible with digital TV. In fact, at average viewing distances, most consumers probably will not be able to discern the difference in resolution between high-definition and standard definition displays. A consumer must be between two and four screen heights from the display to see a difference. At any further distance away, the increased resolution is not apparent (Bracken & Atkin, 2004; Grant, 1993b).

Peripheral Devices. Delivery of most forms of home video is facilitated by one or more devices attached to a TV receiver. These include: videocassette recorders (VCRs), DVD players, digital video recorders (DVRs), cable and satellite set-top boxes, external audio amplifiers and speakers, and other types of set-top boxes.

The most established of these is the VCR, which was first introduced to the home market in the 1970s, diffusing quickly in the 1980s (Secunda, 1990). The rapid diffusion of VCRs and the potential of the device to alter the nature of TV viewing led to numerous research projects on the use and impacts of VCRs (see e.g., Dimmick, 2003; Dobrow, 1990a; Levy, 1989; Lin, 1992, 2002; Noh & Grant, 1997; Rubin & Eyal, 2002).

One of the most important findings of these studies was the identification of the two primary functions of VCRs: time shifting (allowing a viewer to record a program for later viewing) and source shifting (allowing a viewer to consume content on a TV that was formerly not available, such as watching a movie or a travel video; e.g., Rubin & Eyal, 2002; Secunda, 1990). Most of the peripherals just listed allow some type of source shifting, expanding the choices available to TV viewers (Lin, 2001). The notable exception is the DVR, which allows users to record TV programming from almost any source for later playback, as well as allowing users to pause live TV and (for recorded programs) to skip through commercials.

The fact that none of the newer technologies listed earlier offers the same dual functions of time shifting and source shifting as offered by VCRs suggests room in the home video marketplace for devices that allow DVD recording. As of this writing (late 2005), a few such devices have been introduced to the market, but the comparatively high price has limited their market penetration.

Among the home video peripherals listed earlier, the DVR has the greatest potential to revolutionize the TV marketplace. A DVR user has the option of watching programs when they are scheduled, pausing them at will, or automatically recording programs to watch at leisure, with the potential of fast-forwarding through commercials.

The key difference between the DVR and other home video technologies is the need for the DVR to regularly download an electronic program guide that is customized for the city, cable TV service, and/or satellite system connected to the DVR. As a result, three separate business models have emerged to support DVRs. Companies such as TiVo and ReplayTV require consumers to buy equipment and then charge a monthly fee for the program guide (with the alternative of paying a large, up-front "lifetime subscription" fee). Cable TV and satellite companies offering DVRs, in contrast, see the technology as a service, charging a monthly fee with no equipment cost. Finally, a few companies sell equipment only, setting aside a portion of the equipment cost to amortize the cost of providing the electronic program guide.

Home Video Entertainment Content

The majority of home video software is TV programming, with movies running a distant second. This section analyzes factors underlying the availability and consumption of all forms of TV programming.

TV Programming. Just as traditional terrestrial-based broadcast TV networks compete against cable TV services for audience shares, the latter also compete against broadcast satellite (DBS) for subscribers. As of spring 2005, cable TV programs surpassed broadcast TV networks in their prime time (Eastern time 8–11 p.m.) viewing ratings (Nielsen Media, 2005), continuing a trend that started more than 20 years ago (e.g., Krugman & Rust, 1993). Although cable TV channels are available in 66.8% of U.S. homes with TV, another 25% of homes are direct broadcast satellite (DBS) subscribers, with each offering over 100 channels of programs (NCTA, 2005).

In the year 2009, when the Federal Communications Commission (FCC) implements the digital broadcasting standard on over-the-air broadcast services, the competitive landscape will also change among the broadcast TV, cable TV, and satellite TV services. Digital broadcasting technology enables terrestrial broadcasters, which formerly could each transmit only one signal in its assigned spectrum capacity, the technical capability to broadcast several digitally compressed signals through the same spectrum space. By comparison, cable TV and DBS have already been providing digital services for nearly a decade. This abundant media environment reflects a wide diversity in channels and further fragments the audience, even though diversity of channels does not equate to diversity in terms of either substantive content choices or quality.

Home Video Content. Although home video software revenues are about equally divided between sales and rentals, the type of content in each category can be dramatically different. Traditionally, theatrical movies comprised the bulk of video rentals, children's videos, and movies-dominated video sales. Over the past 10 years, however, a new category is taking an increasing share of home video sales revenues: prime-time TV programs. Program producers and syndicators have discovered a new revenue stream from packaging and selling TV series, usually one season at a time. For example, after a stellar network run and equally successful off-network syndication, producers of "Seinfeld" began releasing episodes of that situation comedy on DVD, one season at a time. As of the end of 2004, more than 4 million copies of the first three seasons of "Seinfeld" had been sold, with sales of the remaining six seasons expected to equal or exceed those numbers (Snider, 2005).

The popularity of off-network programming in home video follows a trend that began with independent TV stations and continued with cable TV. Although the new media have the potential to distribute new programs and new types of programs, a disproportionate amount of time and (independently) a disproportionate amount of audience attention are devoted to programming that first appeared in prime-time TV program schedules or in theaters (Abelman & Atkin, 2002; Waterman & Grant, 1991).

THEORETICAL PERSPECTIVES

As the preceding discussion illustrates, the biggest factors in determining the success of any home video technology are usually not technological factors, but some combination of technology, individual users, and organizational factors. The interplay of these different levels of analysis can be examined from many different perspectives. As we have already explicated the technology factors, the following discussion examines this interplay dynamic using the theories stemming from both the economic and audience perspectives.

Economic Perspective

Economics of Distribution. Home video content can either be paid for by advertising support (indirect pricing), subscription/pay-per-view (direct pricing), or, rarely, a combination of the two. One of the keys to understanding why some content is distributed through broadcasting and other advertiser-supported media versus some type of pay-per-view can be found in examining the underlying economics of direct versus indirect pricing.

The factor that makes broadcasting different from most other media is the fact that it costs about the same amount to provide the content to one person as it does to thousands (or, in some cities, millions). In economic terms, virtually all of the costs of broadcasting are fixed costs, with the marginal cost of adding additional viewers being zero. As a result, content that will attract large audiences can be more efficiently distributed through indirect pricing (advertising support), rather than having consumers pay directly for the content. However, content that is of interest to a relatively small audience can be more efficiently distributed using direct-pricing media such as pay-per-view and video rental (Grant, 1993a; Reagan, 2002). Moreover, the production of most home video content has larger variable costs than fixed costs. In turn, the cost of the content (software) makes up a much greater proportion of the variable costs than the costs of the physical distribution of the content, whether by disk, tape, or online (Komiya & Litman, 1990).

The Principle of Relative Constancy. The economic resources of individuals must also be considered in understanding the home video economic landscape. Consider that the average household consumes just over 8 hours per day of TV (with the average *person* watching between 3 and 4 hours of TV a day), and weekly household use totals about 50 hours of TV (Media InfoCenter, 2005d). The average household also pays to watch about four movies a month through either video rental or pay-per-view TV, taking up about 2 hours of that 50 hours per week. The question is why so little TV is directly paid for by the individual user. One factor is certainly the presence of so much advertiser-supported TV programming, available to the viewer at no charge per viewing. But an equally important factor is the economic resources—the money—consumers have to devote to home video.

One perspective on the limits of consumer spending on entertainment media is the "principle of relative constancy" (McCombs & Nolan, 1992), which states that, over time, the proportion of disposable income spent on entertainment media is relatively constant. Study of more than 100 years of history of consumer spending on media generally supports the principle of relative constancy. The implications for the media are quite important—almost any new medium requiring consumer spending on equipment, services, programming, and so on must "take" that spending from other existing media or other leisure options.

It should be noted that consumer spending on media is not quite a zero-sum game. During good economic times, when disposable income increases, consumers spend more on media, making these times the best to introduce new media options. There is also one notable exception to the principle of relative constancy—the introduction of the videocassette recorder in the 1980s in the United States was accompanied by an increase in the proportion of disposable income devoted to media. This "VCR aberration" has been attributed to the fact that the VCR enabled a new set of functional alternatives for consumption of media content, leading to an increase in the amount of disposable income devoted to the media (Noh & Grant, 1997).

Technology Clusters and Media Substitution

These various home entertainment technologies may belong to the same technology clusters, a concept that Rogers (2003) uses to describe groups of functionally similar innovations (e.g., videocassettes, cable TV, remote control devices) that are often adopted collectively or in sequence. Further, in the parlance of diffusion theory, these technology clusters may involve innovations that offer a *relative advantage* over incumbent technologies. In cases where innovations fail to meet user expectations, they may be *reinvented* to better suit those needs. For instance, VCRs were primarily designed for playback purposes, but were used to do more time shifting and home-video library building (e.g., Dobrow, 1990a; Klopfenstein, Spears, & Ferguson, 1991; Levy, 1981; Lin, 2000; Rubin & Eyal, 2002). VCRs also became a tool for creating social viewing opportunities with families and friends (e.g., Krendel, Clark, Dawson, & Troiano, 1993; Lin, 1992). As Rubin and Eyal (2002) suggest, VCRs, DVDs, and DVRs inhabit a similar innovation cluster centered around home video. Digital cable, pay-per-view, and video on demand are also likely to occupy a similar cluster focused on enhanced viewing (Atkin et al., 2003).

The principle of relative constancy (PRC; see Reagan & Lee, chap. 10, this volume) suggests that the media compete for a relatively fixed pool of audience leisure time, among other factors. Lasswell (1948) first applied the concept over a half century ago, when he correctly predicted the displacement of radio by the nascent TV medium. Thus, a new medium like cable or the VCR must derive its sustenance by displacing its traditional media forebears (see e.g., Henke & Donohue, 1989). Lin (1999) notes that "audience members may substitute the use of a functionally similar medium for another when such a substitution need arises and the circumstance presents itself" (p. 80). She posits, for instance, that audiences would use prerecorded video playbacks as a replacement for movie-outing activity because video facilitates better audience control over household leisure time and budget allocations. These substitution mechanisms may not be as readily apparent, however, when media with distinctive origins begin to converge (e.g., the computer and cable) and begin to compete for the delivery of video and other services (see Reagan, 2002).

Uses and Gratifications. In their seminal work, Katz, Blumler, and Gurevitch (1974) define the focus of uses and gratifications (U & G) as: "(1) the social and psychological origins of (2) needs, which generate (3) expectations of (4) the mass media or their sources, which lead to (5) differential patterns of media exposure (or engagement in other activities), resulting in (6) need gratifications…" (p. 20).They further note that such media use is based on the assumption that the audience is active and initiates the choice of medium in response to those needs. In that regard, the U & G approach is an umbrella perspective that includes several different models, enabling "researchers to study mediated communication situations via a single or multiple sets of psychological needs, psychological motives, communication channels, communication content, and psychological gratifications" (Lin, 1996, p. 574).

The uses and gratifications perspective has been employed more than any other perspective to explain individual motives for adoption and use of home

TABLE 9.1
Media Consumption: Hours per Person per Year (2003)

Medium	*Hours per Person per Year (2003)*
Cable and satellite TV	949
Broadcast TV	778
Home video	67
Box office	13
Interactive TV	2

Note: From Media InfoCenter (2005a).

video technologies (Rubin & Eyal, 2002). Simply put, the perspective suggests that the strength of an audience member's motivation for using a new medium—be it for the purposes of diversion, escape, information, or any number of gratifications—determines the nature and extent of his or her media adoption and use. Dupagne (2002) posits that adopters of a more visually compelling medium—like high-definition TV hardware —will be more strongly motivated by a need to be stimulated by program content (e.g., outdoors reality programs).

EMPIRICAL FINDINGS

Economic Perspective

As illustrated in Table 9.1, the majority of time spent viewing TV is devoted to broadcast and cable programming. Despite the relatively small amount of time devoted to watching movies and related home video content, this sector of the entertainment industry commands vast revenues. For comparison, revenues from all forms of broadcast TV in 2003 (including all network and local revenue) totaled just under $40 billion (Media InfoCenter, 2005b), whereas revenues from home video sales and rentals for the same year totaled $24 billion (Dick, 2004).

The fact that so much viewing of motion pictures takes place in the home via DVD, cable TV, or broadcast TV has led to speculation regarding why motion pictures cannot receive their initial release to the home video market. Indeed, a few studios have attempted nontheatrical releases of major motion pictures, with the most notable example being "The Pirates of Penzance" in 1983. Without exception, these releases have failed, and *direct-to-video* has become a term used to describe films that are perceived as not being good enough to open in theaters.

The underlying process in the motion picture industry that is at play is referred to as *intertemporal tiering* (Waterman, 1985). Intertemporal tiering is the process of releasing a media product through a planned series of media over time, starting with the media that provide the highest revenues per user (initial theatrical release), then moving to slightly lower revenues (dollar houses, airlines, and hotel pay-per-view), then to home video, home pay-per-view, pay cable, and, finally, basic cable TVand broadcast syndication.

Technology Clusters and Media Substitution

Recent industry figures suggest that analog broadcast TV can be found in 98% of American households, and penetration levels for other electronic media are as follow: VCRs (92%), cable (60%), DVDs (72%), direct broadcast satellite (27%), and digital TV (10%; see Brown, 2006). As noted earlier, research on the principle of relative constancy notes a VCR aberration to the displacement mechanisms outlined earlier, which suggests that the facilities offered by the VCR were so compelling as to prompt audiences to increase media expenditures in order to adopt the VCR (Noh & Grant, 1997). That said, studies of audience time suggest that the VCR does not so much displace as supplement regular TV viewing (e.g., Henke & Donohue, 1989; Lin, 1994, 2001). VCRs are, however, likely to displace moviegoing (e.g., Childers & Krugman, 1987; Dick, 2004; Henke & Donohue, 1989; Lin, 1993).

Although findings on this point have been mixed with cable TV—with some studies noting positive relationships with overall TV viewing (e.g., Reagan, 2002)—aggregate studies of broadcast viewership indicate that audiences are substituting cable for broadcast TV (e.g., Krugman & Rust, 1993; Lin & Atkin, 2002). Thus, as Reagan (2002) concludes, the literature on VCR displacement dynamics seems rather inconsistent, a finding that suggests that simple conceptions of displacement or complementarity among media uses may overlook some of the nuances in audience media selection and use. Audiences may, instead, be drawn to selectively use media that are functionally similar (Atkin et al., 1998) to media that provide comparable contents or gratifications.

One of the biggest factors affecting selection of home video alternatives is awareness that the choices exist. In a 500-channel universe, it is virtually impossible for a person to know all of the programs that are available. Instead, viewers are usually aware of a small set of programs in which they have a strong interest and, through viewing the channels that offer these programs, become aware of other programs offered on those channels. The result is that most people's TV viewing is clustered around a limited number of channels. Studies of "channel repertoire" indicate that the typical viewer watches only a small fraction of the number of channels available on a regular basis. For example, viewers who can receive 20 channels watch an average of only about 6 channels regularly, while viewers who can receive 120 or more channels watch fewer than 20 of those channels regularly (Media InfoCenter, 2005c). The problem for new programs and new channels is creating awareness among viewers and getting viewers to add a new channel to their repertoire because adding a channel typically means replacing a current channel in that person's channel repertoire (Ferguson & Perse, 1993; Heeter, 1985; Neuendorf, Atkin, & Jeffres, 2001).

Uses and Gratifications

As outlined earlier, the U & G perspective is assuming a central role in explaining audience use behaviors with new media, particularly as they become more widely diffused in the population. Empirical work in this tradition has addressed the underlying motives behind the usage patterns of VCRs, for instance, and the manner in which consumers use VCRs to gratify needs and have a greater degree of control over their viewing (Rubin & Bantz, 1989; Rubin & Rubin, 1989), and the issue of audience activity in VCR use (Krugman & Johnson, 1991; Lin, 1990). Lin (1994, 2001) notes that the basic audience motives for seeking either traditional or new video media contents are similar, finding that new media adopters are typically more active, motivated, and satisfied audience members.

As of this writing, few comparable studies of DVRs have been published. Findings from one of the first such studies, however, demonstrate gratifications similar to those reported for VCR use, with a strong link between the ability to control TV programming and the enjoyment derived from the viewing experience (Ferguson & Perse, 2004). In his investigation of research addressing a range of digital home video technologies, Dupagne (2002) found little support for the notion that audience adoption is related to greater needs for stimulation. Although the nascent state of digital TV penetration limits the amount of uses and gratifications work conducted in that area, preliminary work (e.g., Atkin et al., 2003) indicates that adoption intentions are linked with a greater audience need for program formats that are not widely available on broadcast TV.

In comparative terms, Dimmick (2003) found that VCRs offer relatively broader gratification activities, followed by cable and finally broadcast TV. He concludes that the VCR's ability to supply greater gratification opportunities can, in turn, help account for the home video revolution. Such results confirm a raft of findings indicating that cable audiences are generally more strongly motivated viewers and derive greater satisfaction from TV, relative to their broadcast counterparts (e.g., Jeffres & Atkin, 1996; Lin, 1994). Similarly, preliminary work on expanded (e.g., 500+) channel repertoires, available over the Internet, suggests that adoption of multimedia cable is also related to viewing motivation levels (Lin & Jeffres, 1998).

SOCIAL CHANGE IMPLICATIONS

From a technical perspective, many manufacturers have started to produce devices that combine TVs and computers in a single unit, using the liquid crystal display technology. These prototypical devices have not yet found success in the market largely because of the manner in which the TV set and personal computer are still being used from a functional point of view. Computers are typically used by a single person who is located no more than 2 feet from the viewing screen. In contrast, TVs are watched as often by groups as by individuals, with typical viewing distance of 7 to 10 feet (e.g., Bracken & Atkin, 2004).

Even so, the technology underlying the next generation of TVs is likely to be much more similar to computer technology than tradition TV technology, as microprocessors and memory are built into TV sets in order to enhance functionality with advanced program guides, digital TV decoders, and DVR capabilities. The conversion to digital TV may eliminate the need for much of the electronics related to encoding and decoding TV signals, making it possible to build DVR capability into almost any TV (or any set-top box). This incorporation of DVRs into TV sets will become more common as the cost of hard drives and other data storage falls. The result may well be a TV set with electronics that are almost functionally indistinguishable from those in a computer, although the uses of these functions will be more oriented toward TVentertainment rather than computer-driven tasks (e.g., Rubin & Eyal, 2002).

From an economic perspective, among all of the home video technologies discussed in this chapter, the DVR may have the greatest economic implications. By giving viewers the ability to skip commercials and control when they want to watch TV programs, DVRs undermine the advertising base that provides the economic foundation for broadcast TV and a good portion of cable TV. This problem is not a new one—when VCRs were first introduced, the same fears were expressed, but skipping of commercials during VCR playback ultimately amounted to a small fraction of viewing (Sapolsky & Forrest, 1989). The ease of use of DVRs, combined with the potential to pause live viewing and then skip through commercials, suggests that the challenge represented by the DVR to advertising revenues will be more formidable than that presented by the VCR in the 1980s. As the issue develops, look for advertisers to explore product placement and other advertising opportunities that will not be affected by DVR use.

From a user perspective, it appears that more content is available than ever for home entertainment consumption, and audiences now can have extensive control over what they watch and when they watch it. Given these two developments, it may logically be assumed that the amount of home video programming consumed will increase, but analyses of previous technologies indicate that this may not be the case. Rather, it is more likely that availability of "any program, any time, anywhere" will lead to less overall program consumption. The empowerment that viewers have from using DVRs and other time- and source-shifting devices will decrease the immediacy of consuming most programming (Rubin & Eyal, 2002). Studies of VCR use (Dobrow, 1990b) demonstrate that more than a third of all programming recorded for later viewing is never watched. Once a person can watch a particular program at his or her leisure, the notion of "appointment TV," where a person sets aside time to watch a particular program when it is scheduled by a network, disappears (Abelman & Atkin, 2002). (The primary exceptions to this rule are live sports and breaking news, where audience members want to watch developments as they occur in real time.)

Another significant phenomenon accompanying the use of home video technologies is how the increase in viewer control has increased audience utility for TV (Dimmick, 2003). Not only have these technologies changed the way that families watch TV and movies, but they have also changed the family viewing culture (Andreason, 2000; Ferguson & Perse, 2004). Examples include using videos as a babysitter (Lin, 2000), negotiating family viewing choice (Krendl et al., 1993; Rubin & Eyal, 2002), and planning social activities around a video showing (Lin, 1992, 2000; Morgan, Alexander, Shanahan, & Harris, 1990). Nonetheless, this family viewing phenomenon may be changing as proliferation of home media entertainment options is allowing individual family members to pursue their own favorite home entertainment fare (e.g., Bryant & Bryant, 2000; Jeffres, Atkin, Neuendorf, & Lin, 2004). As a case in point, the popularity of video games and different types of online activities has reduced the amount of time that families spend on watching TV or movies together (see Klopfenstein, 2002; Montgomery, 2000).

In the "brave new world" narrowcasting, the availability of almost any movie or home video fare poses challenges for those who are concerned about the programming to which children their children are exposed (see Krcmar & Stirzhakova, chap. 4, this volume). The debate over standards regarding sexual and violent content on TV is typically focused on broadcast TV (e.g., Abelman & Atkin, 2002), but almost any type of content is available to children who have access to a DVD player. Research has also found that adolescents are disproportionately attracted to R-rated, adult content on home video (Atkin, 2000; Greenberg & Lin, 1989; Montgomery, 2000). This issue, which may become more important with the increased use of the Internet to distribute content for home video consumption, must consider the full range of content and media available to viewers of all ages.

In conclusion, the proliferation in media content and technological capabilities represents agents of perpetual change to the home video entertainment marketplace. Nonetheless, regardless of the number of options, most audiences will always be tuned to a small set of favorite channels and programs (e.g., Jeffres et al., 2004). This phenomenon is related to the fact that the amount of time, money, and interest that audience members have to devote to home video entertainment is limited, no matter what abundance of content can be made available due to advances in technological or programming devices. Finally, the most important question that remains is what this easy access to such advanced digital media environment and growing entertainment repertoire may mean to the American family's home entertainment culture, which plays a significant role in the socialization of our children in society.

Only time can tell what the ultimate social impact of this continuing audience erosion might be. Just as audience fragmentation represents a threat to old economic structures like the broadcast networks, emerging channels like the Internet may give rise to a fragmentation of our popular political and cultural discourse (e.g., Davis, 1999). Later work should explore the impact of this as well as any corollary fragmentation in family viewing, which, based on our experience with new video technologies, seems poised to accelerate as emerging multimedia applications promise even greater audience control and content selectivity. In addition to the sheer diversity offered by new channels—now encompassing such niche favorites as "Queer Eye", "L-word," and poker, as well as skiing channels—emerging channels will also offer audiences greater capabilities to receive and manipulate their contents. This might occur with a digital program transmission that is downloaded to a viewer's iPod, for instance, and later downloaded and edited on a home work station. In the end, however, it remains to be seen whether these new contents and facilities will maximize audience utilities for TV or simply represent a multiplicity of mediocrity.

REFERENCES

Abelman, R., & Atkin, D. (2002). *The televiewing audience*. Cresskill, NJ: Hampton.

Andreason, M. (2000). Evolution in the family's use of television: An overview. In J. Bryant & J. A. Bryant (Eds.), *Television and the American family* (pp. 3–30). Mahwah, NJ: Lawrence Erlbaum Assoicates.

Atkin, D. (2000). The home ecology of children's television viewing. In J. Bryant & J. A. Bryant (Eds.), *How children view television* (pp. 49–74). Mahwah, NJ: Lawrence Erlbaum Associates.

Atkin, D., Jeffres, L., Neuendorf, K. (1998). Understanding Internet adoption as telecommunication behavior. *Journal of Broadcasting & Electronic Media, 42*, 475–490.

Atkin, D., Neuendorf, K., Jeffres, L., & Skalski, P. (2003). Predictors of audience interest in adopting digital television. *Journal of Media Economics, 16*(3), 159–173.

Bracken, C., & Atkin, D. (2004). How screen size affects perception of television. *Visual Communication Quarterly, 9*, 9–14.

Brown, D. (2006). Communication technology timeline. In A. E. Grant & J. H. Meadows (Eds.), *Communication technology update* (pp. 7–45). Boston: Focal Press.

Bryant, J., & Bryant, J. A. (2000). *How children view television*. Mahwah, NJ: Lawrence Erlbaum Associates.

Childers, T., & Krugman, D. (1987). The competitive environment of pay-per-view. *Journal of Broadcasting & Electronic Media, 31*(3), 335–342.

Davis, R. (1999). *The web of politics: The Internet's impact on the American political system.* New York: Oxford University Press.

Dick, S. J. (2004). Home video technology. In A. E. Grant & J. H. Meadows (Eds.), *Communication technology update* (9th ed., pp. 216–224). Boston: Focal.

Dimmick, J. (2003). *Media competition and coexistence: The theory of niche*. Mahwah, NJ: Lawrence Erlbaum Associates.

Dobrow, J. (1990a). *Social and cultural aspects of VCR use*. Hillsdale, NJ: Ablex.

Dobrow, J. (1990b). The re-run ritual: Using VCRs to re-view. In *Social and cultural aspects of VCR use* (pp. 181–194). Hillsdale, NJ: Ablex.

Dupagne, M. (2002). Adoption of high-definition television in the United States: An Edsel in the making? In C. Lin & D. Atkin (Eds.), *Communication technology and social change: Audience adoption and uses* (pp. 279–306). Cresskill, NJ: Hampton.

Dupagne, M. & Seel, P. B. (2006). Digital television. In A. E. Grant & J. H. Meadows (Eds.), *Communication technology update* (10th ed., pp. 49–69). Boston: Focal Press.

Ferguson, D. A., & Perse, E. M. (1993). Media and audience influences on channel repertoire. *Journal of Broadcasting & Electronic Media, 37*, 31–47.

Ferguson, D. A., & Perse, E. M. (2004). Audience satisfaction among TiVo and ReplayTV users. *Journal of Interactive Advertising, 4*(2).

Grant, A. E. (1993a). Prospects for video on demand. *New Telecom Quarterly, 1*(3), 51–55.

Grant, A. E. (1993b). HDTV: An assessment of impacts and applications. *New Telecom Quarterly, 1*(2), 38–45.

Grant, A. E. (2004). The umbrella perspective on communication technology. In A. E. Grant & J. H. Meadows (Eds.), *Communication technology update* (9th ed., pp. 1–6). Boston: Focal.

Greenberg, B. S., & Lin, C. A. (1989). Adolescents and the VCR boom: Old, new, and nonusers. In M. R. Levy (Ed.), *The VCR age: Home video and mass communication* (pp. 73–91). Newbury Park, CA: Sage.

Heeter, C. (1985) Program selection with abundance of choice: A process model. Human *Communication Research, 12*, 126–152.

Henke, L., & Donohue, T. (1989). Functional displacement of traditional TV viewing by VCR owners. *Journal of Advertising Research, 29*(2), 18–23.

Jeffres, L., & Atkin, D. (1996). Predicting use of technologies for communication and consumer needs. *Journal of Broadcasting & Electronic Media, 40*, 318–330.

Jeffres, L. W., Atkin, D. J., Neuendorf, K. A., & Lin, C. A. (2004). The influence of expanding media menus on audience content selection. *Telematics and Informatics, 21*(4), 317–334.

Katz, E., Blumler, J. G., & Gurevitch, M. (1974). Utilization of mass communication by the individual. In J. G. Blumler & E. Katz (Eds.), *The uses of mass communication* (pp. 19–32). Beverly Hills, CA: Sage.

Klopfenstein, B. (2002). The Internet and Web as communication media. In C. Lin & D. Atkin (Eds.), *Communication technology and society: Audience adoption and uses* (pp. 353–378). Mahwah, NJ: Lawrence Erlbaum Associates.

Klopfenstein, B. C., Spears, S. C., & Ferguson, D. A. (1991). VCR attitudes and behaviors by length of ownership. *Journal of Broadcasting & Electronic Media, 35*(4), 525–531.

Komiya, M., & Litman, B. (1990). The economics of the prerecorded videocassette industry. In J. Dobrow (Ed.), *Social and cultural aspects of VCR use* (pp. 25–44). Hillsdale, NJ: Ablex.

Krendl, K. A., Clark, G., Dawson, R., & Troiano, C. (1993). Preschoolers and VCRs in the home: A multiple methods approach. *Journal of Broadcasting and Electronic Media, 37*(3), 293–312.

Krugman, D. M., & Johnson, K. F. (1991). Differences in the consumption of traditional broadcast and VCR movie rentals. *Journal of Broadcasting and Electronic Media, 35*, 213–232.

Krugman, D. M., & Rust, R. T. (1993). The impact of cable penetration and VCR penetration on network viewing: Assessing the decade. *Journal of Advertising Research, 35*, 213–232.

Lasswell, H. (1948). The structure and function of comunication in society. In L. Bryson (Ed.), *The communication of ideas* (pp. 37–51). New York: Harper.

Levy, M. R. (1981). Home video recorders and time shifting. *Journalism Quarterly, 58*(3), 401–405.

Levy, M. R. (1989). *The VCR age: Home video and mass communication.* Newbury Park, CA: Sage.

Lin, C. A. (1990). *Audience activity and VCR use.* In J. Dobrow (Ed.), *Social and cultural aspects of VCR use* (pp. 75–92). Hillsdale, NJ: Ablex.

Lin, C. A. (1992). The functions of the VCR in the home leisure environment. *Journal of Broadcasting and Electronic Media, 36*, 345–351.

Lin, C. A. (1993). Exploring the role of the VCR in the emerging home entertainment culture. *Journalism Quarterly, 70*, 833–842.

Lin. C. A. (1994). Audience fragmentation in a competitive video marketplace. *Journal of Advertising Research, 34*(4), 30–38.

Lin, C. A. (1996). Looking back: The contribution of Blumler and Katz's *Uses of mass communication* to communication research. *Journal of Broadcasting & Electronic Media, 40*, 571–581.

Lin, C. A. (2000). The VCR, home video culture, and new media technologies. In J. Bryant & J. A. Bryant (Eds.), *Television and the American family* (pp. 91–107). Mahwah, NJ: Lawrence Erlbaum Associates.

Lin, C. A. (2001). Audience attributes, media supplementation and likely online service adoption. *Mass Communication & Society, 4*(1), 19–38.

Lin, C. A. (2002). A paradigm for communication and information technology adoption research. In C. A. Lin & D. Atkin (Eds.), *Communication technology and society: Audience adoption and uses* (pp. 447–475). Cresskill, NJ: Hampton Press.

Lin. C. A., & Atkin, D. (2002). *Communication technology and society: Audience adoption and uses* (pp. 23–42). Cresskill, NJ: Hampton.

Lin, C. A., & Jeffres, L. W. (1998). Factors influencing the adoption of multimedia cable technology. *Journalism & Mass Communication Quarterly, 75*, 341–352.

McCombs, M. E., & Nolan, J. (1992). The relative constancy approach to consumer spending for media. *Journal of Media Economics, 5*(2), 43–52.

Media InfoCenter. (2005a). *Media consumption based on hours per person.* Retrieved June 9, 2005, from http://www.mediainfocenter.org/film/competition/consumption.asp

Media InfoCenter. (2005b). *Communication industry revenue.* Retrieved June 9, 2005, from http://www.mediainfocenter.org/compare/revenues/tot_revenues.asp

Media InfoCenter. (2005c). *Channels receivable vs. viewed.* Retrieved June 9, 2005, from http://www.mediainfocenter.org/television/size/r_vs_viewed.asp

Media InfoCenter. (2005d). *Time spent viewing—Households.* Retrieved September 19, 2005, from http://www.mediainfocenter.org/television/tv_aud/time_house.asp

Montgomery, K. C. (2000). Children's media culture in the new millennium: Mapping the digital landscape. In *The future of children: Children and computer technology* (Vol. 10, part 2, pp. 145–167). Los Altos, CA: David and Lucille Packard Foundation.

Morgan, M., Alexander, A., Shanahan, J., & Harris, C. (1990). Adolescents, VCRs, and the family environment. *Communication Research, 17*, 83–106.

National Cable & Telecommunications Association. (2005). *Industry statistics.* Retrieved November 10, 2005, from http://www.ncta.com/Docs/PageContent.cfm?pageID=86.

Neuendorf, K., Atkin, D., & Jeffres, L. (2001). Reconceptualizing channel repertoire in the urban cable environment. *Journal of Broadcasting & Electronic Media, 45*(3), 464–482.

Nielsen Media. (2005). *Annual report on* viewing. New York: Author.

Noh, G. Y., & Grant, A. E. (1997). Media functionality and the principle of relative constancy: An explanation of the VCR aberration. *Journal of Media Economics, 10*(3), 17–31.

Reagan, J. (2002). The difficult world of predicting telecommunication innovations: Factors affecting adoption. In C. A. Lin & D. Atkin (Eds.), *Communication technology and society: Audience adoption and uses* (pp. 65–87). Cresskill, NJ: Hampton.

Rogers, E. M. (2003). *Diffusion of innovations* (5th ed.). New York: The Free Press.

Rubin, A. M., & Bantz, C. R. (1989). Uses and gratifications of videocassette recorders. In J. Salvaggio & J. Bryant (Eds.), *Media use in the information age: Emerging patterns of adoption and use* (pp. 181–195). Hillsdale, NJ: Lawrence Erlbaum Associates.

Rubin, A. M., & Eyal, K. (2002). The videocassette recorder in the home media environment. In C. A. Lin & D. J. Atkin (Eds.), *Communication technology and society: Audience adoption and uses* (pp. 329–349). Cresskill, NJ: Hampton.

Rubin, A. M., & Rubin, R. B. (1989). Social and psychological antecedents of VCR use. In M. R. Levy (Ed.), *The VCR age: Home video and mass communication* (pp. 92–111). Newbury Park, CA: Sage.

Sapolsky, B. S., & Forrest, E. (1989). Measuring VCR "ad-voidance." In M. R. Levy (Ed.), *The VCR age: Home video and mass communication* (pp. 148–167). Newbury Park, CA: Sage.

Secunda, E. (1990). VCRs and viewer control over programming: An historical perspective. In J. Dobrow (Ed.), *Social and cultural aspects of VCR use* (pp. 9–24). Norwood, NJ: Ablex.

Snider, M. (2005, January 6). DVD continues spinning success. *USA Today*. Retrieved June 30, 2005, from http://www.usatoday.com/life/movies/news/2005-01-05-dvd-sales-inside_x.htm

Waterman, D. (1985). Prerecorded home video and the distribution of theatrical feature films. In E. Noam (Ed.), *Video media competition: Regulation, economics and technology* (pp. 221–243). New York: Columbia University Press.

Waterman, D., & Grant, A. E. (1991). Cable television as an aftermarket. *Journal of Broadcasting and Electronic Media, 35*(2), 179–188.

10

Online Technology, Edutainment, and Infotainment

Joey Reagan and Moon J. Lee

This chapter examines the potential social impact of online media. With the rapid changes in access to and content of the Internet, the potential for major changes in how people use media is great. Some people view this as positive, such as enhancing the ability to develop grassroots movements. Others are less sanguine about negative implications, such as the expansion of opportunities for people to become gambling dependent.

The Internet arose from the U.S. defense information network in 1973. This was followed by the NSFNET in 1986 (which still provides a large component of today's Internet), and BITNET and CSNET in the 1980s (a scientific exchange with e-mail), in addition to the Corporation for Research and Educational Networking (CREN), designed for academic networking and supported by member dues (Internet Society, 2004). The domain name system (DNS) came from the University of Wisconsin, and in 1992 the World Wide Web (WWW) was released by CREN (Kristula, 2005). From 1993 on, browsers allowed expansion into commercial and entertainment use (Pizza Hut offered the first online ordering service). A milestone occurred in 2004 when, for the first time, a majority of domain names were located outside of the United States (Zakon, 2005).

Recently, more users are on the Internet and they are more likely to have high-speed access. Early in the new millennium, over 50% enjoyed Internet access at some location—home, work, or school (U.S. Census Bureau, 2005)—and major network web sites, newspaper web sites, and cable news (such as C-SPAN and MSNBC) web sites were most used for news (Pew Research, 2004).

In terms of high-speed Internet connections, access through DSL is available to 11% of the U.S. population, and broadband cable penetration is roughly double that (National Cable, 2005). The Internet puts more emphasis on user control aspects of information consumption (Lee, 2001). For instance, individuals can create or access blogs to share their own news and opinions, play interactive computer games, and talk with friends or strangers. There are online dating services and Usenet News Groups for those who are not well served by traditional news. These uses may be based on human needs associated with Internet use: information, communication, entertainment, and escape (e.g., Charney & Greenberg, 2002; Jansen, Spink, Bateman, & Saracevic, 1998; Katz & Aspden, 1997; Kraut, Kiesler, Boneva, Cummings, Hegelson, & Crawford, 2002; Lin, 2004; Papachrissi & Rubin, 2000; University of California, 2003).

Some argue that the Internet is particularly useful in civil society at mobilizing grassroots social movements such as civic protests and demonstrations (Rhee & Kim, 2004). For example, one of the U. S. presidential candidates, Howard Dean, utilized the Internet as a core part of his campaign and fundraising base (Pew, 2004). However, others (LaRose, Lin, & Eastin, 2003; Lin & Yan, 2002) are concerned about the negative impacts of online media environments, such as Internet addiction (e.g., computer game addiction, online gambling, or online sex addiction) and antisocial activities (e.g., online scams, organized terrorist or cult activities, and criminal activities).

The uses and potential impacts of online entertainment on society bear careful examination by communication researchers. The "Theoretical Explanations" section of this chapter reviews the potential for theories to account for these possible social changes, and the "Empirical Findings" section looks at research about social impacts in these areas:

- Education and edutainment (e.g., hobbies, marketing, shopping).
- Social isolation, well-being, and anxiety.
- Information and infortainment, and blogs (e.g., TV shows, movies, music, video, sports, entertainment news).
- Entertainment (e.g., podcasting, webcasting, etc.).
- Dependency and addiction (games, gambling, porn, etc.).

THEORETICAL EXPLANATIONS

There are few theories that provide good generalizations about the social impact of the Internet because there are so many different uses and types of users. This section provides examples of the limits of theory in the domain of online entertainment media adoption.

Obviously, if people are seeking information or entertainment on the Internet, then theories that assume a "strong audience," like "uses and gratifications" (U & G) theory, apply. Lin (2002b), for example, found that there were two predictors of online service from U & G, "escape" and "information seeking", but that conception fails to fully capture the complexity of this theory. Charney and Greenberg (2002) found that "keeping informed" was such a dominant factor that it accounted for 40% of the variance in Internet use gratification, overwhelming the major predictor, "diversion and entertainment." Moreover, there remains the question of causality. These limitations were noted by LaRose, Mastro, and Eastin (2001), who argued that "Studies of the uses and gratifications of the Internet have tended to repeat the pattern of weak predictions of media behavior common to this body of research" (p. 2). They proposed that "Social-cognitive theory explains behavior in terms of reciprocal causation among individuals, their environments, and their behaviors" (p. 3).

Like LaRose et al. (2001), Knobloch, Carpentier, and Zillman (2003) rejected U & G because of "inconsistencies and troublesome ambiguities," noting that U & G lacks "specificity in its operationalizations" (p. 3). Thus, it ignores the needs that consumers have for news, for instance, but instead uses "what news should be" as the criterion. Further pros and cons associated with uses and gratifications theory—and its applicability in other media adoption domains—are discussed in the chapters by Caplan et al. and Grant (this volume).

Knobloch et al. (2003) examined and discarded various theories that researchers had used to explain online news use. A few of those covered here demonstrate the difficulty of finding a single strong theory to explain the social effects of the Internet. For instance, the "knowledge-gap hypothesis"—which suggests that new media can widen gaps between media "haves and have-nots"—was found deficient because it ignored "interest." Interest in a topic influences selective exposure to information about that topic. So outcomes, like knowledge gaps, will widen whenever there is a difference in interest between population segments, *not* solely because of access differentials.

Diffusion theory (viz. Rogers, 2003) explores adoption dynamics, including the critical point at which people adopt new services, when they are accessible, perceived as sufficiently reliable, and so on. In particular, the perspective focuses on the attributes of an innovation and of those who are relatively earlier to adopt it (e.g., an emerging Internet channel; see chapters by Grant and by Krcmar & Strizhakova, this volume). Innovation theory predicts that innovative individuals would have the education and resources necessary to acquire new media technologies, along with the psychological traits (e.g., innovativeness, need for information) that could motivate such adoption (Lin, 1998). Adoption would also be related to perceptions of an innovation's observable benefits, trialability, relative advantage, complexity, and compatibility with a user's values, needs, and lifestyle. For instance, at some point, enough people will have access to the Internet—or enough people connect to certain sites confidently—that the behavior becomes a social asset or a social problem. Once an innovation like the Internet reaches a critical mass, there may be a point at which its relative advantage will be so great that traditional news sources decline and the supply of news diminishes.

Convergence (e.g., Baldwin, McVoy, & Steinfield, 1996; Lin & Jeffres, 1998) applies as a theoretical concept because the behaviors arise out of the merging of media and related behaviors, such as the use of electronic chats for social discourse. As Atkin (2002) notes, new media are rendering a convergence along economic (e.g., mergers like AOL–Time Warner) and technology domains. Lin (2002a) refers to a telematic channel like the Internet as an Intermedia medium, owing to its ability to operate in mass (e.g., Web) and interpersonal (e.g., e-mail) modalities. Her theory of fluidity provides an explication of this concept, drawing from the Internet's ability to fluidly morph from one such modality to the next and even recoil to an earlier application.

The media substitution hypothesis flows from the principle of relative constancy, discussed in Grant's chapter (this volume), and suggests that new media like the DVD will displace functionally similar existing media. This displacement notion dates back to the 1940s, when Lasswell (1948) correctly predicted that TV would compete with radio for audience leisure, program talent, and advertising revenue. Although this has not held true for all emerging media—the VCR aberration, for example—a new medium must generally derive its sustenance from resources that would otherwise go to preexisting media. More recent work (e.g., Henke & Donohue, 1989; Lin, 1998; Lin & Jeffres, 1998; Perse & Courtwright, 1993) finds evidence of this displacement effect across a range of emerging media.

The concept of *repertoires*—the principle that people select a smaller set of offerings from the total based on their interests or needs (see e.g., Heeter & Greenberg, 1985; Reagan, 1996)—would predict that people do not use all of the Internet's offerings, but rather select a small number of sites to which they return on a regular basis. Jeffres, Atkin, Neuendorf, and Lin (2004), for instance, theorized that Internet users prefer moderate levels of content diversity—in contrast to unbounded content environments—because they are threatened by limitless choice.

Functional similarity (Atkin, 1993), the principle that some media may be more successful because they fit existing media uses and improve on existing technology, is also a sound concept. Drawing from the notion of innovation compatibility, as outlined in diffusion theory, audiences would endeavor to adopt new media that fall within a technology cluster of functionally similar media (or services; Lin & Atkin, 2002). One would expect that Internet services that provide more efficient alternatives for existing behaviors—say, faster news sites for those who already use traditional media outlets—would be more successful.

In *parasocial interaction* (e.g., Horton & Wohl 1956; Rubin, Perse, & Powell 1985), media users identify with the characters in media, which may lead to higher viewership or increased consumption of products related to the character's behavior, and it might apply to electronic interactions. Parasocial interaction represents one of several media uses and gratifications, defined earlier, that can predict new media adoption and use (Charney & Greenberg, 2002; Lin, 1998, 2002b, 2004; Rubin & Eyal, 2002). To some extent, all interaction potentially can engender a relationship. For instance, Grant and Meadows (2002) report that parasocial interaction is a critical variable for any kind of online retailing, as customers "spend more time with these hosts than with many of their closest friends, and the result is a continuing psychological need to interact with these hosts, both by watching and by making purchases from the service" (p. 270). According to that logic, audiences are likely to be more engaged with (and feel greater affinity toward) media vehicles that are better able to satisfy these interaction needs (see Li, chap. 1, this volume). Although often applied in the context of teleshopping, this dynamic explains the success of an interactive web site for Diet Pepsi featuring Ray Charles.

Augmenting parasocial interaction is *media dependency theory* (e.g., Ball-Rokeach, 1976; Fry & McCain, 1983; McDonald, 1983), which examines the relationships between users and the media (rather than personality influences on media use). Grant and Meadows (2002) suggest that audience members develop a relationship with a medium as the user "learns that the medium can help him or her fulfill a specific goal or set of goals" (p. 270). Following that logic, this dependency would magnify any media effect, such that users indicating higher levels of dependency would be more likely to engage in such behaviors as home shopping (see Lin, chap. 1, this volume). Although studied in a variety of contexts, dependency is a social-system-based theory that assumes audiences may come to rely on a medium if it helps fulfill one's (a) cognitive and affective needs, or (b) media-use goals based on past use experiences (Ball-Rokeach & DeFleur, 1985; Grant, Guthrie, & Ball-Rokeach, 1991).

Psychological impacts may also be a fruitful area to explore. Does the change from traditional media to the virtual world affect the way people physically respond in their minds (e.g., Biocca, 1997)? One would predict that, as interaction increases—for example, from watching news to selecting news stories to chatting on a blog to a virtual reality experience in a video game—one might feel that the interaction is more "real" and perhaps more credible or engender other internal responses. Virtual media with greater levels of "presence" would thus prove more compelling vehicles (e.g., Biocca & Nowak, 2002), perhaps even engendering the kinds of parasocial relationships reviewed above, in conjunction with human actors.

Perhaps what one needs is more of a model rather than a theory to follow behavior as it takes place on the Internet. Lin (2002a) provided the *dynamic interactivity* model. This model posits a set of factors that determines uses of new media. For example, "adoption factors" determine whether someone will even use the media in the first place. Technical factors affect whether the interaction is possible (and, of course, there is room for mutual causation, such as adoption that is contingent on such factors as ease of use; technical factors can also influence adoption, as when users reinvent how a technology is used). Although this model has the potential to more generally explain Internet behavior, there is just not enough research available to fully develop the model at this time.

As the preceding review has shown and as the empirical findings show next, these theories do not fully explicate social impacts, especially those occurring across users. That is, no single theory provides broad explanations for the impacts that are found. So one should expect that no theory could or would be developed that accurately predicts technology use, behavior, or impacts. What seems to work in predicting Internet impacts are needs, social factors, and so on, but those are not fully explanatory theories. The fact that there is no clear set of theoretical explanations for the social impacts of the Internet does not mean none of the theories will prove useful in the future. It just means there is a lot of "trench work" to do. More research should examine the characteristics and motivations people take to their Internet experience. Toward that end, it is useful to provide a review of empirical findings on the adoption of online media entertainment services.

EMPIRICAL FINDINGS

Access, Digital Divide, and Knowledge Gap. The first consideration of how technology impacts people is whether people have access. In the realm of mass media, scholars (e.g., Tichenor, Donohue, & Olien, 1980) reported gaps in knowledge between high- and low-status groups, which grew as the volume of information diffused increased. In the new media realm, research on Internet diffusion reveals a link with education, which can enlarge knowledge gaps (Atkin, Jeffres, & Neuendorf, 1998). Here the major research thrusts involve the "digital divide" and the related issue of a "knowledge gap" that arise when people have either a lack of access or a lack of skills or knowledge in the use of the technology (see e.g., Donohue, Tichenor, & Olien, 1995; Hoffman, Novak, & Schlosser, 2000; Speight, 1999). For instance, Bonfadelli (2002) uncovered a gap in Internet use by education, one that had implications of the knowledge gap. The social consequence is a loss of economic and informational power.

However, that effect is mediated by the internal motivations that people bring to their use of technologies (see e.g., Hoffman et al., 2000; LaRose & Eastin, 2002). A study by Althaus and Tewksbury (2000), for instance, found that students used the Web primarily for entertainment, although use of online media as a news source was related to newspaper reading.

Education and Learning. A benefit of technologies may be through intentional learning (distance education) and incidental and lifelong learning. Recent articles have noted that the Internet is useful for accessing resources and sharing work (Gordon, Gomez, Pea, & Fishman, 1996), "widening opportunities" for lifelong learning and virtual campuses (Beller & Or, 1998), tying together multiple campuses to enhance learning (Wheeler, Valacich, Alavi, & Vogel, 1995), and providing a flexible learning environment (Bryant, Campbell, & Kerr, 2003). Chester and Gwynne (1998) revealed that in the relative anonymity of the electronic classroom students can develop community identities and more commitment with a related reduction in antisocial behavior, which leads to a better learning environment. Online interactive technologies are not only useful for in-class learning (Woods, Shimon, Karp, & Jensen, 2004), but can directly serve people outside of the formal learning environment with entertaining information content as well. This mixing of education and entertainment is "edutainment." For instance, a PBS web site may feature a Sesame Street videogame that teaches kids how to read (see Krcmar & Strizhakova, chap. 4, this volume, for a discussion of online applications for children) is an example of this edutainment phenomenon.

Online technology has also made a difference in the field of health and medicine, from doctors seeking to convey health information to those trying to sell herbal remedies (see Whitten, chap. 12, this volume). Although "information therapy" has been encouraged since 1992 (Lindner, 1992), it was not until recently that use of the Internet as a medium for this therapy was tested. For example, D'Alessandro, Kreiter, Kinzer, and Peterson (2004) found that "information prescriptions," training patients to use the Internet, and providing online information increased the use of high-quality information sources. Yet there is the risk that a wealth of misleading and downright dangerous information is being purveyed on the Internet. Chapman (1996) observed that stories about and coverage of health may be misleading. A specific example is a human growth hormone that is being distributed (illegally in the United States) by thousands of sites supplying inaccurate information about the drug. One e-marketer was responsible for "5 deaths, 18 heart attacks, 26 strokes, and 43 seizures" (Perls, 2004, p. 683).

Social Interaction. As noted earlier, the human need for communication may be associated with Internet use. That use may be positive or negative. For example, dependency on games and pornography is noted later in this chapter, and some studies have found that heavier Internet use negatively impacts users. In some cases, instant messaging may lead to more anxiety (Gross, Juvonen, & Gable, 2002), and heavier users of the Internet may experience social withdrawal (Li & Chung, 2006), depression (Shapira, Goldsmith, Keck, Kholsab, & McElroy, 2000), and feelings of loss of control (Armstrong, Phillips, & Saling, 2000). Several researchers are concerned that overuse of the Internet will lead to fewer social interactions offline and increases in loneliness and depression, but some of these negative effects are small and not necessarily alarming (Kraut et al., 1998).

Conversely, Shotton's (1991, 1989) work found that heavier computer use led to increased self-esteem and less stress. Others posited that people might seem more appealing to each other when their communication is mediated by technology—relative to face-to-face communication—while amount of Internet use did not affect adolescent psychological well-being (McKenna, Green, & Gleason, 2002).

Walther and Boyd (2002) examined computer-mediated communication and found that bulletin boards and chat rooms (i.e., both in real time and with delayed messaging) provided support for emotions and self-esteem. They also acknowledged the Internet's limits on providing actual physical support without in-person intervention (such as buying groceries for a shut-in). Atkin, Jeffres, Neuendorf, Lange, and Skalski (2005) found that interest in a diversity of communication sources and current events were the most powerful predictors of chat room use.

Information. The potential for the Internet to be a potent information technology is exemplified by the use of online news sources and the recent phenomenon of blogs. For example, in Australia, there were over 5.5 million unique hits recorded for the top five newspapers, with over 150 million pages read in the month of November 2004 (Nielsen NetRatings, 2004). In the United States, although 30% to 40% of the electorate receive campaign information from traditional sources, those numbers are declining—down 10% for network news from 2000. Usage of other news sources is increasing, including the Internet, which was used by 4% of the electorate. Traditional media are also well represented online, as 11% of people get their news from web sites of news organizations (Pew Center, 2004; see also Jeffres, chap. 7, this volume).

Blogs and Infotainment. An interesting trend on the Internet is the development of blogs. Although initially blogs were simply individual journals or diaries, they have developed into alternative sources of infotainment. Blogs have the highest credibility among blog users for news; as with all online sources being perceived as more credible than traditional sources, TV news was considered the lowest credibility news source (Johnson & Kaye, 2004). Gilmor (2003) has a positive view of weblogs, arguing that they democratize society by decentralizing and augmenting news gathering. However, this optimism must be tempered by the fact that most blogs overtly espouse a position or belief that colors their content.

The possible expansions of information sources might be dampened, nevertheless, by a highly discussed phenomenon, whereby much of news has been merged with entertainment, called "infotainment." This phenomenon can portend great social influence because it reflects an important system factor that examines the fabric that sustains our social structure—namely, how people become informed about politics, their community, and their lives. Some (e.g., Dholakia, Mundorf, & Dholakia, 1996) define infotainment as the merging of the new technical advances (information technologies) with entertainment, but this chapter uses the definition that relates to the potential social impact of the merging of informative content (news, documentary, and other usable information) with entertainment content (see e.g., Brants, 1998; Brants & Neijens, 1998).

Two recent reports in *Broadcasting and Cable* and *Multichannel News* (Rosenberg, 2004; Grebb, 2004, respectively) highlight this phenomenon. Rosenberg reported that there really is no line anymore between news and entertainment, with "Americans ages 21–29 [using] Comedy Central's *Daily Show* and NBC's *Saturday Night Live* as primary sources for election news, almost equaling the percentage watching nightly newscasts" (p. 32).

Studies of traditional media find infotainment in many forms, in news and information, drama, and other programs (Surette & Otto, 2002), and virtually "all of the informational programs examined contained some elements of entertainment" (Brants & Neijens, 1998, p. 1). As the Internet and other interactive technologies become major sources of information for the public, the question of whether such information will be substantive or ephemeral arises. About a decade ago, the Pew Center documented the changing use of the Internet for information (Pew Research Center, 1996), with almost three fourths of online users getting some news from the Internet. Some 12% of adults obtained political information from the Internet, and 40% got news about science, health, and technology from the Internet.

More recent work (e.g., Lin, 2004; Salwen, Garrison, & Driscoll, 2005) suggests that the Internet does not appear to compete directly with traditional sources because those who use the Internet for news also use traditional sources just as much as they did before, although online news users tended to be younger and less interested in local news. The biggest change in Internet news use was for "infotainment," where "the largest increase in online activities has occurred among users who go online to get entertainment-related information (30% do this at least weekly, compared to 19% in 1995) and for financial information (22%, compared to 14% in 1995)" (Brants & Neijens, 1998, p. 14). More recent work (e.g., Pew Research Center, 2004) suggests that these trends have continued, although at a more moderate pace. Having reviewed the utilities associated with personal communication technologies, it is useful now to focus on some of the perceived downsides of Internet use.

Dependency and Addiction. The psychiatric community has discussed TV addiction (see e.g., McIlwraith, 1998), but the Internet is the first of the interactive technologies that has been labeled by the psychiatric community as a disorder, called "Internet behavioral dependence" (see e.g., Hall & Parsons, 2001). The symptoms include typical addictive symptoms, such as a need for longer amounts of time online, withdrawal when reducing Internet use, and use of the Internet for mood modification. Some studies categorize as many as 60% of Internauts as dependent, but Hall and Parsons' review of previous work arrives at a 6% estimate. Griffiths (1999) found that those addicted to the Internet are "an exceedingly tiny minority." Young (1997) suggests that the Internet may engender dependency and addiction much like drugs and eating disorders, and that heavy users primary need social support, sexual fulfillment, and the creation of a persona.

There is a concern that students may have poorer study habits or grades, absence from classes, or be disciplined because of excessive Internet use (Young, 1996). Other dependency problems relate to pornography (Freeman-Longo & Blanchard, 1998), gambling—among high school and college students (Mitka, 2001)—depression (LaRose, Lin, & Eastin, 2003), teacher relations, studies, peer relations (Yang & Tung, 2005), social withdrawal (Li & Chung, 2006), depression (Shapira, et al., 2000), and loss of control, productivity, and sleep (Armstrong et al., 2000). Petrie and Gunn (1998) discovered that heavy Internet use was negatively correlated with extroversion and positively correlated with self-reported depression, concluding that heavy Internet users are introverted and more likely to be suffering from depression or low self-esteem (Griffiths, 1995). Those with addictive symptoms report an increased sense of loneliness (Morahan-Martin & Schumacher, 2000).

Some assume that individuals' cognitive (or affective) styles may be related to Internet use (Tan & Lo, 1991). Adventurous individuals are less likely to express feelings of disorientation when reading hypertext than nonadventurous individuals (Lee, 2001). Amichai-Hamburger, Fine, and Goldsmith (2004) discovered that an individual's need for closure had a significant effect on the person's preference of a web site with interactive features like hyperlinks. This finding indicates that needs for sensation or closure may influence how much and what type of stimulation users would seek through their access to online media environments. Of course, for the majority, excessive Internet use seems to be incidental or temporary in the sense that it does not disrupt their daily routine and well-being.

Zuckerman (1983) suggested that individuals have different optimal levels of desired stimulation and that this tendency is related to their preferences for certain media types (Donohew, Lorch, & Palmgreen, 1991), such as videogame addiction (Chiu, Lee, & Huang, 2004) or violent web site content (Slater, 2003). However, empirical data seem to suggest mixed results. Lin and Yan (2002), studying a sample of Taiwanese high school adolescents, found a positive relationship between sensation seeking and Internet addiction. However, Armstrong et al. (2000) found that low self-esteem was related to excessive use of the Internet rather than sensation-seeking tendencies. Still another study resulted in Internet-dependent subjects scoring lower on sensation seeking than did nondependent subjects (Lavin, Marvin, McLarney, Nola, & Scott, 1999).

When examining specific dependencies like gambling addiction, some research shows that these are not addictive behaviors that are caused by the technologies. The ability to "blame" the technologies is limited by most studies' methods, which tend to be self-report surveys—and many of those involve online volunteer samples. Charlton (2002) notes that most research on computer addiction tends to be anecdotal. More important, internal characteristics tend to lead to the reported addictions and dependencies (e.g., Griffiths, 1995). So the technology merely provides an outlet for those who are prone to develop psychological and/or behavioral dependency. In the case of gambling dependency, the Griffiths, Richard, and Wood (2000) review of the research on addiction found that risk factors for gambling, videogame, and Internet dependence were internal characteristics such as arousal or depression, and some researchers attribute the behavior to being "overzealous" rather than addicted (Kraut et al., 1998). Still, they note that access to technologies facilitates engaging in the dependent behavior.

SOCIAL CHANGE IMPLICATIONS

With so many positive and negative findings, and with so many research programs and theoretical explanations still evolving, it is difficult to make affirmative generalizations about the impact of the Internet on society much less what theories explain that impact. Scholarly work (e.g., LaRose & Eastin, 2002) suggests that the social change implications for online entertainment content are as multifaceted as the Internet medium and can be summarized as both positive and negative. The Internet's primary impact is to offer access to information and entertainment as well as goods and services at any location where people can go online. For the gambler, the Internet offers easy access to gaming sites. For the person suffering from a disease, the Internet may offer access to coping information. For the workaholic, the Internet offers constant access to work-related correspondence, even from remote locations. Bearing in mind that some research areas overlap (e.g., shopping addiction and e-commerce), a set of generalizations can be offered for later research.

The first is that the Internet simply offers easier access to entertainment content and related products. This was summarized by Tyler (2002) when he reviewed an entire issue of the *Journal of Social Issues* (Vol. 58, No. 1, 2002) devoted to the social impact of the Internet, noting that there is not much social change, just that the Internet has created a "new way of doing old things" (p. 195). Others (e.g., Atkin et al., 2005) suggest that once distinctive Internet behaviors are becoming normalized over time.

Second, individual characteristics and motivations are most important. In the same *Journal of Social Issues* already cited, the opening article points out that "One theme stands out: These effects depend on how the unique qualities of Internet communication modes interact with the particular characteristics and goals of the individuals, groups, and communities using them" (Bargh, 2002, p. 1).

Previous studies have shown that Internet use is affected by many factors such as social and demographic characteristics, individuals' attitudes toward the Internet, and their support groups such as family or peers (Atkin et al., 1998; Lin, 1998; Rogers, 1995; Rhee & Kim, 2004; Zhu & He, 2002). Until more research like this is done, conclusions about how or why the Internet seems to have an impact will not be possible.

Third, there is a lack of clear theory, which results in a need for research development. Sawhney (1997), reviewing a book on infotainment technologies, concluded that, "Although the chapters … are individually insightful, they do not coalesce into a meta-framework of any sort. This lack of coherence, however, is understandable as this area of research is still underdeveloped and these studies are first steps in a new direction" (p. 443). Moreover, it is also unlikely that there could be a theory that is sufficiently comprehensive in scope to explain—as well as predict—the vast array of potential behavioral outcomes and social impacts stemming from Internet use.

Fourth, there is a need to study these impacts related to other technologies. Many suggest that we cannot study the Internet in isolation because people's uses both lead to and depend on the use of other technologies (e.g., Lin, 2002a). By considering the present chapter's observations in the context of others in this volume, we can hopefully begin the process of contextualizing Internet use in the context of broader sets of converging media domains.

How does this relate back to the "Theoretical Implications" section? Overall, there are some good theories on adoption. For example, diffusion and functional similarity will help us predict points at which new services become viable. The problem is that the theories are not general and are poor on predicting impacts before they occur. As noted already, these theories "do not coalesce into a meta-framework of any sort."

More positively, however, some specific impacts, when they are found, can be predicted to get stronger or weaker depending on some theories. For example, as access becomes easier or more efficient, the impact should get stronger (based on functional similarity and convergence); as motivations, interests, and other internal factors become stronger, the impact should broaden (based on repertoires and uses and gratifications); as relations with the Internet become stronger, the impact should increase (based on dependency and parasocial interaction).

So the theories should not be discarded or ignored. They could benefit from better frameworks and more research. Obviously, the "trench work" needs to continue to deal with the inconsistencies discovered in the empirical research and to search for more theoretical generalizations. Finally, Lin's (2002a) model of dynamic interactivity can be used as a guide for research, one that prompts empiricists to examine multiple factors along with the mutual causes involved (or lack thereof). In attempting to define new media like the Internet as a "fluid" channel—which can easily morph from mass (e.g., Web) to interpersonal (e.g., e-mail) modalities—this conception provides a sound point of departure for theoretical explication of converging, multifunctional online media. In particular, new theories will need to be grounded theories—that is, based on a consistent body of generalizable research.

REFERENCES

Althaus, S. L., & Tewksbury, D. (2000). Patterns of Internet and traditional news media use in a networked community. *Political Communication*, *17*, 21–45.

Amichai-Hamburger, Y., Fine, A., & Goldstein, A. (2004). The impact of Internet interactivity and need for closure on consumer preference. *Computers in Human Behavior, 20*, 103–117.

Armstrong, L., Phillips, J. G., & Saling, L. L. (2000). Potential determinants of heavier internet usage. *International Journal of Human-Computer Studies, 53,* 537–550.

Atkin, D. J. (1993). Adoption of cable amidst a multimedia environment. *Telematics and Informatics, 10*, 51–58.Atkin, D., Jeffres, L., & Neuendorf, K. (1998). Understanding Internet adoption as telecommunications behavior. *Journal of Broadcasting & Electronic Media, 42*, 475–490.

Atkin, D. J. (2002). Convergence across media. In C. Lin & D. Atkin (Eds.), *Communication technology and society: Audience adoption and uses* (pp. 37–51). New York: Harper.

Atkin, D. J., Jeffres, L., Neuendorf, K., Lange, R., & Skalski, P. (2005). Why they chat : Predicting adoption and use of chat rooms. In M. B. Salwen, B. Garrison, & P. D. Driscoll (Eds.). *Online news and the public* (pp. 303–322). Mahwah, NJ: Lawrence Erlbaum Associates.

Baldwin, T. F., McVoy, D. S., & Steinfield, C. (1996). *Convergence: Integrating media, information and communication.* Newbury Park, CA: Sage.Ball-Rokeach, S. J. (1985). The origins of individual media-system dependency: A social framework. *Communication Research, 12*, 485–510.

Ball-Rokeach, S. J., & DeFluer, M. L. (1976). A dependency model of mass media effects. *Communication Research, 3*(1), 1–21.

Bargh, J. A. (2002). Beyond simple truths: The human-internet interaction. *Journal of Social Issues, 58,* 1–8.

Beller, M., & Or, E. (1998). The crossroads between lifelong learning and information technology a challenge facing leading universities. *Journal of Computer Mediated Communication, 4.* Retrieved January 20, 2005, from http://www.ascusc.org/Journal of Computer Mediated Communication/vol4/issue2/beller.html.

Biocca, F. (1997). The cyborg's dilemma: Progressive embodiment in virtual environments. *Journal of Computer Mediated Communication, 3.* Retrieved September 11, 2005, from http://jcmc.indiana.edu/vol3/issue2/biocca2.html.

Biocca, F., & Nowak, K. (2002). Plugging your body into the telecommunication system: Mediated embodiment, media interfaces, and social virtual environments. In C. A. Lin & D. Atkin (Eds.), *Communication technology & social change: Audience adoption and uses* (pp. 379–408). Cresskill, NJ: Hampton.

Bonfadelli, H. (2002). The Internet and knowledge gaps: A theoretical and empirical investigation. *European Journal of Communication, 17*, 65–84.

Brants, K. (1998). Who's afraid of infotainment? *European Journal of Communication, 13*, 315–335.

Brants, K., & Neijens, P. (1998). The infotainment of politics and political communication. *European Journal of Communication, 15*, 149–164.

Bryant, K., Campbell, J., & Kerr, D. (2003). Impact of Web based flexible learning on academic performance in information systems *Journal of Information Systems Education, 14,* 41–50.

Chapman, S. (1996). The commodification of prevention. *British Medical Journal (International edition), 312,* 730–731.

Charlton, J. P. (2002). A factor-analytic investigation of computer "addiction" and engagement. *British Journal of Psychology, 93,* 329–344.

Charney, T., & Greenberg, B. S. (2002). Uses and gratifications of the internet. In C.A. Lin & D. J. Atkin (Eds.), *Communication technology and social change: Audience adoption and uses* (pp. 379–408). Cresskill, NJ: Hampton.

Chester, A., & Gwynne, G. (1998). Online teaching: Encouraging collaboration through anonymity. *Journal of Computer Mediated Communication, 4.* Retrieved December 15, 2004, from http://www.ascusc.org/Journal of Computer Mediated Communication/vol4/issue2/chester.html.

Chiu, S.-L., Lee, J.-Z., & Huang, D.-H. (2004). Video game addiction in children and teenagers in Taiwan. *CyberPsychology & Behavior, 7*, 571–581.

D'Alessandro, D. M., Kreiter, C. D., Kinzer, S. L., & Peterson, M. W. (2004). A randomized controlled trial of an information prescription for pediatric patient education on the internet. *Archives of Pediatrics & Adolescent Medicin, 158*, 857–863.

Dholakia, R. R., Mundorf, N., & Dholakia, N. (1996). *New infotainment technologies in the home: Demand side perspectives*. Hillsdale, NJ: Lawrence Erlbaum Associates.

Donohew, L., Lorch, E. P., & Palmgreen, P. (1991). Sensation-seeking and the targeting of televised anti-drug PSAs. In L. Donohew, H. E. Sypher, & W. J. Bukoski (Eds.), *Persuasive communication and drug abuse prevention* (pp. 209–226). Hillsdale, NJ: Lawrence Erlbaum Associates.

Donohue, G. A., Tichenor, P. J., & Olien, C. N. (1975). Mass media and the knowledge gap: A hypothesis reconsidered. *Communication Research, 2*, 3–23.

Freeman-Longo, R. E., & Blanchard, G. T. (1998). *Sexual abuse in America: Epidemic of the 21st century*. Brandon, VT: Safer Society Press.

Fry, D. L., & McCain, T. A. (1983). Community influentials' media dependency in dealing with a controversial local issue. *Journalism Quarterly, 60*, 458–463, 542.

Gillmor, D. (2003). Moving toward participatory journalism. *Nieman Reports, 57*, 79.

Gordon, D. N., Gomez, L. M., Pea, R. D., & Fishman, B. J. (1996). Using the World Wide Web to build learning communities in K–12. *Journal of Computer Mediated Communication, 2*. Retrieved December 15, 2004, from http://www.ascusc.org/Journal of Computer Mediated Communication/vol2/issue3/gordin.html.

Grant, A., & Meadows, J. (2002). Electronic commerce: Going shopping with QVC and AOL. In C. A. Lin & D. J. Atkin (Eds.), *Communication technology and social change: Audience adoption and uses* (pp. 247–274). Cresskill, NJ: Hampton.

Grant, A. E., Guthrie, K., & Ball-Rokeach, S. J. (1991). Television shopping: A media system dependency perspective. *Communication Research, 18*(6), 773–798.

Grebb M. (2004, December 20). Special report: Story of the year, seriously funny; "Daily Show" provides voice of dissent for those who differ with status quo. *Multichannel News*, p. 30.

Griffiths, M. (1999). Internet addiction: Fact or fiction? *The Psychologist: Bulletin of the British Psychological Society, 12*, 246–250.

Griffiths, M., Richard, T., & Wood, A. (2000). Risk factors in adolescence: The case of gambling, videogame playing, and the Internet. *Journal of Gambling Studies, 16*, 199–225.

Griffiths, M. D. (1995). Technological addictions. *Clinical Psychology Forum, 76*, 14–19.

Gross, E. F., Juvonen, J., & Gable, S. L . (2002). Internet use and well-being in adolescence. *Journal of Social Issues, 58*, 75–90.

Hall, A. S., & Parsons, J. (2001). Internet addiction: College student case study using best practices in cognitive behavior therapy. *Journal of Mental Health Counseling, 23*, 312–318.

Heeter, C., & Greenberg, B. (1985). Cable and program choice. In D. Zillman & J. Bryant (Eds.), *Selective exposure to communication* (pp. 203–224). Hillsdale, NJ: Lawrence Erlbaum Associates. Hoffman, D. L., Novak, T. P., & Schlosser, A. E. (2000). The evolution of the digital divide: How gaps in Internet access may impact electronic commerce. *Journal of Computer Mediated Communication, 5*. Retrieved December 15, 2004, from http://jcmc.indiana.edu/vol5/issue3/hoffman.html.

Henke, L., & Donohue, T. (1989). Functional displacement of traditional TV viewing by VCR owners. *Journal of Advertising Research, 29*(2), 18–23.

Horton, D., & Wohl, R. R. (1956). Mass communication and para-social interaction: Observations on intimacy at a distance. *Psychiatry, 19*, 215–229.

Internet Society. (2004). *A brief history of the Internet and related networks*. Retrieved December 15, 2004, from http://www.isoc.org/internet/history/cerf.shtml.

Jansen, B. J., Spink, A., Bateman, J., & Saracevic, T. (1998, November). Searchers, the subjects they search, and sufficiency: A study of a large sample of Excite searches. *Proceedings of WebNet98 Conference of the WWW* (p. 52). Orlando, FL.

Jeffres, L., Atkin, D., Neuendorf, K., & Lin, C. (2004). The new "wired nation" and menu of abundance. *Telematics & Informatics, 21,* 317–334.

Johnson, T. J., & Kaye, B. K. (2004). Wag the blog: How reliance on traditional media and the Internet influence perceptions of credibility of weblogs among blog users. *Journalism & Mass Communication Quarterly, 1*, 600–620.

Katz, J., & Aspden, P. (1996, October). *Motivations for and barriers to internet usage: Results of a national public opinion survey*. Paper presented at the 24th annual Telecommunications Policy Research conference, Solomons, MD.

Knobloch, S., Carpentier, F. D., & Zillmann, D. (2003). Effects of salience dimensions of information utility on selective exposure to online news. *Journalism and Mass Communication Quarterly, 80*, 91–108.

Kraut, R., Kiesler, S., Boneva, B., Cummings, J., Hegelson, V., & Crawford, A. (2002). Internet paradox revisited. *Journal of Social Issues, 58*, 49–74.

Kraut, R., Patterson, M., Landmark, V., Kiesler, S., Mukopadhyay, T., & Scherlis, W. (1998). Internet paradox: A social technology that reduces social involvement and psychological well-being? *American Psychologist, 53*, 1017–1031.

Kristula, D. (2005). *The history of the Internet*. Retrieved January 11, 2005, from http://www.davesite.com/webstation/net-history2.shtml.

LaRose, R., & Eastin, M. S. (2002). Is on-line buying out of control? Electronic commerce and consumer self- regulation. *Journal of Broadcasting and Electronic Media, 45*, 549–564.

LaRose, R., Lin, C. A., & Eastin, M. S. (2003). Unregulated Internet usage: Addiction, habit, or deficient self-regulation? *Media Psychology, 5*, 225–253.

LaRose, R., Mastro, D., & Eastin, M. S. (2001). Understanding Internet usage. *Social Science Computer Review, 19*, 395–404.

Lavin, M., Marvin, K., McLarney, A., Nola, V., & Scott, L. (1999). Sensation seeking and collegiate vulnerability to internet dependence. *CyberPsychology & Behavior, 2*, 425–430.

Lasswell, H. (1948). The structure and function of communication in society. In L. Bryson (Ed.), *The communication of ideas* (pp. 37–51). New York: Harper.

Lee, M. J. (2001). *Effective tailored-communication in learning from hypertext: Introducing expanding hypertext based on individuals' sensation seeking and working memory capacity*. Unpublished doctoral dissertation, University of Florida, Gainesville.

Li, S., & Chung, T. (2006). Internet function and Internet addictive behavior. *Computers in Human Behavior, 22*, 1067–1071.

Lin, C. A. (1998). Exploring personal computer adoption dynamics. *Journal of Broadcasting and Electronic Media, 42*, 95–112.

Lin, C. A. (2002a). A paradigm for communication and information technology adoption research. In C. A. Lin & D. J. Atkin (Eds.), *Communication technology and society: Audience adoption and uses* (pp. 447–475). Cresskill, NJ: Hampton.

Lin, C. A. (2002b). Perceived gratifications of online media service use among potential users. *Telematics and Informatics, 19*, 3–19.

Lin, C. A. (2004). Webcasting adoption: Technology fluidity, user innovativeness, and media substitution. *Journal of Broadcasting & Electronic Media, 48*, 446–465.

Lin, C. A., & Atkin, D. (2002). *Communication technology and society: Audience adoption and uses*. Cresskill, NJ: Hampton.

Lin, C. A., & Jeffres, L. W. (1998.) Factors influencing the adoption of multimedia cable technology. *Journalism and Mass Communication Quarterly, 75,* 341–352.

Lin, S. H., & Yan, G. G. (2002). Internet addiction disorder, online behavior and personality. *Chinese Mental Health Journal, 16*, 501–502.

Lindner, K. (1992). A piece of my mind: Encourage information therapy [comment]. *Journal of the American Medical Association, 267*, 2592.

McDonald, D. G. (1983). Investigating assumptions of media dependency research. *Communication Research, 10*, 509–528.

McIlwraith, R. D. (1998). "I'm addicted to television": The personality, imagination, and TV watching patterns of self-identified TV addicts. *Journal of Broadcasting & Electronic Media, 42,* 371–386.

McKenna, K. Y. A., Green, A. S., & Gleason, M. E. J. (2002). Relationship formation on the Internet: What's the big attraction? *Journal of Social Issues, 58*, 9–31.

Mitka, M. (2001). Win or lose, Internet gambling stakes are high. *Journal of the American Medical Association, 285*, 1005–1006.

Morahan-Martin, J. M., & Schumacher, P. (2000). Incidence and correlates of pathological Internet use among college students. *Computers in Human Behavior, 16*, 13–29.

National Cable and Telecommunications Association. (2005). *2004 Year-end industry overview*. Retrieved December 15, 2004, from http://www.ncta.com/pdf_files/NCTA YearEndOverview04.pdf.

Nielsen NetRatings. (2004). *New insights in on-line news*. Retrieved December 15, 2004, from http://www.ncta.com/pdf_files/NCTAYearEndOverview04.pdf.

Papachrissi, Z., & Rubin, A. M. (2000). Predictors of Internet use. *Journal of Communication, 46*, 80–97.

Perls, T. T. (2004). Anti-aging quackery: Human growth hormone and tricks of the trade-more dangerous than ever. *Journals of Gerontology: Series A: Biological Sciences and Medical Sciences, 59A*, 691.

Petrie, H., & Gunn, D. (1998). *Internet "addiction": The effects of sex, age, depression and introversion*. Paper presented at the conference of the British Psychological Society, London.

Perse, E. M., & Courtright, J. A. (1993). Normative images of communication media: Mass and interpersonal channels in the new media environment. *Human Communication Research, 19*, 485–503.

Pew Research Center for the People and the Press. (1996). *News attracts most Internet users one-in-ten voters online for campaign '96*. Retrieved December 15, 2004, from http://people-press.org/reports/display.php3?ReportID=117.

Pew Research Center for the People and the Press. (2004). *Cable and Internet loom large in fragmented political news universe perceptions of partisan bias seen as growing, especially by Democrats*. Retrieved December 15, 2004, from http://people-press.org/reports/display.php3?ReportID=200.

Reagan, J. (1996). The repertoire of information sources. *Journal of Broadcasting & Electronic Media, 40*, 112–121.

Rhee, K. Y., & Kim, W. (2004). The adoption and use of the Internet in South Korea. *Journal of Computer-Mediated Communication, 9*. Retrieved December 15, 2004, from http://jcmc.indiana.edu/vol9/issue4/rhee.html.

Rogers, E. M. (1995). *Diffusion of innovations* (4th ed.). New York: The Free Press.

Rogers, E. M. (2003). *The diffusion of innovations* (5th ed.). New York: The Free Press.

Rosenberg, H. (2004, November 1). The fact is, the joke is the news. *Broadcasting and Cable*, p. 32.

Rubin, A. M., & Eyal, K. (2002). The videocassette recorder in the home media environment. In C. Lin & D. Atkin (Eds.), *Communication technology and society: Audience adoption and uses* (pp. 329–349). Cresskil, NJ: Hampton.

Rubin, R. B., Perse, E. M., & Powell R. A. (1985). Loneliness, parasocial interaction, and local television news viewing. *Human Communications Research, 12*, 155–180.

Salwen, M., Garrison, B., & Driscoll, P. D. (2005). *Online news and the public*. Mahwah, NJ: Lawrence Erlbaum Associates.

Sawhney, H. (1997). New infotainment technologies in the home: Demand-side perspectives. *Journal of Broadcasting & Electronic Media, 41*, 442–444.

Shapira, N. A., Goldsmith, T. D., Keck, P. E., Khoslab, U. M., & McElroy S. L. (2000). Psychiatric features of individuals with problematic internet use. *Journal of Affective Disorders, 57*, 267–272.

Shotton, M. A. (1989). *Computer addiction? A study of computer dependency*. New York: Taylor & Francis.

Shotton, M. A. (1991). The costs and benefits of "computer addiction." *Behaviour & Information Technology, 10*, 219–230.

Slater, M. D. (2003). Alienation, aggression, and sensation seeking as predictors of adolescent use of violent film, computer, and website content. *Journal of Communication, 53*, 105–121.

Speight, K. (1999). Gaps in the worldwide information explosion: How the Internet is affecting the worldwide knowledge gap. *Telematics and Informatics, 16*, 135–150.

Surette, R., & Otto, C. (2002). A test of a crime and justice infotainment measure. *Journal of Criminal Justice, 30*, 443–453.

Tan, B. W., & Lo, T. W. (1991). The impact of interface customization on the effect of cognitive style on information system success. *Behavior and Information Technology, 10*, 297–310.

Tichenor, P. J., Donohue, G. C., & Olien, C. (1980). *Community conflict and the press*. Beverly Hills, CA: Sage.

Tyler, T. R. (2002). Is the Internet changing social life? It seems the more things change, the more they stay the same. *Journal of Social Issues, 58*, 195–205.

University of California, Center for Communication Policy. (2003). *UCLA Internet Report 2001, University of California, Los Angeles, CA*. Retrieved January 7, 2005, from www.ccp.ucla.edu.

U.S. Census Bureau. (2005). *Table 5A: Use of a computer at home*. Retrieved December 15, 2004, from http://www.census.gov/population/socdemo/computer/ppl-175/tab05A.pdf.

Walther, J. B., & Boyd, S. (2002). Attraction to computer-mediated social support. In C. A. Lin & D. Atkin (Eds.), *Communication technology and society: Audience adoption and uses* (pp. 153–188). Cresskill, NJ: Hampton.

Wheeler, B. C., Valacich, J. S., Alavi, M., & Vogel, D. (1995). A framework for technology-mediated inter-institutional telelearning relationships. *Journal of Computer Mediated Communication, 1*. Retrieved January 20, 2005, from http://jcmc.indiana.edu/vol1/issue1/wheeler/essay.html .

Woods, M. L., Shimon, J. M., Karp, G. G., & Jensen, K. (2004). Using webquests to create online learning opportunities in physical education. *Journal of Physical Education, Recreation & Dance, 75*, 41–48.

Yang, S. C., & Tung, C. (2005). Comparison of Internet addicts and non-addicts in Taiwanese high school. *Computers in Human Behavior,* in press.

Young, K. S. (1996, August). *Internet addiction: The emergence of a new clinical disorder*. Paper presented at the 104th annual meeting of the American Psychological Association, Toronto.

Young, K. S. (1997, August). *What makes the Internet addictive: Potential explanations for pathological internet use*. Paper presented at the 105th annual conference of the American Psychological Association, Chicago. Retrieved January 20, 2005, from http://www. netaddiction.com/articles/habitforming.htm.

Zakon, R. H. (2005). *Hobbes' Internet timeline v8.0*. Retrieved December 15, 2004, from http://www.zakon.org/robert/internet/timeline/.

Zuckerman, M. (1983). Sensation seeking: A biosocial dimension of personality. In A. Gale & J. A. Edwards (Eds.), *Correlates of behavior: Individual differences of psychopathology* (p. 3). London: Academic Press.

Zhu, J. H., & He, Z. (2002). Diffusion, use and impact of the Internet in Hong Kong: A chain process model. *Journal of Computer Mediated Communication, 7*. Retrieved January 20, 2005, from http://jcmc.indiana.edu/vol7/issue2/hongkong.html.

VI

CONSUMER SETTING

11

Interactive Media Technology and Electronic Shopping

Carolyn A. Lin

Since the Internet entered our social consciousness a little over a decade ago, it has evolved into a multifaceted communication medium. In fact, the Internet now mirrors a technology appliance, a medium that is essential for many people to receive their news, information, and entertainment, in addition to communicating with others. One of the most beneficial utilities for both marketers and consumer alike is the Internet's ability to be an interactive direct-response marketing channel to move goods and services. At present, sales revenues of offline teleshopping still greatly outnumber those of online teleshopping. Based on the most recent available statistics, the sales revenue for offline teleshopping, including telephone marketing, TV, and radio, totaled an estimated $529.1 billion (Direct Marketing Association, 2004). This compares with online retail sales revenue of $53 billion for the same fiscal year, which is scarcely 10% of its offline counterpart (Evans, 2004).

Nonetheless, online sales revenues continued to climb at a steady rate. The National Retail Federation (Mangalindan, 2005), an association for online retailers, reported that online retail sales in 2004 reached $89 billion or 4.6% of all retail sales in the United States. That number was projected to reach $172 billion for the year 2005. Advertising expenditures for online marketing would also rise to $14.7 billion according to one industry estimate in 2005, which would reflect an increase of 23% from the year before; the average annual growth rate for online advertising is projected at 15% (C/Net, 2005).

To fully understand the relations between different types of retail shopping modalities, Hoffman, Novak, and Chatterjee (1996) suggested that online shopping should be studied in conjunction with other similar modes of offline shopping channels. This chapter introduces the teleshopping phenomenon both off- and online. It also reviews social scientific theories that were used to study this phenomenon and explains relevant study findings. Social change implications associated with this form of personalized shopping behavior are also discussed.

BACKGROUND

Teleshopping is a form of direct-response marketing sponsored by a wide variety of advertisers, including those who also engage in traditional mass media advertising and who also market their products in traditional retailer stores. It entails a form of direct-to-consumer marketing communication from the marketers (or advertisers), and it draws a direct-to-marketer response from the consumers. This form of marketing communication involves the use of telephone, TV, radio, and the Internet venues.

The use of telephones as a direct marketing tool has been referred to as telemarketing. The term *telemarketing* was coined by AT&T in the early 1980s, when it began marketing its long-distance telephone services (Romano, 1998). The advantage of using the telephone to market goods and services is that it allows one-on-one interpersonal communication and instantaneous feedback. According to the most recent consumer survey (Direct Marketing Association, 2004), telemarketing had the highest direct consumer order rate (at 5.78%), compared with all other direct marketing outlets (including direct mail). In 2003, residential consumer telemarketing generated $315.3 billion in sales revenue (Direct Marketing Association, 2004).

Television ads are used to prompt the consumer to order a number of different goods and services by calling an 800 number, visiting a designated web site or sending in an order via mail. Informercials, the program-length commercials dedicated to promote a particular advertiser's product (e.g., Bowflex), are also a form of televised direct-response marketing modality, in addition to TV shopping channels (e.g., QVC and HSN), which directly market a number of different products from different advertisers . Direct-response TV commercials and informercials produced about $154.1 billion in 2003 (Direct Marketing Association, 2004).

Radio ads are also utilized to market an array of goods and services to the consumers directly by prompting the listeners to call an 800 number, visit a designated web site, or send in an order via mail. This form of direct-response radio marketing spawned $59.7 billion of revenues in 2003 (Direct Marketing Association, 2004). Although radio is the oldest form of electronic advertising and marketing channel, it is also an emerging form of hybrid direct-marketing medium owing to the development of podcasting. Podcasting is a form of online audio broadcast produced and disseminated by Internet users; users can "tune in" to a wide variety of podcasting directories to find these grassroots-based user-generated online broadcast "programs" by using either a digital audio player or an MP-3 player to download the audio files.

The Internet is a multimodal direct-response marketing medium. It encompasses the use of e-mail outlets, banner ads, rich-media ads (i.e., advertising messages that contain enhanced visual displays and interactive mechanisms), pop-up ads, web-page-embedded ads, instant messaging (IM), blogs (i.e., a form of online journal posted by journalists and individual online users), and podcasts to deliver targeted advertising and marketing messages. Web-page-embedded ads can appear in a dedicated advertising web page (e.g., dietcoke.com), in a third-party web page (e.g., Yahoo! homepage), or in a dedicated corporate webpage (e.g., Best Buy). Web-driven direct marketing sales revenues in 2004 arrived at $52.5 billion (Evans, 2004) and are projected to grow more rapidly than its offline teleshopping counterparts.

Of these different teleshopping channels, both telemarketing and Internet marketing have been criticized for their intrusion on consumer privacy, while online personal data security has been a serious social concern. To curb telemarketing fraud, Congress passed the Telemarketing and Consumer Fraud and Abuse Prevention Act in 1994 (15 U.S. C. §§. 1601-). The Federal Trade Commission (FTC) subsequently adopted the Telemarketing Sales Rule (FTC, 1994) to implement this act. In 2003, the FTC adopted a new rule (FTC, 2004), which allows the public to register their telephone numbers at the National Do Not Call Registry. The Federal Communications Commission (FCC) also issued its own rules to coordinate this FTC policy (FCC, 2004).

To protect consumers from receiving unsolicited commercial e-mail messages, Congress enacted the Controlling the Assault of Non-Solicited Pornography and Marketing (CAN-SPAM) Act of 2003 (U.S. Code, 2003). Under this act, the FTC has proposed rules (FTC, 2004) to protect consumers from receiving unsolicited commercial e-mails, and the FCC (2005) has also implemented rules on banning unsolicited commercial e-mails to be delivered to consumer wireless devices. See chapter 14 by Lee (this volume) to learn more about the issue of privacy.

Even as consumers seemed to be resistant to the invasion of privacy from electronic marketers, just under 50% of them still initiated a purchase through the telephone (41%) and/or the Internet (45%), according to a recent consumer study of 1,000 consumers (American Telemarketing Association, 2002). According to another study of 2,500 households, consumer-direct marketing channels—which include both nonelectronic (i.e., catalogs and direct mail) and electronic channels (i.e., interactive TV and online)—will continue to grow and may account for 12% of the overall retail sales revenues in our economy by the year 2010 (Pastore, 2000).

The question remains, however, of what might propel consumers to engage in this type of in home shopping activity, given that it is considered antisocial by some and parasocial by others. The following section helps answer this question by reviewing a set of relevant social-scientific theoretical explanations.

THEORETICAL EXPLANATIONS

There was limited research that investigated consumer perceptions, attitudes, and behaviors toward unsolicited direct marketing via teleshopping and nonteleshopping channels, even as privacy concerns have been increasing in recent years

(Nowak & Phelps, 1997). Within the limited domain of in-home teleshopping literature, consumer motives have been explained by motivational factors including perceived value (Dholakia & Uusitalo, 2002) and perceived psychological gratifications of shopping (Salomon & Koppelman, 1992). Behavior-based studies have often linked actual shopping patterns with consumers' demographic characteristics (Donthu & Gilliland, 1996). Consumer personality traits were examined by the diffusion of innovations theory (Rogers, 2003). Dysfunctional shopping behavior was explicated by media dependency theory (Ball-Rokeach & DeFleur, 1985; DeFleur & Ball-Rokeach, 1989) and consumer addiction theory (Faber, O'Guinn, & Krych, 1987; Rook, 1987).

Consumer Motivations

Utilitarian Versus Hedonic Value. According to Babin, Darden, and Griffin (1994), there are two basic categories of consumer perceived value of shopping. These two categories—utilitarian value and hedonic value—can influence what consumers' attitudes about shopping are, which can in turn dictate subsequent consumer action and behavior.

Utilitarian value refers to the type of shopping motivation that is task-oriented, deliberate, and efficiency-driven (Engel, Blackwell, & Miniard, 1993). Hence, consumers who have a work-mentality attitude and utility-driven perception about their shopping action and experience tend to treat such shopping activity as necessary errands, obligations, and routines. For instance, the more mundane types of shopping, such as grocery, office supplies, and so on, could prompt a negative perception about the tedious nature of shopping and evoke negative feelings about such shopping activity.

By contrast, hedonic value is associated with perceived fun and enjoyment of the shopping activity and experience as well as the attitude of "play" or "recreation" instead of "mission" or "performance (Fischer & Arnold, 1990; Sherry, 1990). Therefore, consumers who find shopping activity stimulating, gratifying, and relaxing tend to cast such activity as an escape, therapy, and a good way to manage negative moods. For example, shopping is seen as a recreational activity for many consumers; these consumers often use shopping activity to alleviate their stress, to socialize, and to enjoy the ambiance of the shopping-locale surroundings.

Psychological Gratifications. Aside from those studies that focused on the dualistic views of utilitarian versus hedonic value, other researchers have examined in-home shopping behavior with a broader set of psychological factors involving different dimensions of consumer motives. The limited number of studies that explore consumer motives for teleshopping often utilize the theory of uses and gratifications as their base for exploring these psychological motivation factors. The uses and gratifications theory presupposes that audiences are motivated by a set of cognitive and affective needs when they decide consume media content and that their actual media-use behavior is goal-directed to help find fulfillment for these cognitive and affective needs (Katz, Blumler, & Gurevitch, 1974).

The range of media-use motives usually includes the following general dimensions: diversion, escape, surveillance, social identity, informational learning, companionship, interpersonal communication, and parasocial interaction (Lin, 1999). In the context of in-home shopping, the psychological motives for such shopping behavior could include, for instance, information, conversation, and social escapism (Korgaonkar & Wolin, 1999), in addition to diversion, role playing, social interaction, and recreational functions (Salomon & Koppelman, 1992).

This need-fulfillment motivation paradigm is conceptualized within the context of the basic human psychological needs hierarchy, according to Maslow's (1943) theory of human motivation. In essence, our cognitive needs as part of our needs for self-actualization can be expressed in our desire to know, to experience, and to express ourselves. Hence, a shopping activity that can serve as an experiential functional alternative for mood management, knowledge acquisition, and social interaction can then help fulfill such cognitive needs for self-actualization (Lin, 1996).

Consumer Attributes and Behavior

Consumer characteristics—in terms of their tendency and ability to tolerate and accept the risk of buying newer products and/or unfamiliar brands—can separate those who are teleshopping adopters and nonadopters. These might be considered alongside other inhibiting factors, including the lack of human interaction and physical contact with the products in the shopping experience. Researchers have conceptualized this type of consumer risk-taking attributes within the theoretical framework of consumer innovativeness, where novelty-seeking, venturesomeness, and creativity are also considered part of one's innovative consumer traits (Hirschman, 1980; Midgley & Dowling, 1978). The concept of individual innovativeness has its origin in the theory of diffusion of innovations. This theory suggests that individuals who adopt new products, technologies, or ideas earlier than others in society are also considered to be more innovative in nature (Rogers, 2003).

This concept of consumer innovativeness has been utilized to study offline in-home teleshoppers in conjunction with demographic profiles, where individual consumers' judgment of risk-taking, attitude toward in-store shopping, and evaluation of perceived benefits were also examined (e.g., Darian, 1987; James & Cunningham, 1987). Similarly, a few pioneering studies have also applied this theoretical concept to online teleshopping behavior. Some of these studies, in addition to explicating the cognitive and behavior risk-taking tendencies, also linked online shopping to other psychological traits of the consumers, including impulsive consumption (e.g., Donthu & Garcia, 1999; LaRose & Eastin, 2002). Because in-home shopping is often done in private, consumers who engage in heavy teleshopping behavior could also become socially isolated or further reduce their social mobility.

Consumer Shopping Dependency

Consumer dependency on shopping activity, which can range from impulsive buying to shopping addiction, has been studied from both sociological and psychological perspectives. The sociological perspective addresses individual behavior as an outcome of a social system. The theory of media dependency is a social-system-based theory that assesses the relations between audience behavior and media systems. It assumes that audiences can develop a dependency relationship with a medium, if the medium has been helping to fulfill their cognitive and affective needs or meet their media-use goals due to past experiences of consuming certain media content over time (Auter, 1992; Ball-Rokeach & DeFleur, 1985).

In the case of in-home teleshopping, research has explored the relationships between TV shopping programming and consumer dependency (Grant, Guthrie, & Ball-Rokeach, 1991). This research explicates how certain consumers rely on TV shopping programs for meeting their goals of utilitarian and hedonic needs. As the consumers come to rely on the TV shopping channels more and more to gratify these needs, they become more dependent on these channels for fulfilling these needs. This dependency relationship was also linked to how consumers develop a parasocial relationship with the program hosts (or sales personalities) featured in these TV shopping programs. This parasocial relationship reflects how consumers perceive these program hosts as "friends" who visit and interact (either on the air or in the consumers' minds) with them through their TV set on a regular basis. It often evolves as a result of frequent "visits" and interactions between these program hosts and consumers/viewers while they watch these shopping programs.

This type of pseudosocial relationship has also been observed within the teleshopping venue from a different scenario. In particular, this research involves studying the parasocial relationship that is formulated, in part, due to observation of the interactions between those consumers who phoned in to converse with the TV-shopping program show hosts by those consumers who did not engage in such on-the-air interactions. The cognitive and affective outcomes of this unique form of parasocial relationship—one obtained from watching interactions between other consumers and TV shopping sales personalities—are described as the effects of broadcast teleparticipation (Skumanich & Kintsfather, 1998).

The concept of impulsive shopping behavior appears to be a preliminary form of compulsive shopping, which could lead to addictive shopping behavior. According to Rook and Hoch (1985), there are five characteristics that distinguish impulsive and nonimpulsive shopping behavior: (a) an unprecedented spontaneous desire to act, (b) a temporary out-of-control feeling due to psychological disequilibrium, (c) psychological conflict between obtaining instant gratification and resisting irrational urges, (d) a lowering of product-utility evaluation criteria, and (e) a disregard for negative consumption consequences. Hoch and Loewenstein (1991) contend that impulsive and compulsive purchase behaviors are a result of the interaction between consumer willpower, resistance to shopping stimuli, and desire for instant gratification. Faber et al. (1987) describe how compulsive shopping behavior can reflect the characteristics of addictive behavior when it demonstrates (a) an impulse to engage in a certain behavior, (b) a denial of any harmful behavioral consequences, and (c) inability to control or modify the harmful behavior.

Impulsive shopping behavior is also manifested in consumers who conduct in-home teleshopping on a regular basis (Donthu & Gilliland, 1996). For instance, dedicated TV shopping channels that air informercials or live-shopping programs could present an environment that is conducive to impulsive purchasing. This is because while watching these types of programs, consumers could become engrossed in the program content and lower their will power to resist their shopping impulses. Similarly, conducting online shopping activity, an activity that is often done by an individual sitting in front of a computer alone, could also lead to unregulated consumption. This could be particularly true when consumers perceive themselves as interacting with the online shopping sites with a recreational mind set (Donthu & Garcia, 1999). Common to both off- and online teleshopping are the ease of purchasing a product through either type of shopping modalities and the instant gratification one could get from such quick action, as all it usually requires is a phone call or a mouse click that provides an individual's credit card number to the advertisers.

EMPIRICAL FINDINGS

Consumer Motivations

There was little social-scientific empirical research evidence associated with offline teleshopping, although this shopping modality generates over $500 billion in our economy. Early studies reported that consumers who shopped through in-home teleshopping channels were often motivated primarily by utilitarian motivations. For instance, Cox and Rich (1964) found that consumers were motivated by the convenience of home shopping that helped them save time, in addition to liking the availability of a wide assortment of products. These study findings were supported by Darian (1987). James and Cunningham (1987), investigating consumer motives for purchases through TV shopping channels, discovered that consumers who made frequent purchases through TV shopping channels tend to have a greater convenience motivation than those who do not shop with these channels.

By contrast, Grant, Meadows, and Handy (1996) contend that hedonic motivations were equally important forces behind making purchases from watching

TV shopping programs. Their findings reveal that the percentage of consumers who watched these programs without making a purchase was nearly the same as the percentage of consumers who did make a purchase. They characterized these nonpurchasing consumers as passive watchers who enjoyed parasocial interactions provided by these shopping programs, as well as the recreational values of these programs. This explanation is consistent with the "shopping as recreation" behavioral pattern, where consumers place a heavier emphasis on the entertainment or hedonic aspect of their shopping experience.

Nowak's (1992) study also confirmed this recreational aspect of teleshopping, whereas the behavior of viewing TV shopping channel commercials and the initiation of making purchases from such channels were found to be related to the consumer emotion of pleasure. In a similar vein, psychological gratifications related to the fulfillment of shopping motives—such as diversion, role playing, social interaction, and recreational functions—were also found to be important motivational factors for engaging in teleshopping activities (Salomon & Koppelman, 1992). These findings are consistent with the theoretical assumptions of the uses and gratifications perspective.

Early studies found, then, that utilitarian motivations were the primary reasons behind in-home teleshopping, and the later studies discovered that hedonic motivations were comparable (if not stronger) motives for this type of shopping modality. The discrepancies in this empirical evidence may be a result of the following three developments in the way teleshopping channels communicated with their consumers. First, TV shopping programs added a much wider range of products, including a number of more upscale items, to attract an even larger audience base. Second, the quality of the talents who appear as hosts on TV shopping programs has steadily improved to help establish a more loyal consumer following. Third, the production value and quality of these programs have also become more polished and professional over time, such that these programs can be looked on as having recreational content.

Donhu and Garcia (1996) studied consumers who made purchases through watching informercials. Their findings indicate that informercial shoppers were motivated by the hedonic benefits of this shopping venue, which provided a type of in-home recreational activity. Utilitarian benefits were also found to be a significant motivator for these informercial shoppers, as they sought to purchase name brand products at a lower price and product varieties that were not available or easily accessible through other retail outlets.

In comparing the perceived values between infomercials and traditional commercials, Hetsroni and Asya (2002) found that, although consumers indicated informercials contain greater hedonic values than traditional commercials, they also thought that traditional commercials were better in presenting the "joy of life" type of hedonistic values—including happiness, adventure, youthful spirit, and leisure. Moreover, these consumers also were motivated by the voyeuristic aspects—such as beauty (e.g., cosmetic informercials) and sex (e.g., 1-900 sex lines)—of informercials, which are typically more fully produced in an informercial than in a 30-second traditional commercial. More importantly, the ratio between functional (or utilitarian) and hedonistic benefits of informercials was 3.3 to 1. Hence, utilitarian values may have weighed more heavily in motivating informercial shopping than did hedonistic values.

These findings are also consistent with those associated with consumer shopping through telemarketing channels. Where interpersonal interactions between the consumer and the sales representative—either through inbound (i.e., call-ins to place an order) or outbound (i.e., call-outs to solicit an order) channels—also carry a practical and recreational dimension, this helps determine the transaction outcome (Romano, 1998). By and large, the empirical evidence of all of these offline in-home teleshopping studies appears to be consistent with those of the general shopping literature, where consumers' recreation and task-value orientations often go in tandem (Berkowitz, Walton, & Walker, 1979; Gehrt & Carter, 1992; Settle, Alreck, & McCorkle, 1994).

This type of work versus play duality among consumers is similar to online shopping motivation, but online in-home shoppers do have a broader set of motivations than their offline counterparts.

For instance, early Internet shopping studies have indicated that browsing online to research and compare product information provides both cognitive and informational gratification, in addition to a stimulating and enjoyable hedonic experience (which includes diversion, role playing, social interaction, and recreational functions; e.g., Hoffman & Novak, 1996; Salomon & Koppelman, 1992).

Similar findings were reported by Korangaonkar and Wolin (1999), who distinguished between those consumers who made web purchases and those who did not (during the past year) based on their information, conversation, social escapism, interactive control, and economic motives. Menon and Kahn (2002) contend that pleasure and pleasant feelings in the initial online product search process could lead to additional browsing for products and subsequent shopping behavior. Novak, Hoffman, and Duhachek (2003) confirmed these findings by suggesting that, although goal-oriented consumers were gratified by the amount of time, effort, and money saved through shopping online, experiential consumers were rewarded by their enjoyment of their online shopping activity.

As the Internet shopping phenomenon evolved and became more widespread, consumer behavior also started to diverge. For instance, Donthu and Garcia (1999) found that Internet shoppers were primarily motivated by utilitarian values. As such, Internet shoppers were convenience seekers who sought product variety and products that would meet their needs, but not necessarily price advantages or recreational values, as compared with their offline teleshopping counterparts. These findings were confirmed by other researchers (Chiang & Dholakia, 2003; Mathwick, Malhotra, & Rigdon 2002). Chen and Chang (2003) nonetheless revealed a contradictory finding that indicates that price comparison (or shopping), due to the ease of information search, was a significant online shopping motive. Grant and Meadows (2002) also suggest that the primary advantages of online shopping were in the areas of convenience and information gathering (e.g., product features and prices) in a setting free of any salesperson pressure. By the same token, according to Li (2004), the main attraction of online shopping is that it offers consumers the convenience to customers of shopping within a comfortable home setting.

Consumer Attributes and Behaviors

According to Stark (1998), informercials were able to attract viewership that was as large as the average rating of such cable flagship channels as MTV or CNN. A more recent study (Direct, 2002) reported that 62.7% of TV viewers 16 years or older were exposed to some form of direct-marketing TV advertising, including long-form direct spots (i.e., informercials), short-form direct spots (i.e., regular-length TV commercials), and live-form shopping programs. Of these viewers, 28.6% did purchase a product advertised on an infomercial.

In terms of demographic characteristics, early studies found that direct-marketing TV shoppers were prone to be older and female (James & Cunningham, 1987); otherwise, they were older and better educated (Darian, 1987). As teleshopping became more mature and pervasive over time, these unique demographic factors seemed to dissipate over time, as later studies reported that TV shoppers were not significantly different from noneshoppers in terms of their age, education, income, or gender (Donthu & Gilliland, 1996).

Nonetheless, what distinguishes nonshoppers from shoppers with regard to their personality attributes as consumers remains consistent, in that teleshoppers were less risk averse and more willing to try new and different products. Early studies provided evidence supporting this negative relationship between risk tolerance and in-home shopping technology adoption (Eastlick, 1991; Lin, 1994; Shim & Mahoney, 1991). Later, Donthu and Gilliland's (1996) informercial study also demonstrated how shoppers were less risk-averse compared with nonshoppers. Eastlick and Lotz (1999) examined a broader set of consumer innovativeness traits and reported that interactive teleshoppers tended to possess the characteristics of personal innovativeness and that those who demonstrated the strongest innovativeness and opinion-leadership tendencies were also the most likely to make frequent purchases from TV shopping programs.

Consumer characteristics associated with online shopping were by and large found to be similar to those of in-home offline teleshoppers. Even as the Internet diffusion curve has reached the level of critical mass, and about two thirds of the population considered themselves Internet users, online shopping is still more likely to be exercised by those Internet users who are more venturesome and less concerned about financial risks, product quality, privacy, security, and other risks associated with electronic shopping in cyber space (Donthu & Garcia, 1999; Ratnasigham, 1998).

More recently, Li (2004) explored online shopping in Taiwan and reported that Taiwanese Internet shoppers typically demonstrated more adventuresome attributes than nonadopters, given that the Internet shopper and nonshopper groups were similar in their technology adoption behavior. By expanding the investigation of online shopping phenomenon to a webcasting environment where the direct marketing messages were displayed within live webcasts, video clips, or video streams, Lin (2004) found that adopters of webcasting-based shopping services possessed stronger individual innovativeness traits than nonadopters.

When consumers are willing to be venturesome in their shopping behavior and choose to use electronic interactive shopping channels to make purchases, they may intentionally or inadvertently displace the noninteractive shopping channels. Muldoon (1996) predicted that online catalog shopping may displace offline print catalog shopping, as online catalogs hold distinct and relative advantage over print catalogs, which include the convenience of reviewing product functions and ease of placing an order online. This prediction is similar to the assumption of the media substitution hypothesis, which posits that audience members may substitute the use of a functionally equivalent new medium (e.g., online catalog) for a more conventional old medium (e.g., print catalog).

Although this media substitution behavior may occur for innovative consumers who adopt new technology products (e.g., the digital video recorder such as TiVo) to replace older technology products (e.g., the analog videocassette recorder), this type of consumer behavior is consistent with consumer evaluation of the "relative advantage" of the new versus old technologies—a concept stemming from the diffusion of innovations theory (Rogers, 2003). Lin (2001) explicated the media substitution hypothesis and concluded that an old media modality can be displaced, supplemented, and/or complemented by a new one when the new media modality possess greater benefits in the areas of "superior content, technical benefits and cost efficiency" (p. 20).

Eastlick and Liu (1997) compared browsers and purchasers of online shopping outlets and revealed that consumer shopping frequency online was not predictive of TV shopping channel purchases, and online shopping behavior did not displace consumer loyalty toward the offline teleshopping modality. Their findings were contrary to earlier work (Korgaonkar & Smith, 1986; Shim & Mahoney, 1991) that pointed out that offline teleshopping experience helped push and enhance consumer engagement in online shopping activities. Kaufman-Scarborough and Lindquist (2002) compared consumers' browsing and purchasing behavior associated with the Internet, informercials, regular TV commercials, TV shopping channels, and print catalogs. Their findings suggest that the displacement function between off- and online teleshopping channels was not obvious or definite, as existing online shoppers may also browse product information online and then make their purchases offline.

Consumer Shopping Dependency

Grant et al. (1991) explored consumer motives associated with TV shopping channel use from the perspective of consumers' relationship with these shopping shows and their hosts. They found that greater frequencies of parasocial interactions between the consumers and the hosts of these TV shopping programs were related to an increase in the viewing of these programs, in addition to an increase in the number of purchases. Gumpery and Drucker (1992) further explored the relationship between the consumers and these TV shopping programs by examining how consumers perceived the on-air visits between other consumers who spoke with these program hosts on the air. They found that a pseudosocial relationship was constituted between those consumers/viewers who called in to the shopping programs and those who did not, where the noncallers perceived those callers as their "friends" in the electronic-shopping social circle.

Skumanich and Kintsfather (1998) characterized this particular social phenomenon as the broadcast teleparticipation effect. Their study found support for a consumer–consumer parasocial relationship that was based on the hypothesis of broadcast teleparticipation effects, in addition to a consumer–program host parasocial relationship that was proposed by Grant et al. (1991). The implications from their findings indicate that when consumers see other consumers having a purchase-related social interaction with the program hosts on the air, they feel a sense of "fellowship" and vicariously experience that pseudosocial relationship. As such, a dependency relationship can be developed between teleshopping purchases and informercial exposures, which can become intensified and lead to further repeated and continued use over time. Those parasocial interaction behaviors that demonstrate media/genre-specific dependency then raise the concerns of a potential deepening social isolation for those consumers who are already less socially mobile.

Rook and Fisher (1995) contend that buying impulses exist in all consumers and can be triggered by shopping stimuli that present themselves. They suggest that having purchase impulses does not mean acting on them, and normative influences (such as social approval) can help moderate impulse buying behavior. According to James and Cunningham (1987), those who patronize a television shopping channel more often tend to pay less attention to the convenience factor and are typically more socially isolated. When consumers shop in social isolation, especially in an in-home environment, they may become more vulnerable—compared to in-store shoppers—to compulsive shopping impulses due to the convenience of making purchases from home. Donthu and Gilliland (1996), for instance, revealed that informmercial shoppers are more impulsive in their purchase behavior than nonshoppers. Study evidence has shown that dependency on media shopping programs can drive consumers to become compulsive home shoppers—a social consequence attributed to unregulated extensive viewing and frequent purchases induced by compulsive consumption impulses (Wesson, 1990).

Nataraajan and Goff (1992) differentiate between the concepts of compulsive shopping from compulsive buying. They contend that compulsive shopping (including browsing and searching) can help a consumer "escape from the realities of his/her environment, and also allow for a temporary alteration of the individual's self-concept toward a desired state…the thrill associated with the very act of shopping…is an important source of satisfaction" (pp. 6–7). By comparison, compulsive buying (or actual exchange of goods and services) signifies "the manifestation of the nature of compulsiveness in the act of buying," and that compulsive shopping usually precedes compulsive buying (pp. 7–8). Hassay and Malcolm (1996) confirmed this conceptual clarification by illustrating that compulsive buying is propelled by the process or act of buying exchanges instead of the acquisition the accumulation of the items, as compulsive buyers did not purchase more products than noncompulsive buyers from teleshopping channels.

As the online shopping environment provides a continuous flow of stimuli (Hoffman & Novak, 1996; Novak, Hoffman, & Duhachek, 2003; Novak, Hoffman, & Yung, 2000), this type of consumption search process then can help facilitate an immersive experience, where a consumer may lose a sense of time or reality in his/her online shopping and buying activity. Preliminary empirical studies seemed to have corroborated this assumption. Donthu and Garcia (1999) reported that online shoppers appeared to be more impulsive than nonshoppers. LaRose and Eastin (2002) examined potentially dysfunctional shopping activity online by testing the concept of self-regulation or self-judgment, which stems from social cognitive theory (Bandura, 1991). Their findings suggest that consumer inability in regulating their online-use activity was related to increased online buying activity.

This concept of deficient self-regulation of consumer shopping impulses was further confirmed by a subsequent study (Kim & LaRose, 2004) that indicated that online buying behavior could become excessive or dysfunctional when consumers were incapable of self-monitoring or self-judging their own behavior against a set of rational behavioral standards.

As such, this type of dysfunctional online shopping behavior could occur regardless of whether the consumers were originally motivated by utilitarian motives—such as convenience—or hedonic motives such as recreation. Negative social outcomes have been evidenced in studies that addressed compulsive use of credit cards online, among other dysfunctions (McVeigh & Hill, 2001).

SOCIAL CHANGE IMPLICATIONS

Consumers have been drawn to teleshopping, either on or offline, for both utilitarian as well as hedonic motives. The types of utilitarian benefits perceived by in-home shoppers often include such convenience factors as time saving, time flexibility, reduced physical effort, reduced aggravation, and even opportunity for impulsive buying (Darian, 1987). By contrast, hedonistic benefits seen by teleshoppers usually present consumers an opportunity to find psychological gratifications—by fulfilling their needs for excitement, stimulation, pleasure, relaxation, parasocial interaction and/or companionship—often through the use of interesting and attractive images, graphics, sounds, and videos (Childers, Carr, Peck, & Carson, 2001).

Although the consumer welfare uniquely associated with electronic shopping venues—including such functional efficiency and advantages as convenience, lower prices, greater product variety—seems socially desirable, these marketing strengths also contain undesirable social consequences. First, shopping through teleshopping channels typically means conducting purchases in an isolation environment that doesn't involve any true social interactions characterizing exchanges in an in-store retail setting (Grant & Guthrie, 1991; Urbaczewski & Jessup, 1998). For those consumers who choose to conduct their shopping activities in isolation due to their social immobility, there could be several harmful implications.

As a case in point, consumers who lack sufficient social interactions may become even more socially detached. Those consumers who enjoy parasocial interactions with sales personnel through a mediated shopping channel may develop a preference for such pseudosocial relationships over other social contacts. In more severe cases, consumers may come to depend on these parasocial interactions for fulfilling their social companionship needs, which could lead to a harmful dissociation from social reality and real social relations.

Second, the pleasure and emotional fulfillment provided by the stimuli originating from teleshopping channels (especially those that contain high entertainment values) can often break up the mundane aspect of a consumer's daily life and provide a good escape for the individual from an unpleasant social reality. If the consumers become reliant on receiving mood management from these teleshopping channels, this could help reduce self-monitoring of shopping activity and drive these individuals to compulsive shopping or buying behaviors.

Third, those consumers who have become obsessed with seeking instant gratifications from their teleshopping shopping activities, to meet a set of different psychological needs, could see their self-control ability erode and spiral downward. Without reversing the loss of self-efficacy or self-consciousness to modify their harmful behavior, these consumers can transit from the state of deficient self-regulation to a true psychological disorder characterized by an addictive condition. This transformation process—starting from need gratification, proceeding to a compulsion for seeking gratification, and ending in an addiction for gratification—is similar to how individuals develop other types of addictive behaviors, including cigarette smoking and alcoholism, among others.

Fourth, the teleshopping industry, especially with the recent rapid growth in online commerce, has emerged to become an even stronger force in our economy. As this direct-to-consumer shopping modality becomes more and more pervasive in society, the concerns about consumer privacy and data security have also become more critical. This is because the ease, convenience, and ubiquity of Internet shopping and the ability of Internet shopping sites to offer nearly every type of product or service—ranging from daily staples, to recreational items, booking services (e.g., travel), banking and investing, bill paying, and medicines—are both a blessing and a curse. On the one hand, teleshopping has made many consumers' lives better by providing them a cost- and time-efficient way to shop. On the other hand, consumers often are unaware of the danger of exposing their personal identification and financial account information to teleshopping-channel operators who commit marketing fraud or whose database systems are not properly secured.

The most financially vulnerable segment of teleshopping fraud has been seniors or older adults. In 2004, consumers ages 50 and older suffered $152,000,000 in fraud, and 85% of all consumer fraud complaints the FTC received were filed by people in this age group as well (Federal Trade Commission, 2005). Of the fraud categories, Internet auction and Internet services/computer complaints occupied the top and third spots, respectively. This is not surprising because it is difficult to enforce regulatory and legal remedies to curb marketing fraud committed in a virtual environment that knows no national boundaries and can be easily manipulated to change the site identification.

In summary, from a sociocultural perspective, the teleshopping industry can be seen as providing a valuable social service to those consumers who enjoy the ease of in-home shopping for functional and/or recreational purposes. It can also be regarded as a source of antisocial and dysfunctional social behaviors that carry damaging social consequences stemming from drawing less socially mobile consumers into deeper social detachment, driving more impulsive consumers to become more compulsive, and turning more compulsive consumers into shopping addicts.

From an economic perspective, the nimbleness of the teleshopping industry as a marketing force has helped position the industry as a major player in consumer economy. At this stage of the online market development, the majority of consumers still rely on offline outlets to meet the bulk of their shopping needs (Fallows, 2004). Moreover, online marketing channels remain an alternative means for consumer product distribution, due to delivery channel efficiency considerations (Jones & Vijayasarathy, 1998; Strader & Shaw, 1999). Nonetheless, as the channels of consumer economy continue to evolve and to incorporate e-commerce components—whether using the online channel as a primary or ancillary marketing outlet—the increased role of online shopping seems to be an inevitable market trend for consumers. This suggests that the significance of trust, privacy, and security concerns (Ratnasigham, 1998; Strader & Ramaswami, 2002) associated with electronic transactions will loom even larger if economic losses caused by consumer fraud is not properly addressed and controlled.

In conclusion, the role of teleshopping as a mediated-marketing modality in our consumer culture will continue to expand. The impact of online commerce will fall on all consumers—either through consumers making direct purchases online of retail goods and services or through indirect transactions from a third party (e.g., a stock brokerage conducting an electronic trade on behalf of its customer). Hence, as market forces are driven to increasingly rely on new media technologies, consumers will also need to develop technology savvy to take advantage of what these technologies can offer rather than become the victims of technology-driven frauds.

REFERENCES

American Telemarketing Association. (2002). *Consumer Study 2002*. Retrieved April 20, 2004, from http://www.ataconnect.org/IndustryResearch/ConsumerStudy2002.html.

Auter, P. J. (1992). TV that talks back: An experimental validation of a parasocial interaction scale. *Journal of Broadcasting & Electronic Media, 36*(2), 173–181.

Bandura, A. (1991). Social cognitive theory of self-regulation. *Organizational Behavior and Human Decision Processes, 50*, 248–287.

Babin, B. J., Darden, W. R., & Griffin, M. (1994). Work and/or fun: Measuring hedonic and utilitarian shopping value. *Journal of Consumer Research, 20*, 644–656.

Ball-Rokeach, S. J., & DeFleur, M. L. (1985). The origins of individual media-systemdependency A sociological framework. *Communication Research, 12*, 485–510.

Berkowitz, E. N., Walton, J. R., & Walker, O. C., Jr. (1979). In-home shoppers: The market for innovative distribution systems. *Journal of Retailing, 55*(3), 15–33.

Chen, S.-J. & Chang, T.-Z. (2003). A descriptive model of online shopping process: Some empirical results. *International Journal of Service Industry Management, 14*(5), 556–569.

Chiang, K.-P., & Dholakia, R. R. (2003). Factors driving consumer intention to shop online: An empirical investigation. *Journal of Consumer Psychology, 13*(1 & 2), 177–183.

Childers, T. L., Carr, C. L., Peck, J., & Carson, S. (2001). Hedonic and utilitarian motivations for online retail shopping behavior. *Journal of Retailing, 77*, 531–535.

C/Net. (2005, August 8). *Online advertising set to skyrocket, study says*. Retrieved August 8, 2005, from http://news.com.com/Online+advertising+set+to+skyrocket%2C+study+says/2100–1024 3–5823583.html.

Cox, D. F., & Rich, S. U. (1964). Perceived risk and consumer decision-making—The case of telephone shopping, *Journal of Marketing Research, 1*, 32–39.

Darian, J. C. (1987). In-home shopping: Are there consumer segments? *Journal of Retailing, 63*(2), 163–186.

Defleur, M. L., & Ball-Rokeach, S. J. (1989). Media system dependency theory. In M. DeFleur & S. Ball-Rokeach (Eds.), *Theories of mass communication* (pp. 292–327). New York: Longman.

Dholakia, R. R., & Uusitalo, O. (2002). Switching to electronic stores: Consumer characteristics and the perception of shopping benefits. *International Journal of Retail & Distribution Management, 30*(10), 459–469.

Direct. (2002, October 21). Plenty of Americans catching DRTV Ads: Study. *Direct*. Primedia Business Magazines & Media Inc. Retrieved April 5, 2005, from http://directmag.com/news/marketing_plenty_americans_catching/index.html, on April 5, 2005.

Direct Marketing Association. (2004). *The DMA 2004 Response Rate Report*. Retrieved August 6, 2005, from http://www.the-dma.org/cgi/registered/research/2004responserate_report_ executivesummary.pdf.

Donthu, N., & Garcia, A. (1999). The Internet shopper. *Journal of Advertising Research, 39*(3), 52–58.

Donthu, N., & Gilliland, D. (1996). The informercial shopper. *Journal of Advertising Research, 36*(2), 69–76.

Eastlick, M. A. (1991). Catalog shoppers as potential adopter of videotex. In R .L. King (Ed.), *Retailing: Reflections, insights and forecasts* (pp. 9–13). Richmond, VA: Academy of Marketing Science.

Eastlick, M. A., & Liu, M. (1997). The influence of store attitudes and other nonstore shopping patterns on patronage of television shopping programs. *Journal of Direct Marketing, 11*(3), 14–24.

Eastlick, M. A., & Lotz, S. (1999). Profiling potential adopters and non-adopters of an interactive electronic shopping medium. *International Journal of Retail & Distribution Management, 27*(6), 209–245.

Engel, J. F., Blackwell, R. D., & Miniard, P. W. (1993). *Consumer behavior.* Chicago: Dryden.

Evans, P. E. (2004, January 8). "Market Forecast: Retail 2004–2008." *Jupiter Research.* Retrieved January 16, 2006, fromhttp://www.jupiterresearch.com/bin/item.pl/research:vision/107/id=94865.

Faber, R. J., O'Guinn, T. C., & Krych, R. (1987). Compulsive consumption. In M. Wallendorf & P. Anderson (Eds.), *Advances in consumer research* (Vol. 14, pp. 132–135). Provo, UT: Association for Consumer Research.

Fallows, D. (2004). *Many Americans use the Internet in everyday activities, but traditional offline habits still fulminate.* Pew Internet & America Life Project. Retrieved August 11, 2004, from http:// www.pewinternet.org.

Federal Communications Commission. (2004). *Subpart L Restrictions on telephone solicitation.* 47 C.F.R. §64.1200.

Federal Communications Commission. (2005, April 5). *CAN-SPAM unwanted text messages and e-mail on wireless phones and other mobile devices.* Retrieved April 10, 2005, from http://www.fcc.gov/cgb/consumerfacts/canspam.html.

Federal Trade Commission. (1994). *Telemarketing sales rule of 1994.* 16 CFR 310. Billing Code 6750-01-P.

Federal Trade Commission. (2004a). *Definitions, implementation, and reporting requirements under the Can-Spam Act.* 16 CFR Part 316.

Federal Trade Commission. (2004b). *Telemarketing Sales Rules of 1994 Amended.* 16 CFR Part 310, §§. 310.4(b)(3)(iv). Billing Code: 6750-01-p.

Federal Trade Commission. (2005, July 27). *FTC testimony: Identifying and fighting consumer fraud against older Americans.* Retrieved August 5, 2005, from http://www.ftc.gov/opa/2005/07/seniortest.htm

Fischer, E., & Arnold, S. J. (1990). More than a labor of love: Gender roles and Christmas shopping. *Journal of Consumer Research, 17*, 333–345.

Gehrt, K. C., & Carter, K. (1992). An exploratory assessment of catalog shopping orientations. *Journal of Direct Marketing, 6*(1), 29–39.

Grant, A. E., Guthrie, K., & Ball-Rokeach, S. J. (1991). Television shopping: A media system dependency perspective. *Communication Research, 18*(6), 773–798.

Grant, A. E., & Meadows, J. H. (2002). Electronic commerce: Going shopping with QVC and AOL. In C. A. Lin & D. J. Atkin (Eds.), *Communication technology and society: Audience adoption uses* (pp. 255–278). Cresskill, NJ: Hampton.

Grant, A. E., Meadows, J. H., & Handy, S. L. (1996). The passive audience for interactive technology. *New Telecom Quarterly, 4*(1), 48–51.

Gumpert, G., & Drucker, S. J. (1992). From the agora to the electronic shopping mall. *Critical Studies in Mass Communication, 9*, 186–200.

Hassay, D. N., & Malcolm, C. S. (1996). Compulsive buying: An examination of the consumption motive. *Psychology & Marketing, 13*(8), 741–752.

Hetsroni, A., & Asya, I. (2002). A comparison of values in informercials and commercials. *Corporate Communication, 7*(1), 34–45.

Hirschman, E. C. (1980). Innovativeness, novelty seeking, and consumer creativity. *Journal of Consumer Research, 7*, 283–295.

Hoch, S. J., & Lowenstein, G. F. (1991). Time-inconsistent preferences and consumer self-control. *Journal of Consumer Research, 17*, 492–507.

Hoffman, D. L., & Novak, T. P. (1996). Marketing in hypermedia computer-mediated environments: Conceptual foundations. *Journal of Marketing, 60*, 50–68.

Hoffman, D. L., Novak, T. P., & Chatterjee, P. (1996). Commercial scenarios for the Web: Oppurtunities and challenges. *Journal of Computer Mediated Communicate, 1*(3). Retrieved March 14, 1999, from http://www.ascusc.org/jcmc/vol1/issue3/hoffman.html.

James, E. L., & Cunningham, I. (1987). A profile of direct marketing television shoppers. *Journal of Direct Marketing, 1*(4), 1–3.

Jones, M. J., & Vijayasarathy, L. R. (1998). Internet consumer catalog shopping: Findings from an exploratory study and directions for future research. *Internet Research, 8*(4), 322–330.

Katz, E., Blumler, J. G., & Gurevitch, M. (1974). Utilization of mass communication by the individual. In J. Blumler & E. Katz (Eds.), *The uses of mass communications: Current perspectives on gratifications research* (pp. 19–32). Beverly Hills, CA: Sage.

Kaufman-Scarborough, C., & Lindquist, J. D. (2002). E-shopping in a multiple channel environment. *Journal of Consumer Marketing, 19*(4/5), 333–350.

Kim, J., & LaRose, R. (2004). Interactive e-commerce: Promoting consumer efficiency or impulsivity? *Journal of Computer Mediated Communication, 10*(1). Retrieved April 2, 2005, from http://jcmc.indiana.edu/vol10/issue1/kim_larose.html.

Korgaonkar, P. K., & Smith, A. E. (1986). Psychographic and demographic correlates of electronic in-home shopping and banking service. In T. A. Shimp, S. Shrama, G. John, J. A. Quelch, J. H. Lindgren,W. Dillion, M. P. Gardener, & R. F. Dyer (Eds.), *1986 AMA Educators' Proceedings, Series No. 52* (pp. 167–170). Chicago: American Marketing Association.

Korgaonkar, P. K., & Wolin, L. D. (1999). A multivariate analysis of web usage. *Journal of Advertising Research, 39*(2), 53–68.

LaRose, R., & Eastin, M. S. (2002). Is online buying out of control? Electronic commerce and consumer self-regulation. *Journal of Broadcasting & Electronic Media, 46*(4), 549–564.

Lin, C. A. (1994). Exploring potential factors for home videotext adoption. *Advances in Telematics, 2,* 111–121.

Lin, C. A. (1996). The contribution of the uses and gratifications perspective to communication research. *Journal of Broadcasting & Electronic Media, 40,* 574–581.

Lin, C. A. (1999). Predicting online service adoption likelihood among potential subscribers: A motivational approach. *Journal of Advertising Research, 39*(2), 79–89.

Lin, C. A. (2001). Audience attributes, media supplementation and likely online service adoption. *Mass Communication and Society, 4*(1), 19–38.

Lin, C. A. (2004). Webcasting adoption: Technology fluidity, user Innovativeness, and media substitution. *Journal of Broadcasting & Electronic Media, 48*(3), 446–465.

Li, S. (2004). Internet shopping and its adopters: Factors in the adoption of internet shopping. In P. Lee, L. Leong, & C. So (Eds.), *Impact and issues in the new media: Toward intelligent societies* (pp. 81–102). Cresskill, NJ: Hampton.

Mangalindan, M. (2005, May 24). Online retail sales are expected to rise to $172 this year. *The Wall Street Journal,* p. D5.

Maslow, A. H. (1943). A theory of human motivation. *Psychological Review, 50*, 370–396.

Mathwick, C., Malhotra, N. K., & Rigdon, E. (2002). The effect of dynamic retail experiences on experiential perceptions of value: An Internet and catalog comparison. *Journal of Retailing, 78*, 51–60.

McVeigh, T., & Hill, A. (2001, October 28). Net shopping hooks army of addicts. *The Observer*. Retrieved January 12, 2005, http://observer.guardian.co. uk/uk_news/story/0,6903,582206,00.html

Menon, S., & Kahn, B. (2002). Cross-category effects of induced arousal and pleasure on the Internet shopping experience. *Journal of Retailing, 78*, 31–40.

Midgley, D. F., & Dowling, G. R. (1978, March). Innovativeness: The concept and its measurement. *Journal of Consumer Research, 4*, 229–242.

Muldoon, K. (1996). *How to profit through catalog marketing.* Lincolnwood, IL: NTC Business Books.

Nataraajan, R., & Goff, B. G. (1992). Manifestations of compulsiveness in the consumer-marketplace domain. *Psychology & Marketing, 9*(1), 31–44.

Novak, T. P., Hoffman, D. L., & Duhachek, A. (2003). The influence of goal-directed and experiential activities on online flow experiences. *Journal of Consumer Psychology, 13*(1 & 2), 3–16.

Novak, T. P., Hoffman, D. L., & Yung, Y.-F. (2000). Measuring the customer experience in online environments: A structural modeling approach. *Marketing Science, 19*(1), 22–42.

Nowak, G. J., & Phelps, J. E. (1997). Direct marketing and the use of individual-level consumer information: Determining how and when rivacy matters. *Journal of Direct Marketing, 11*(4), 94–108.

Nowak, G. (1992). TV viewer characteristics and results beyond response. *Journal of Direct Marketing, 6*(2), 181.

Pastore, M. (2000, June 28). Direct-to-consumer sales seen increasing. *Clickz.Com*. Retrieved April 10, 2005, from http://www.clickz.com/stats/sectors/retailing/article.php/6061_404851.

Ratnasingham, P. (1998). The importance of trust in electronic commerce, electronic commerce. *Internet Research, 8*(4), 313–320.

Rogers, E. (2003). *Diffusion of innovations* (5th ed.). New York: The Free Press.

Romano, C. (1998). Telemarketing grows up. *Management Review, 87*(6), 31–34.

Rook, D. W. (1987). The buying impulse. *Journal of Consumer Research, 14*, 189–199.

Rook, D. W., & Fisher, R. J. (1995). Normative influence on impulsive buying behavior. *Journal of Consumer Research, 22*, 305–313.

Rook, D. W., & Hoch, S. J. (1985). Consuming impulses. In E. C. Hirschman & M. B. Holbrook (Eds.), *Advances in consumer research* (pp. 23 27). Provo, UT: Association for Consumer Research.

Salomon, I., & Koppleman, F. S. (1992). Teleshopping or going shopping? An information acquisition perspective. *Behavior & Information Technology, 11*(4), 189–198.

Settle, R. B., Alreck, P. L., & McCorkle, D. E. (1994). Consumer perceptions of mail/phone order shopping media. *Journal of Direct Marketing, 8*(3), 30–45.

Sherry, J. F., Jr. (1990). A sociocultural analysis of a midwestern flea market. *Journal of Consumer Research, 17*, 13–30.

Shim, S., & Mahoney, M. (1991). Electronic shoppers and nonshoppers among Videotex users. *Journal of Direct Marketing, 3*, 29–38.

Skumanich, S., & Kintsfather, D. P. (1998). Individual media dependency relations within television shopping programming: A causal model reviewed and revised. *Communication Research, 25*(2), 200–220.

Stark, S. D. (1998). *Glued to the set: The 60 television shows and events that made us who we are today*. New York: Delta.

Strader, T. J., & Shaw, M. J. (1999). Consumer cost differences for traditional and Internet markets. *Internet Research, 9*(2), 82–92.

Strader, T. J. & Sridhar, N. R. (2002). The value of seller trustworthiness in consumer-to-consumer online markets. *Communications of the ACM, 45*(12), 45–49.

Urbanczewski, A., & Jessup, L. (1998). A manager's primer on electronic commerce. *Business Horizons, 41*(5), 5–17.

U.S. Code (2003). *Controlling the assault of non-solicited pornography and marketing act of 2003.* 15 U.S.C. 7701–7713.

U.S. Code (1994). *Telemarketing Consumer Fraud and Abuse Prevention Act of 1994*, 15 U.S.C. 1601–1608.

Wesson, C. (1990). *Women who shop too much: Overcoming the urge to splurge*. New York: St. Martin's.

12

Interactive Media Technology and Telemedicine

Pamela Whitten

> Anne was a 39-year-old divorced mother of one who had been diagnosed with terminal cancer. During the final weeks of her life, she was moved into a nursing home for 24-hour care. A low-end piece of telemedicine equipment, a videophone, was put into her room. With this technology, Anne was able to receive ongoing end-of-life services from her hospice agency. In this case, telemedicine provided continuity of care for this patient. However, telemedicine went further than simply providing health-related support services for Anne. In this case, a second videophone was put in the home of her 5-year-old son who was living more than an hour away with Anne's ex-husband. Every day, Anne would see and talk to her son several times a day. She was even able to attend his sixth birthday party via the telemedicine equipment. Although Anne only lived less than 2 months once admitted to the nursing home, the nurses reported a miraculous final quality of life. She got up and out of bed every day, became physically stronger, and enjoyed her final days. We have not yet begun to understand the health, cost, and social benefits of telemedicine.

The health field is employing telecommunication technologies in exciting and innovative ways. Telemedicine, the use of telecommunication technologies to deliver health services and education over a distance, has garnered growing interest over the past decade. However, the application of telemedicine actually dates back more than 50 years in North America (e.g., telepsychiatry applications by Wittson, Affleck, & Johnson, 1961).

Telemedicine has been defined in a variety of ways, both narrow and broad in scope. Park (1974) defined telemedicine narrowly, with a focus on the technology, when he described it as the use of interactive or two-way TV to provide health care. Others describe telemedicine in terms of diagnosing or treating a patient who is at a different location (Willemain & Mark, 1971). Bennet, Rappaport, and Skinner (1978) coined the term telehealth, which includes education, administration, and patient care. The Office for the Advancement of Telehealth (OAT) currently defines telehealth as " the use of electronic information and telecommunications technologies to support long-distance clinical care, patient and professional health-related education, public health and health administration" (Office for the Advancement of Telehealth, n.d.).

BACKGROUND

Telemedicine initially emerged as a potential way to address two significant problems in health delivery. First, access was a challenge in many parts of the United States. For example, patients residing in rural areas would have to travel long distances to see a health provider such as a specialty physician. A second driving force behind initial telemedicine deployment involved cost and resource efficiencies. Health care administrators were seeking solutions to address the high costs associated with delivering many types of health services.

The use of telecommunication technologies for medical diagnosis, care, and education has commonly involved the use of interactive video for synchronous delivery of care. *Interactive video* refers to real-time videoconferencing that occurs between two or more sites. The quality of the interactions depends on the equipment and transmission speeds employed. A handful of specialty applications employ asynchronous or store-and-forward solutions, including radiology, pathology, and often dermatology. Store-and-forward connections permit audio clips, video clips, still images, or data to be held now and transmitted or received at a later date. The recipient is free to choose the actual time when he or she reviews the content of the transmission.

Telemedicine techniques have developed over the past four decades. Wittson and colleagues (1961) were the first to employ telemedicine for medical purposes in 1959, when they set up telepsychiatry consultations via microwave technology between the Nebraska Psychiatric Institute in Omaha and the state mental hospital 112 miles away. In the same year, Montreal, Quebec, was the site for Jutra's (1959) pioneering teleradiology work. The advent of teleradiology allowed for the asynchronous transmission of radiographic images to experts located at distant sites.

The 1970s witnessed a flurry of telemedicine activity, as several major projects developed in North America and Australia, including the Space Technology Applied to Rural Papago Advanced Health Care (STARPAHC) project of the National Aeronautics and Space Administration (NASA) in southern Arizona, a project at Logan Airport in Boston, and programs in northern Canada (Dunn, Conrath, Action, & Bain, 1980). These initial projects provided valuable insight into technological constraints and acceptance issues that paved the way for future telemedicine endeavors such as telehome health.

Although data are limited, early reviews and evaluations of these programs suggest the equipment was reasonably effective at transmitting the information needed for most clinical applications and users were mostly satisfied (Conrath,

Buckingham, Dunn, & Swanson, 1975; Dongier, Tempier, Lalinec-Michaud, & Meunier, 1986; Fuchs, 1974; Murphy & Bird, 1974). Interestingly, with the exception of one simple program at Memorial University Hospital of Newfoundland, no telemedicine programs survived past 1986. When external sources of funding were withdrawn, the programs simply folded.

The 1960s, 1970s, and 1980s exhibited a series of telemedicine pilot and demonstration projects. However, the 1990s proved to be a period of rapid growth. In the early 1990s, new, fairly inexpensive, and commonly available digital technologies enabled video, audio, and other imaging information to be digitized and compressed. This facilitated the transmission of information over telephone lines with relatively narrow bandwidths, instead of through more expensive satellites or relatively unavailable private cable or fiber-optic lines. In 1990, there were four active telemedicine programs. By 1997, there were almost 90 such programs, and by 1998, there were 200 documented telemedicine programs (Federal Telemedicine Update, 2002). Today, so many health systems employ some form of telecommunication technology to deliver health services or education that it is no longer possible to quantify the number of telemedicine programs.

A wide range of applications and contexts have emerged where telemedicine is proving to be a useful solution. Telepsychiatry, telemedicine for behavioral health, has stood out as perhaps the most common form of synchronous tertiary care. Home telecare, via interactive nursing visits and/or data monitoring, is often considered to be the fastest growing segment of telemedicine (Whitten & Gregg, 2001). Teleradiology, the electronic transmission of radiological images, is even considered mainstream today due to its ubiquity and cost effectiveness (Maheu, Whitten, & Allen, 2001). The deployment and adoption of telemedicine have followed a bumpy and uneven growth trajectory. Some telemedicine programs, such as the project at Marquette General Health System in Michigan's Upper Peninsula, have survived and thrived past their period of grant funding. To date, the vast majority of telemedicine research projects have simply tested feasibility. However, there is a growing body of research that seeks to study the social impacts of telemedicine through a specific theoretical lens. We now turn to a sampling of theoretical explanations addressing telemedicine.

THEORETICAL EXPLANATIONS

Communication theorists account for only a handful of the research being conducted in the field of telehealth. However, several interesting approaches have made significant contributions to the field. Gagnon et al. (2003) sought to assess the predictors of physicians' intentions to use telemedicine in their clinical practice based on the theory of interpersonal behavior. They concluded that those doctors who perceived professional and social responsibilities as key expressed a stronger intention to use telemedicine.

Communicational theory (focusing on linguistic pragmatics) was used by Schall and colleagues (2003) to compare telemedical applications with traditional care. The researchers sought to demonstrate that the structures in electronic medical communication showed communicative structures with unique features. They employed a semiotic model that focused on the roles of partners in an interaction, the respective message, and their relations. They specifically focused on the channel, sender, addressee, and other structural roles and found that electronic medical communication does have special features not discovered in traditional communication.

Other scholars have sought to explain and integrate agency and structure. For example, Lehoux, Sicotte, Denis, Berg, and Lacroix (2002) employed Anthony Giddens' structuation theory to analyze the views, communications needs, and referral strategies of physicians performing teleconsultations. In structuration theory, humans and social structure are not seen as two separate concepts or constructs, but rather as two ways of considering social action. There is a duality of structures so that one side is composed of situated actors who undertake social action and interaction in concordance with the rules, resources, and social relationships that are produced and reproduced in social interactions. Structuration seeks to understand the ways in which social systems are produced and reproduced through social interactions.

The Technology Acceptance Model (TAM) is another theoretical perspective employed to understand telemedicine. The Technology Acceptance Model (Davis, 1989) seeks to explain users' intention to use a new system through two beliefs: perceived usefulness and perceived ease of use. Chau and Hu (2002) investigated technology acceptance by individual professionals by examining physicians' decisions to accept telemedicine technology. Results from their study suggest several areas where individual "professionals" might subtly differ in their technology acceptance decision making, as compared with end users and business managers in ordinary business settings. Specifically, physicians appeared to be fairly pragmatic, largely anchoring their acceptance decisions in the usefulness of the technology, the compatibility of the technology with their practices, and peers' opinions about using the technology.

Diffusion theory is the most popular communication-based theory by far that has been applied to the study of telemedicine. Specifically, this theory argues that diffusion "is the process by which an innovation is communicated through certain channels over time among the members of a social system" (Rogers, 1995, p. 10). In a study conducted in New Mexico, researchers employed this theory to better understand the dynamic interactions between the characteristics of telehealth and the social system in which it is applied (Helitzer, Heath, Maltrud, Sullivan, & Alverson, 2003). Walker and Whetton (2002) argued that diffusion theory suggests that organizational structures and cultures will affect health professionals' perceptions of telehealth as the introduction of telehealth services impacts existing work practices.

In an earlier study, scholars extended Atwell's theory of knowledge barriers to explain why actual diffusion and adoption of telemedicine remained low (Tanriverdi & Iacono, 1999). The theory of knowledge barriers (Attewell, 1992) suggests that the burden of learning technical knowledge creates knowledge barriers that inhibit diffusion. Tanriverdi and Iacono (1999) argued that the acquisition of technical knowledge must be supplemented with economic, organizational, and behavioral barriers to fully explain diffusion challenges.

Other scholars have suggested that traditional diffusion theory is unable to fully explain the deployment and adoption of telemedicine. Whitten and Collins (1997) argued that a linear approach fails to account for the diffusion of decentralized and continually reinvented innovations such as telemedicine. Instead, they offer that a communicative focus that privileges the role of participatory conversation is used to examine and explain the invention, diffusion, and reinvention of telemedicine.

Whitten (2005) has recently been working on explaining the success or failure of telemedicine through the use of social embeddedness theory developed by Granovetter (1985). Granovetter posits that embeddedness is a multidimensional construct, indicating that people (actors) within complex social networks (such as health care) face unique resources and constraints relative to those not embedded in such relations. Delving further into the notion of embeddedness, Moody and White (2001) propose the theory of structural cohesion, which suggests that there is a minimum number of actors (disciplines within health care) that, if removed from the group, would disconnect the group. Whitten (2005) argues that understanding which disciplines or stakeholders (e.g., doctors, patients, insurance providers, legal and regulatory agents, information technology [IT] specialists) "break down" the telemedicine intervention—when they exit—will help us explain the success or failure of telemedicine. So, for example, if insurance providers refuse to reimburse for telemedicine, this may cause physicians to be less enthusiastic about providing services over this particular modality.

Theoretical explanations for telemedicine's successes and failures are still in their infancy. However, this does not mean that important data have not been collected to help us understand perceptions and outcomes of this phenomenon. We now turn to a summary of empirical findings.

EMPIRICAL FINDINGS

Telemedicine researchers are interested in perceptions regarding telehealth, as well as understanding both clinical and cost outcomes. A summary of research results for each of these categories is provided here. However, the reader should note that it is highly likely that there is a recursive (or two-way) relationship among these three categories.

Numerous articles purport to demonstrate the clinical effectiveness of telemedicine. A cursory review of many of the articles that address clinical outcomes in general reflects an almost universal tendency to make positive claims about the impact of telemedicine on patient outcomes. Taken at face value, the lengthy list of publications that document positive outcomes appears to substantiate this claim. For example, a study examining clinical efficacy of telemedicine in emergency radiotherapy (i.e., the treatment of disease with radiation, especially by selective irradiation with x-rays or other ionizing radiation and by ingestion of radioisotopes) for malignant spinal cord compression reported that not only were image quality, transmission time, and cost-benefit satisfactory, but also there was a significant increase in the number nonambulant patients who became ambulant (Hashimoto et al., 2001). Similarly, in the area of home-based interventions, where home health nurses use telemedicine to monitor patient symptoms, studies report encouraging data such as improved mean arterial pressure in hypertension (or high blood pressure) patients (Rogers et al., 2001) and improved health-related outcomes for newly injured spinal cord patients (Phillips et al., 2001).

A number of other studies also have documented positive outcomes from the delivery of behavioral health care via telemedicine. An evaluation of a prison telepsychiatry service, which employed telemedicine to provide psychiatric services for inmates, concluded that telepsychiatry is an effective way to deliver mental health services to a prison population (Zaylor, Nelson, & Cook, 2001). In a study of the use of telemedicine for discharge planning for inpatients in a tertiary psychiatric hospital, D'Souza (2002) found that a higher proportion of patients in the control group were readmitted over a 12-month period and reported medication side effects than in the telemedicine group. In the primary care setting, Hunkeler et al. (2002) published the finding that nurse telehealth care, which enables nurses to provide services via telemedicine, improves clinical outcomes for antidepressant drug treatment.

A handful of studies, however, do not claim unequivocal evidence for telemedicine health benefits. As an example, in the area of teledermatology, a randomized controlled trial compared results of real-time, store-and-forward, and conventional consultations (Loane et al., 2000). Although real-time teleconsults were deemed clinically feasible, store-and-forward consults were considered inefficient, as significantly more store-and-forward consults resulted in a request for a subsequent appointment by the dermatologist.

Published review articles on telemedicine clinical outcomes all draw similar conclusions—namely, that irrefutable evidence regarding telemedicine's positive impact on clinical outcomes still eludes us. These review papers document a spattering of potential evidence for clinical effectiveness in a range of services. Roine, Ohinmaa, and Hailey's (2001) review of more than 150 articles meeting preestablished review criteria concluded that potential effectiveness could only be attributed to teleradiology, telepsychiatry, transmission of echocardiographic images, and consultations between primary and secondary health providers. A recent systematic review that examined more than 1,300 papers making claims about telemedicine outcomes found only 46 publications that actually assessed at least some clinical outcomes (Hailey, Roine, & Ohinmaa, 2002). In this case, Hailey et al. (2002) found the most convincing evidence of the efficacy and effectiveness of telemedicine was for teleradiology, telepsychiatry, tele-home services, and a few medical consultations.

Currell, Urquhart, and Wainwright (2002) performed a literature review to analyze the effects of telemedicine as an alternative to face-to-face care. They searched for publications on randomized trials, controlled before-and-after studies, and interrupted time series that compared telemedicine to traditional patient care. Seven trials including more than 800 patients were critiqued. The reviewers concluded that establishing systems for patient care using telecommunications technologies is feasible; however, the studies provide inconclusive results regarding clinical benefits and outcomes. In an analysis of 25 articles meeting review criteria, Hersch and colleagues (2001) found home-based telemedicine for chronic disease management, hypertension, and AIDS ranked highest in demonstrated efficacy of telemedicine for clinical benefits. They also documented that there was reasonable evidence that telemedicine is comparable to face-to-face medicine in emergency medicine and is beneficial in surgical and neonatal intensive care units, as well as for patient transfers in neurosurgery.

As is the case with clinical benefits, bibliographic database searches (e.g., Telemedicine Information Exchange, Pub Med) yield hundreds of publications making claims regarding the cost effectiveness of telemedicine. These publications vary in quality and nature and encompass a wide range of topics within the rubric of economic outcomes, including cost of equipment, staff, or maintenance; comparative cost of traditional versus telecare; and potential savings by provider or patient.

Once again, a quick review of many of the articles that address economic outcomes reflects an almost universal tendency to make positive claims about the cost benefits of telemedicine. For example, a cost measurement study for a home-based telehospice service documented cost per visit for traditional care at $126 and telehospice visits at $29 (Doolittle, 2002). Another study found that the utilization of home uterine monitoring systems and telephone nursing by pregnant women with preterm labor had a mean cost of $7,225 for the telemedicine group and a mean cost of $21,684 for the control group (Morrison et al., 2001). A remote community in Canada used store-and-forward consults between the on-site nurse and hospital physicians (Jong et al., 2001). Factoring in the cost of replacing the system biennially, the health authority projected a savings of $20,877 per year on patient transportation with the patient estimated to save $8,008.

A pilot study employed cost minimalization analysis to evaluate a telemedicine chronic wound consultation clinic and documented that the average cost of a consultation was $126 versus $246 for an equivalent face-to-face consultation (Pringle-Specht, Wakefield, & Flanagan, 2001). Despite the encouraging picture presented by cost-benefit studies such as the ones just illustrated, several rigorous review articles have been published that paint a less rosy picture regarding the documented cost benefits of telemedicine.

Whitten et al. (2002) published a systematic review in the *British Medical Journal*. This paper identified 612 articles specifically addressing costs associated with the telemedicine intervention through an extensive database search of scholarly journals in the area. Of these 612 articles, 557 did not include any type of cost data. Of the 557 articles that lacked cost data, only 2% made claims of negative economic outcomes. From the remaining 55 articles with cost data, only 24 included criteria qualifying them for a full systematic review (e.g., complete explanation of methods for study designs that could provide a common baseline for analysis).

Although most of the 24 studies provided clear goals, only 3 provided a formal research hypothesis for empirical testing. Most studies failed to make clear the perspective and boundaries—that is, research approach and scope utilized in the analysis—and employed inappropriate economic analytical techniques (e.g., unique measures that preclude comparison across studies). None of these studies used conventional cost-utility analysis—specifying the costs and benefits per intervention—to determine "value for money."

Without cost-utility analysis, results cannot be generalizable across different treatments, making it "impossible to assess the extent to which telemedicine represents a sensible priority for health care investment" (Whitten et al., 2002, p. 1437). The short duration engaged by most of these projects (only 20% lasted longer than 12 months) further compounded the problems of drawing long-term conclusions regarding telemedicine's cost-effectiveness. In general, Whitten and colleagues found no uniformity of analysis, and the studies reviewed were often plagued by poor design and inadequate technical quality.

In a guest commentary, Wootton (2002) points out that other review articles have drawn similar conclusions regarding the questionable conclusions of the cost effectiveness of telemedicine. However, Wootton (2002) documents an important point as well: "It is important to understand that a lack of evidence for the cost-effectiveness of telemedicine is not the same as evidence that telemedicine is not cost effective" (p. 12).

Perhaps the most commonly researched aspect of telemedicine services involves perceptions and satisfaction. The heaviest emphasis has been placed on patient perceptions. For the most part, these studies report results from small-scale or feasibility studies with less than 100 patients (Whitten & Mair, 2000). The vast majority of studies report extremely high patient satisfaction. These studies typically have involved real-time medical consultations for a variety of services. In two different projects providing services across multiple specialty areas, satisfaction ratings were quite high. Harrison, Clayton, and Wallace (1997), for example, found that 84% of their participants involved in outpatient consultations in England felt "positive" about their session, which was the same percentage reported by Huston and Burton (1997) for their Kentucky-based project. Itzak, Weinberger, Berkovitch, and Reis (1998), in their study of primary care teleconsulting in Israel, reported 100% of patients satisfied with their care via this medium.

It also appears that the positive perceptions patients have regarding telehome health cuts across service types and disease contexts. Meyer (2002) reported a 90% patient satisfaction rate with a telehome health project launched through the Veterans Administration Home and Community Care Service Line in South Georgia, Florida, and Puerto Rico. Researchers of a telehome health project in Michigan's Upper Peninsula found levels of satisfaction with telehome health for diabetic, chronic obstructive pulmonary disease (COPD), and congestive heart failure (CHF) patients to display no statistically significant differences (Whitten, 2003).

Patients have agreed that there are definite advantages to telemedicine, which include reduced waiting times to be seen, enhanced access to care, reduced costs to the health care system, impressions that examinations are more thorough, and excitement with the use of this new technology. Disadvantages that have been noted include nervousness about use of the new technology, difficulty in talking to health care providers via the TV system, a tendency to be less candid when talking to the provider via this medium, and an experience of emotional distance between themselves and their provider (Allen & Hayes, 1995; Allen, Roman, Cox, & Cardwell, 1996; Baer et al., 1995; Blackmon, Kaak, & Ranseen, 1997; Clarke, 1997; Couturier, Tyrrell, Tonetti, Rhul, & Franco, 1998; Dongier et al., 1986; Doolittle, Yaezel, Otto, & Clemens, 1998; Harrison et al., 1997; Huston & Burton, 1997; Jones et al., 1996; Kunkler, Rafferty, Hill, Henry, & Foreman, 1998; Pedersen & Holand, 1995).

Other researchers have documented specific components that impact the level of patient satisfaction. For example, Chae and colleagues (2001) found that location of the patient and the patient's perspective on the quality of the communication encounter served as predictors of patient satisfaction. Qualitative research data provide more specific evidence of patient attitudes toward home health services. During interviews, telehospice patients in Michigan expressed frustration that subsequently *more* visits were not conducted via the telehealth technology (Whitten, Doolittle, & Mackert, 2004).

A preponderance of provider satisfaction studies conclude that overall satisfaction or acceptance is positive, if not high (Allen, Hayes, Sadasivan, Williamson, & Wittman, 1995; Allen et al., 1996; Brennan et al., 1999; Chan, Woo, Hui, & Hjelm, 2001; Doolittle et al., 1998; Elford et al., 2001; Hui & Woo, 2002; Kobb, Hoffman, Lodge, & Kline, 2003; Lee et al., 1998; Lowitt et al., 1998; Mark, Willemain, Malcom, Master, & Clarkson, 1976; Nesbitt, Hilty, Kuenneth, & Siefkin, 2000; Nordal, Moseng, Kvammen, & Lochen, 2001). Studies have demonstrated some variations: Nurses are more enthusiastic than doctors (Higgins, Conrath, & Dunn, 1984), referring physicians are more satisfied than the referred specialists (Zollo, Kienzle, Loeffelholz, & Sebille, 1999), satisfaction increases with time and use (Higgins et al., 1984; Cunningham, Marshall, & Glazer, 1978), providers do not universally advocate telemedicine for all situations (Whitten, Collins, & Mair, 1998), and only 45% of physicians felt that telemedicine could adequately assess the conditions of patients (Bratton, 2001).

Studies that go further than simple provider satisfaction with telemedicine reveal richer details. Patient–provider relations (Itzak & Weinberger, 1998) and patient care may be improved (Ricci, Callas, & Montgomery, 1997). In early studies, doctors were wary of technical glitches in diagnoses (Cunningham et al., 1978; Fuchs, 1974), but these concerns appeared to have been mitigated by the advance of technology, as many providers now have reported confidence in their diagnoses (Campbell & Martel, 1999; Endean, Mallon, Minion, Kwolek, & Schwarez, 2001; Kvedar, Menn, Baradagunta, Smulders-Meyer, & Gonalez, 1999). A survey of nurses found that telenurses experienced work satisfaction equal to that of hospital-based nurses (Shlacta-Fairchild, 2002).

Given the overwhelming evidence of positive provider satisfaction and perceptions, it may appear contradictory that generally satisfied providers could be critical gatekeepers. However, a review concludes that we still lack scientific documentation on the impact of telemedicine on patient–physician relations, quality and quantity of medical work, and provider satisfaction (Kjaer & Karlsen, 2002). Examining the process of gaining acceptance generates some insights—namely, that provider acceptance is neither unconditional nor automatic. The direct involvement of health care providers in the development of telemedicine systems positively impacts acceptance and use (Nielsen, Jackson, Kosman, Kiley, & Thompson, 2000) as does initial and follow-up training in the use of the system (Dansky & Bowles, 2002; Mazmanian et al., 1996). Furthermore, many studies do not reflect the entire population of health care providers, but only willing participants who may be part of a funded research grant project, pilot study, or project development. In other words, there is an inherent selection bias whereby the providers who are surveyed or interviewed already had an accepting, if not embracing, attitude toward telemedicine before the study was commenced.

Empirical evidence regarding the effectives of telemedicine is mixed. However, extensive research is still relatively new, and it appears that there is strong potential for this type of innovative approach to providing health care. The actual implications of telemedicine for health care in general are important considerations and are covered in the next section through a discussion of potential consequences.

SOCIAL CHANGE IMPLICATIONS

The jury is still out regarding issues of deployment and adoption for telemedicine. To more clearly see the interrelated role of the communication process and the extent to which it simultaneously creates and re-creates the telemedicine and general health environment, one must examine the current and long-term consequences and constraints (Whitten & Collins, 1997).

One obvious potential consequence concerns whether telemedicine will lessen the cost of medicine because of such factors as decreased travel time and expenses. Or, instead, will telemedicine increase the overall cost of health care as more and more people have access to specialty services that are currently unavailable in their communities? Could telemedicine add to the problem of excessive and unnecessary tests?

Yet another potential economic consequence to be monitored concerns how telemedicine will be deployed. It is possible that telemedicine will become a strategic marketing tool used to by health organizations to garner additional market share. One potential social implication impacted by such economic consequences could be a widening of the gap for those with access to the best medical services.

There is an interesting array of social consequences that must also be considered. Will technologies employed in telemedicine change interpersonal relationships between providers and patients? It is possible that our society could evolve into one that no longer expects a patient and provider to be physically located in the same room. How will this type of setup impact relationships and affect such health outcomes as treatment compliance? It may well be that enhanced access will trump some relational issues. Yet important communication research has taught that, in many cases, health care providers' interpersonal skills were judged by patients to be as important as their clinical skills (Matthews, Sledge, & Lieberman, 1987).

We may find that telemedicine offers a forum for patients and providers to communicate and interact in ways that do not compromise interpersonal rapport, openness, trust, and comfort. If this is the case, telemedicine may ultimately enhance health outcomes, as more patients have access to health services they could not obtain without the use of communication technologies. We also hear arguments from technological pessimists that telemedicine may carry negative repercussions for health outcomes due to mistakes, abuse, or even through the development of poor relationships that could develop through mistrust. To date, studies have not documented these negative potential effects of telemedicine. However, more research effort is needed to better understand the impact of telemedicine on communication and relational issues.

Ethical issues related to patient privacy are also highly significant matters to consider when implementing health care through telemedicine. This is because patient information is inputted, delivered, received, and stored electronically. In other words, the public "airwaves" carry personal histories, images, and patient concerns. The Health Information Portability and Accountability Act (HIPAA) of 1996 includes a number of provisions to aid in the protection and security of personal medical information. However, understanding and implementing these protections have been confusing for health care providers. To date, what it means to be HIPAA compliant is still being explored. Ethical consequences such as privacy, confidentiality, and intimacy are key issues to consider.

Sociological consequences may also emerge from telemedicine deployment. Perhaps quality of life will increase for rural residents as access to health services begins to mirror what is available in more urbanized areas, where health care facilities are more readily available. In addition, widespread adoption of telemedicine may change our expectations for health. Patients may expect faster access to a wider range of services. What about the health of our society? Is it possible that we could become healthier and live longer with telemedicine?

Health organizations in the United States are challenged to find effective and efficient strategies to deliver quality care for all Americans. Telemedicine has emerged as a possible solution to address access, cost, and quality of care challenges. Early evidence indicates that consumers are happy to use telemedicine when it meets a health care need. Much more work is needed to better understand why some health providers embrace this solution and others avoid it. As data-driven services such as electronic medical records (EMRs) merge into traditional and telemedicine services, additional research will be needed to explore and understand how health services should be socially engineered to maximize the employment of communication technologies.

REFERENCES

Allen, A., & Hayes, M. P. A. (1995). Patient satisfaction with tele-oncology: A pilot study. *Telemedicine Journal, 1*(1), 41–46.

Allen, A., Hayes, J., Sadasivan, R., Williamson, S. K., & Wittman C. (1995). A pilot study of the physician acceptance of tele-oncology. *Journal of Telemedicine and Telecare, 1*, 34–37.

Allen, A., Roman L., Cox, R., & Cardwell, B. (1996). Home health visits using a cable television network: User satisfaction. *Journal of Telemedicine and Telecare, 2*(Suppl. 1), 92–94.

Attewell, P. (1992). Technology diffusion and organizational learning: The case of business computing. *Organization Science, 3*(1), 1–19.

Baer, L., Cukor, P., Jenike, M. A., et al. (1995). Pilot studies of telemedicine for patients with obsessive-compulsive disorder. *American Journal of Psychiatry, 152*, 1383–1385.

Bennet, A. M., Rappaport, W. H., & Skinner, E. L. (1978). *Telehealth handbook* (Publication No. PHS 79-3210). Washington, DC: U.S. Department of Health, Education, and Welfare.

Blackmon, L. A., Kaak, H. O., & Ranseen, J. (1997). Consumer satisfaction with telemedicine child psychiatry consultation in rural Kentucky. *Psychiatric Services, 48*(11), 14644–14666.

Boulanger, B., Kearney, P., Ochoa, J., Tsuei, B., & Sands, F. (2001). Telemedicine: A solution to the follow-up of rural trauma patients? *Journal of the American College of Surgeons, 192*, 447–452.

Bratton, R. L. (2001). Patient and physician satisfaction with telemedicine for monitoring vital signs. *Journal of Telemedicine and Telecare, 7*(Suppl. 1), 72–73.

Brennan, J. A., Kealy, J. A., Gerardi, L. H., Shih, H., Allegra, J., Sannipoli, L., et al. (1999). Telemedicine in the emergency department: A randomized controlled trial. *Journal of Telemedicine and Telecare, 5*, 18–22.

Campbell, T., & Martel, R. F. (1999). Realtime remote consultation in the outpatient clinic: Experience at a teaching hospital. *Journal of Telemedicine and Telecare, 5*(Suppl. 1), 70–71.

Chae, Y. M., Heon, L. J., Hee, H. S., Ja Kim, S., Hong Jun, K., & Uk Won, J. (2001). Patient satisfaction with telemedicine and in home health services for the elderly. *International Journal of Medical Informatics, 61*(2 3), 167 173.

Chan, W. M., Woo J., Hui, E., & Hjelm, N. M. (2001). The role of telenursing in the provision of geriatric outreach services to residential homes in Hong Kong. *Journal of Telemedicine and Telecare, 7,* 38–46.

Chau, P. Y. K., & Hu, P. J. (2002). Examining a model of information technology acceptance by individual professionals: An exploratory study. *Journal of Management Information Systems, 18*(4), 191–229.

Clarke, P. H. J. (1997). A referrer and patient evaluation of a telepsychiatry consultation-liaison service in South Australia. *Journal of Telemedicine and Telecare, 3*(Suppl. 1), 12–14.

Conrath, D. W., Buckingham, P., Dunn, E. V., & Swanson, J. N. (1975). An experimental evaluation of alternative communication systems as used for medical diagnosis. *Behavioral Science, 20*(5), 296–305.

Couturier, P., Tyrrell, J., Tonetti, J., Rhul, C., & Franco, A. (1998). Feasibility of orthopaedic teleconsulting in a geriatric rehabilitation service. *Journal of Telemedicine and Telecare, 4*(Suppl. 1), 85–87.

Cunningham, N., Marshall, C., & Glazer, E. (1978). Telemedicine in pediatric primary care: Favorable experience in nurse-staffed inner-city clinic. *Journal of the American Medical Association, 240*, 2749–2751.

Currell, R., Urquhart, C., Wainwright, P., et al. (2002). Telemedicine versus face to face patient care: Effects on professional practice and health care outcomes (Cochrane Review). Available in *The Cochrane Library*, Issue 3. Oxford: Oxford Update Software.

Dansky, K. H., & Bowles, K. H. (2002, April). Lessons learned from a telehomecare project. *Caring*, pp. 18–22.

Davis, F. D. (1989). Perceived usefulness, perceived ease of use, and user acceptance of information technology. *MIS Quarterly, 13*(3), 319–340.

Dongier, M., Tempier, R., Lalinec-Michaud, M., & Meunier, D. (1986). Telepsychiatry: Psychiatry consultation through two-way television: A controlled study. *Canadian Journal of Psychiatry, 31*, 32–34.

Doolittle, G.C. (2000). A cost measurement study for a home-based telehospice service [abstract]. *Journal of Telemedicine and Telecare, 6*(Suppl. 1), 193–195.

Doolittle, G. C., Yaezel, A., Otto, F., & Clemens, C. (1998). Hospice care using home based telemedicine systems. *Journal of Telemedicine and Telecare, 4*(Suppl. 1), 58–59.

D'Souza, M. (2002). Improving treatment adherence and longitudinal outcomes in patients with serious mental illness by using telemedicine [abstract]. *Journal of Telemedicine and Telecare, 8*(Suppl. 2), 113–115.

Dunn, E., Conrath, D., Action, H., & Bain, H. (1980). Telemedicine links patients in Sioux lookout with doctors in Toronto. *Canadian Medical Association Journal, 22*(4), 484–487.

Elford, D. R., White, H., St. John, K., Maddigan, B., Ghandi, M., & Bowering, R. (2001). A prospective satisfaction study and cost analysis of a pilot child telepsychiatry service in Newfoundland. *Journal of Telemedicine and Telecare, 7,* 73–81.

Endean, E. D., Mallon, L. I., Minion, D. J., Kwolek, C. J., & Schwarez, T. H. (2001). Telemedicine in vascular surgery: Does it work? *American Surgeon, 67,* 334–341.

Federal Telemedicine Update. (2002). *Federal telemedicine update*. Retrieved October 30, 2002, http://www.federaltelemedicine.com

Fuchs, M. (1974). Provider attitudes toward STARPAHC, a telemedicine project on the Papago Reservation. *Medical Care, 17*, 59–68.

Gagnon, M. P., Godn, G., Gagne, C., Fortin, J. P., Lamothe, L., Reinharz, D., & Cloutier, A. (2003). An adaptation of the theory of interpersonal behaviour to the study of telemedicine adoption by physicians. *International Journal of Medical Information, 71*(2–3), 103–115.

Granovetter, M. (1985). Economic action and social structure: A theory of embeddedness. *American Journal of Sociology, 91*, 481–510.

Hailey, D., Roine, R., & Ohinmaa, A. (2002). Systematic review of evidence for the benefits of telemedicine. *Journal of Telemedicine and Telecare, 8*(Suppl. 1), 1–7.

Harrison, R., Clayton, W., & Wallace, P. (1997). Can telemedicine be used to improve communication between primary and secondary care? *British Medical Journal, 313*(7069), 1377–1381.

Hashimoto, S., Shirato, H., Kaneko, K., et al. (2001). Clinical efficacy of telemedicine in emergency radiotherapy for malignant spinal cord compression [abstract]. *Journal of Digital Imaging, 14*(3), 124–130.

Helitzer, D., Heath, D., Maltrud, K., Sullivan, E., & Alverson, D. (2003). Assessing or predicting adoption of telehealth using the diffusion of innovations theory: A practical example from a rural program in New Mexico. *Telemedicine Journal of E Health, 9*(2), 179–187.

Hersch, W. R., Helfand, M., Wallace, J., et al. (2001). Clinical outcomes resulting from telemedicine interventions: A systematic review. *BMC Medical Information Decision Making, 1*(1), 5.

Higgins, C. A., Conrath, D. W., & Dunn, E. V. (1984). Provider acceptance of telemedicine systems in remote areas of Ontario. *Journal of Family Practice, 18,* 285–289.

Hui, E., & Woo, J. (2002). Telehealth for older patients: The Hong Kong experience. *Journal of Telemedicine and Telecare, 8*(Suppl. 3), 39S–41S.

Hunkeler, E. M., Meresman, J. F., Hargreaves, W. A., et al. (2000). Efficacy of nurse telehealth care and peer support in augmenting treatment of depression in primary care [abstract]. *Archives of Family Medicine, 9*(8), 700–708.

Huston, J. L., & Burton, D. C. (1997). Patient satisfaction with multispecialty interactive teleconsultations. *Journal of Telemedicine and Telecare, 3,* 205–208.

Itzak, B., Weinberger, T., Berkovitch, E., & Reis, S. (1998). Telemedicine in primary care in Israel. *Journal of Telemedicine and Telecare, 4*(Suppl. 1), 11–14.

Jones, D. H., Crichton, C., Macdonald, A., Potts, S., Sime, D., Toms, J., & McKinlay, J. (1996). Teledermatology in the highlands of Scotland. *Journal of Telemedicine and Telecare, 2*(Suppl. 1), 7–9.

Jong, M. K. K., Horwood, K., Robbins, C. W., et al. (2001). A model for remote communities using store and forward telemedicine to reduce health care costs [abstract]. *Canadian Journal of Rural Medicine Winter 6*(1), 15–20.

Jutra, A. (1959). Teleroentgen diagnosis by means of videotape recording. *American Journal of Roentgenology, 82*, 1099–1102.

Kjaer, N. K., & Karlsen, K. O. (2002). [Telemedicine and general practice future of present: Telemedicine, a way to strengthen the gatekeeper role?]. *Ugeskrift for Laeger, 164*(45), 5262–5266. (In Danish)

Kobb, R., Hoffman, N., Lodge, R., & Kline, S. (2003). Enhancing elder chronic care through technology and care coordination: Report from a pilot. *Telemedicine Journal of E Health, 3*(9), 189–195.

Kunkler, I. H., Rafferty, P., Hill, D., Henry, M., & Foreman, D. (1998). A pilot study of tele-oncology in Scotland. *Journal of Telemedicine and Telecare, 4*(2), 113–119.

Kvedar, J. C., Menn, E. R., Baradagunta, S., Smulders-Meyer, O., & Gonzalez, E. (1999). Teledermatology in a capitated delivery system using distributed information architecture: Design and development. *Telemedicine Journal, 5,* 357–366.

Lee, J. K., Renner, J. B., Saunders, B. F., Stamford, P. P., Bickford, T. R ., Johnston, R. E., et al. (1998). Effect of real-time teleradiology on the practice of the emergency department physician in a rural setting: Initial experience. *Academic Radiology, 5,* 533–558.

Lehoux, P., Sicotte, C., Denis, J. L., Berg, M., & Lacroix, A. (2002). The theory of use behind telemedicine: How compatible with physicians' clinical routines? *Social Science Medicine, 6,* 889–904.

Loane, M. A., Bloomer, S. E., Corbett, R., et al. (2000). A randomized controlled trial to assess the clinical effectiveness of both real-time and store-and-forward teledermatology compared with conventional care [abstract]. *Journal of Telemedicine and Telecare, 6*(Suppl. 1), S1–S3.

Lowitt, M. H., Kessler, I. I., Kauffman, C. L., Hooper, F. J., Siegel, E., & Burnett, J. W. (1998). Teledermatology and in-person examinations: A comparison of patient and physician perceptions and diagnostic agreement. *Archives of Dermatology, 134,* 471–476.

Maheu, M., Whitten, P., & Allen, A. (2001). *E-Health, telehealth & telemedicine.* San Francisco: Jossey-Bass.

Mark, R. G., Willemain, T. R., Malcolm, T., Master, R. J., & Clarkson, T. (1976). *Nursing home telemedicine project: Volume 1.* Springfield, VA: National Technical Information Service, U.S. Department of Commerce.

Matthews, D. A., Sledge, W. H., & Liebermann, P. B. (1987). Evaluation of intern performance by medical inpatients. *American Journal of Medicine, 83,* 938–944.

Mazmanian, P., McCue, M., Parpart, C. F., Marks, T., Fisher, E., Hampton, C.,Krick, S., & Kaplowitz, L. (1996, July). Evaluating telemedicine [abstract]. *International Conference on AIDS,* p. 172.

Meyer, M. (2002). "Virtually Healthy" integrating technology and coordination: A two year success story. *Telemedicine Journal and e Health, 8*(2), 218.

Moody, J., & White, D. R. (2001). Structural cohesion and embeddedness: A hierarchical concept of social groups. *American Sociological Review, 68*(1), 103–127.

Morrison, J., Bergauer, N. K., Jacques, D., et al. (2001). Telemedicine: Cost-effective management of high-risk pregnancy [abstract]. *Managed Care, 11,* 42–46, 48–49.

Murphy, L. H., & Bird, K. T. (1974). Telediagnosis: A new community health resource. *American Journal of Public Health, 64,* 113–119.

Nesbitt, T. S., Hilty, D. M., Kuenneth, C. A., & Siefkin, A. (2000). Development of a telemedicine program: A review of 1,000 videoconferencing consultations. *West Journal of Medicine, 173,* 169–174.

Nielsen, P. E., Jackson, K. B., Kosman, K., Kiley, K. C., & Thomson, B. A. (2000). Standard obstetric record charting system: Evaluation of a new electronic medical record. *Obstetetrics and Gynecology, 96,* 1003–1008.

Nordal, E. J., Moseng, D., Kvammen, B., & Lochen, M. L. (2001). A comparative study of teleconsultations versus face-to-face consultations. *Journal of Telemedicine and Telecare, 7,* 257–65.

Office for the Advancement of Telehealth. (n.d.) *Welcome.* Retrieved February 25, 2005, from http://telehealth.hrsa.gov/welcome.htm

Park, B. (1974). *An introduction to telemedicine: Interactive television for delivery of health services.* New York: Alternative Media Center, New York University.

Pedersen, S., & Holand, U. (1995). Tele-endoscopic otorhinolaryngoligical examination: Preliminary study of patient satisfaction. *Telemedicine Journal, 1,* 47–52.

Phillips, V. L., Vesmarovich, S., Hauber, R., et al. (2001).Telehealth: Reaching out to newly injured spinal cord patients [abstract]. *Public Health Reports, 116* (Suppl. 1), 94–102.

Pringle-Specht, J. K., Wakefield, B., & Flanagan, J. (2001). Evaluating the cost of one telehealth application connecting an acute and long term care setting [abstract]. *Journal of Gerontological Nursing, 1,* 34–39.

Ricci, M. A., Callas, P. W., & Montgomery, W. L. (1997). The Vermont telemedicine project: Initial implementation phases. *Telemedicine Journal, 3,* 197–205.

Rogers, E. (1995). *Diffusion of innovations* (4th ed.). New York: The Free Press.

Rogers, M. A., Small, D., Buchan, D. A., et al. (2001). Home monitoring service improves mean arterial pressure in patients with essential hypertension: A randomized, controlled trial [abstract]. *Annals of Internal Medicine, 134*(11), 1024–1032.

Roine, R., Ohinmaa, A., & Hailey, D. (2001). Assessing telemedicine: A systematic review of the literature. *Canadian Medical Association Journal, 165*(6), 765–771.

Schall, T., Roeckelien, W., Mohr, M., Kampshoff, J., Lange, T., & Nerlich, M. (2003). A communication-theory based view on telemedical consultation. *Studies in Health Technolology and Information, 97,* 115–130.

Schlachta-Fairchild, L. (2002, April). Telehealth: Practice in home care, synopsis of the 2000 US telenursing role study. *Caring,* pp. 10–13.

Tanriverdi, H., & Iacono, C. S. (1999). Diffusion of telemedicine: A knowledge barrier perspective. *Telemedicine Journal, 5*(3), 223–244.

Vassallo, D. J., Hoque, F., Farquharson, R., et al. (2001). An evaluation of the first year's experience with a low-cost telemedicine link in Bangladesh [abstract]. *Journal of Telemedicine and Telecare, 7*(3), 125–138.

Walker, J., & Whetton, S. (2002). The diffusion of innovation: Factors influencing the uptake of telehealth. *Journal of Telemedicine and Telecare, 8*(Suppl. 3), 73–75.

Whitten, P. (2003, September). *Analysis of consumer acceptance across three Michigan telehome health projects.* Paper presented at the American Telemedicine Association Fall Forum: Remote Telemonitoring & Home Telehealth, Ft. Lauderdale, FL.

Whitten, P. (2005, October). *Explaining telemedicine success and failure: Mandate for an interdisciplinary approach.* Colloquium presented at Purdue University, Lafayette, IN.

Whitten, P., & Collins, B. (1997). The diffusion of telemedicine: Communicating an innovation. *Science Communication, 19*(1), 21–40.

Whitten, P., Collins B., & Mair, F. (1998). Nurse and patient reactions to a developmental home telecare system. *Journal of Telemedicine and Telecare, 4,* 152–160.

Whitten, P., Doolittle, G., & Mackert, M. (2004). Telehospice in Michigan: Use and patient acceptance. *American Journal of Hospice & Palliative Care, 21*(3), 191–195.

Whitten, P., & Gregg, J. L. (2001). Telemedicine: Using telecommunication technologies to deliver health services to older adults. In M. L. Hummert & J. Nussbaum (Eds.), *Communication, aging and health: Multidisciplinary perspectives* (pp. 3–22). Mahwah, NJ: Lawrence Erlbaum Associates.

Whitten, P., & Mair, F. (2000). Telemedicine and patient satisfaction: Current status and future directions. *Telemedicine Journal and e-Health, 6*(4), 417–423.

Whitten, P. S., Mair, F. S., Haycox, A., et al. (2002). Systematic review of cost effectiveness studies of telemedicine interventions. *British Medical Journal,* 324(7351), 1434–1437.

Willemain, T. R., & Mark, R. G. (1971). Models of health care systems. *Biomedical Science Instrument, 8*, 9–17.

Wittson, C. L., Affleck, D. C., & Johnson, V. (1961). Two-way television group therapy. *Mental Hospital, 12*, 2–23.

Wootton R. (2002). Guest commentary: Systematic review of cost-effectiveness studies of telemedicine interventions. *Telehealth Practice Report*, *7*(4), 1–12.

Zaylor, C., Nelson, E. L., & Cook, D. J. (2001). Clinical outcomes in a prison telepsychiatry clinic [abstract]. *Journal of Telemedicine and Telecare, 7*(Suppl. 1),47–49.

Zollo, S., Kienzle, M., Loeffelholz, P., & Schille, S. (1999). Telemedicine to Iowa's correctional facilities: Initial clinical experience and assessment of program costs. *Telemedicine Journal, 5,* 291–301.

VII

LEGAL AND REGULATORY SETTING

13

Digital Media Technology and Fair Use

Jeremy Lipschultz

The diffusion of new media technologies has happened in a series of cycles. Widespread adoption in the U.S. consumer market of analog audiocassette tape and videotape recorders (VTRs) in the 1960s and 1970s, analog videocassette recorders (VCRs) in the 1980s and 1990s—and, more recently, personal computer (PC) and digital video recorders (DVRs) and disks (DVDs)—presented challenging legal issues for commercial industries, lawmakers, and the courts.

In this chapter, the intellectual property rights focus is on copyright law rather than the problematic concerns related to expensive and time-consuming patents on ideas and their use in inventions (Stallman, 2002). Although patents are limited in duration based on their application, copyrights are extended in length, easy to obtain, and not associated with ideas; rather, they are related to the specific manner of expression. Article I, Section 8 of the U.S. Constitution (see http://www.house.gov/Constitution/Constitution.html) encouraged the protection of *general* intellectual property rights by empowering Congress to promote science and the arts "by securing for limited times to authors and inventors the exclusive right to their respective writings and discoveries." The constitutional language groups the work of authors and inventors at the same time as it distinguishes each as unique. Thus, many have observed that the phrase "intellectual property" is less meaningful than the more specific language of copyrights and patents.

BACKGROUND

The first copyright law was passed in 1790, and Congress consistently supported ongoing revision (Bowles, 2005). The Copyright Law of 1976 was the most elaborate attempt to that point to protect creative works, as this was a period of time in which the concept was bolstered by international treaties designed to protect against illegal exports. However, in a legal sense, copyright may collide with the core values of the First Amendment, as well as the broad cultural interest in the free flow of ideas. In short, to the extent that a society protects property rights of communication content, these provisions tend to stifle sharing of ideas. Especially in the case of electronic media and new digital media, access to information is an ongoing challenge.

Although the Copyright Act of 1976 granted exclusive rights of intellectual property, it also wrote into law a fair use doctrine—one that may allow for limited uses of content for specific purposes, such as, for example, classroom teaching. Protection of property rights through copyright statute law has been traditionally balanced by a desire of the courts to maintain individual "fair use" rights—legal free copies that must be judged in court based on the following criteria: "(1) the purpose and character of the use, including whether such use is of a commercial nature or is for nonprofit educational purposes; (2) the nature of the copyrighted work; (3) the amount and substantiality of the portion used in relation to the copyrighted work as a whole; and (4) the effect of the use upon the potential market for or value of the copyrighted work" (17 U. S. C. § 107). Fair use must be determined on a case-by-case basis, in which a court would judge the purpose and proportion of use and economic impact of the distribution.

The sharing culture of the 1960s matured with the baby-boomer generation and their children, and blended with the technological innovation of the Internet and World Wide Web to allow users to freely copy and distribute music and video files (Hilt & Lipschultz, 2005). The clash between sociocultural norms of free access to content and statutes, such as the Sonny Bono Copyright Extension Act of 1998 (17 U.S.C. §§ 302–305) and the Digital Millennium Copyright Act of 1998 (17 U.S.C.§ 101), meant that there has been an increase in the amount of social and subsequent legal conflict.

The Digital Millennium Copyright Act of 1998 brought U.S. law into an updated position with respect to international treaties and the World Trade Organization (WTO) efforts designed to offer copyright holders "the basic property right to control distribution of copies of their creations" (Bowles, 2005, p. 120). In point of fact, evidence from the 1980s and Betamax video recorders suggested that "the desire of copyright holders for monopoly protection and the desire of consumers for wider access to information often conflicts" (Lloyd & Mayeda, 1986, p. 60; see also Lin, 1985). Thus, the digital revolution appears to have served to accelerate a social conflict that had its roots many years earlier.

THEORETICAL EXPLANATIONS

Copyrights and Fair Use

Copyrights and fair use exist within a larger context of law. It involves common law rights inherited from the case law of England via the colonial past, constitutional language, and various statutes, such as specific copyright acts. In the Internet age, international law also has become important, specifically treaties that attempt to protect intellectual property of patents and copyright across national borders. To the extent that such law may restrict a free trade of ideas, there is a natural tension between it and our historical value of free expression (Levy, 1985).

From John Milton's *Aereopagitica* in 1644 to the First Amendment to the U.S. Constitution, there is a clear sense of the value of free speech in a free society (Lipschultz, 2000). In the search for truth, the desire for individual enlightenment, and the need for social progress, it has been theorized as necessary that there be broad freedom to share and cultivate ideas (Emerson, 1970). However, in the modern age, ideas often are disseminated through commercial mass media, which have an economic desire to protect property for its potential to create current and future profits. *Infringement* is the legal term to describe when someone has had access to an original work and substantially copied it through "piracy, bootleg or plagiarism" (Bowles, 2005, p. 128). The image of a pirate, it has been argued by some, is probably an exaggeration. However, theft of property, as defined by law, is a serious matter. In the most extreme cases, theft of media content, particularly when it involves redistribution and reselling, has led to prosecutions, fines, and jail time.

Although there have been attempts to carve individual protection for "personal use," it is fair use that has some basis in the law. However:

> Fair use is notoriously vague and unpredictable in application. The statute provides a nonexclusive list of purposes for which copyrighted materials may be used without the copyright owner's permission: criticism, comment, news reporting, teaching, scholarship, and research. It then establishes a nonexclusive, four-factor standard for fair use. The Supreme Court has interpreted the statute in several key cases, which has added some flesh to the statutory skeleton, but left many issues unresolved. (Tussey, 2001, p. 1144)

Digital Media Copyrights and Fair Use

The Internet presented particularly tricky legal issues about protection of copyrights for computer-originated and computer-transmitted material:

> The combination of the Internet and digital technology presents copyright law with what has been described as a digital dilemma. On one hand, digital technology makes it possible to make an unlimited number of perfect copies of music, books, or videos in digital form, and through the Internet individuals may distribute those digital works around the world at the speed of light. (Shih Ray Ku, 2002, pp. 263–264)

Other questions surround the copyright protection of computer-generated materials. To qualify for copyright protection, material must be fixed in a tangible medium and must be original. Therefore, if an online document is a copy of material that already exists on paper, it is unclear whether the online document should be considered "original" and therefore protected. Additionally, it remains to be determined whether material created and transmitted electronically is "fixed in a tangible medium."

Shih Ray Ku (2002), however, argued that computer technologies require users to fund distribution networks costs, and in this environment the theory of copyright breaks down:

> To the extent that there are costs associated with disseminating digital content, they are borne by the users of information through the purchase of computer equipment and connections to the Internet. As such, the Internet and digital technology eliminate one of the principal problems created by the public good nature of information the public's failure to internalize the cost of distributing intellectual works. Copyright, therefore, is no longer necessary to create property rights artificially in digital works to eliminate free riding. (p. 268)

It is also important to note, however, that courts have considered the transformative nature of some creative work, and this may spawn fair use rights in certain contexts. Elkin-Koren (1996) suggested that this is yet another area in which fair use provisions are murky and need clarification:

> It is arguable that the appropriate balance between the owner's monopoly over derivative uses of her work and users' freedom to manipulate, should be resolved under the fair use doctrine. The scheme of providing a broad right to control derivative use, subject to fair use exemption, puts users at a disadvantage. Under fair use, users' rights are less stable. Rights would depend on judicial determination. Users would have to act under the threat that their acts would be found infringing. The prospects of potential litigation may have an inhibiting effect on users. (p. 282)

In the end, the Supreme Court has left for future case-by-case review the judgment about protection of copyright ownership versus fair use.

However, this shift poses significant concerns for those interested in free flow of information:

> DMCA has upset the balance between the interests of copyright holders and the need for public access to protected materials. Through its anti-circumvention provision, the statute prevents people from engaging in actions that traditionally have been considered fair use. The DMCA also creates a chilling effect by requiring Internet service providers to remove content even if the reproduction of such materials is permissible under existing copyright law. The statute, therefore, has raised serious concerns about free speech, privacy, academic freedom, learning, culture, democratic discourse, competition, and innovation. (Yu, 2004, pp. 912–913)

Although copyright law protects tangible products, such as writings, music, multimedia, and recordings, it does not protect ideas or facts.

The Sonny Bono Copyright Term Extension Act (CETA) extended the length of time covered to the life of the author *and* an additional 70 years (17 U.S.C. 3). The U.S. Supreme Court, in hearing a challenge to the law, decided that existing copyrights on, for example, Disney's Mickey Mouse should be extended for another 20 years to 2023. Without an extension, many 20th-century icons would have become available in the public domain for general use on Internet web sites. In *Eldred v. Ashcroft* (2003), the U.S. Supreme Court upheld the extension on the grounds that it did not violate the copyright clause or the First Amendment right of free speech. The 7–2 decision dismissed concerns about free expression.

However, Justices Stevens and Breyer wrote dissenting opinions, which did raise concerns about the large proportion of unused content protected under the latest extension. Justice Breyer wrote:

> This statute will cause serious expression-related harm. It will likely restrict traditional dissemination of copyrighted works. It will likely inhibit new forms of dissemination through the use of new technology. It threatens to interfere with efforts to preserve our Nation's historical and cultural heritage and efforts to use that heritage, say, to educate our Nation's children. It is easy to understand how the statute might benefit the private financial interests of corporations or heirs who own existing copyrights. But I cannot find any constitutionally legitimate, copyright-related way in which the statute will benefit the public. Indeed, in respect to existing works, the serious public harm and the virtually nonexistent public benefit could not be more clear. (p. 266)

EMPIRICAL FINDINGS

Traditional analog radio, TV, music, and movies produced some of the earliest concerns over copying media content. Universal and Disney studios brought a key case in 1976, and it took 8 years to work through the lower courts and end in a split decision against them. In *Sony Corp. v. Universal City Studios* (1984), the U.S. Supreme Court rejected the claims by TV program producers that the manufacturers of consumer recorders could be held liable for ontributory copyright infringement. Justice John Paul Stevens authored a 5–4 decision in which the Court found that the primary use of the Betamax recorder was for time shifting of over-the-air programs, and there was little evidence of economic harm.

In fact, the decision utilized the testimony of Fred Rogers of the PBS children program *Mister Rogers' Neighborhood* that time shifting seemed to benefit busy families. In the Sony case, other TV program producers sued the equipment manufacturers rather than consumers. Justice Harry Blackmun's dissent, however, argued that video recorders could also be used for library-building of personal collections that might adversely impact the economics of the TV industry. In full, the U.S. Constitution had granted Congress a general power "to promote the Progress of Science and useful Arts, by securing for limited Times to Authors and Inventors the exclusive Right to their respective Writings and Discoveries" (U.S. Constitution, Article I, 8).

BOX 13.1. Supreme Court supports use of home recording devices.

SONY CORPORATION OF AMERICA ET AL. v. UNIVERSAL CITY STUDIOS, INC., ET AL.
No. 81–1687
SUPREME COURT OF THE UNITED STATES

464 U.S. 417; 104 S. Ct. 774; 78 L. Ed. 2d 574; 1984 U.S. LEXIS 19; 52 U.S.L.W. 4090; 220 U.S.P.Q. (BNA) 665; 224 U.S.P.Q. (BNA) 736; 55 Rad. Reg. 2d (P & F) 156
January 18, 1983, Argued
January 17, 1984, Decided

* * *

Justice Stevens,
... In a case like this, in which Congress has not plainly marked our course, we must be circumspect in construing the scope of rights created by a legislative enactment, which never contemplated such a calculus of interests. In doing so, we are guided by Justice Stewart's exposition of the correct approach to ambiguities in the law of copyright:

"The limited scope of the copyright holder's statutory monopoly, like the limited copyright duration required by the Constitution, reflects a balance of competing claims upon the public interest: Creative work is to be encouraged and rewarded, but private motivation must ultimately serve the cause of promoting broad public availability of literature, music, and the other arts. The immediate effect of our copyright law is to secure a fair return for an 'author's' creative labor. But the ultimate aim is, by this incentive, to stimulate artistic creativity for the general public good. 'The sole interest of the United States and the primary object in conferring the monopoly,' this Court has said, 'lie in the general benefits derived by the public from the labors of authors.' When technological change has rendered its literal terms ambiguous, the Copyright Act must be construed in light of this basic purpose."

Justice Blackmun,

It may be tempting, as, in my view, the Court today is tempted, to stretch the doctrine of fair use so as to permit unfettered use of this new technology in order to increase access to television programming. But such an extension risks eroding the very basis of copyright law, by depriving authors of control over their works and consequently of their incentive to create. Even in the context of highly productive educational uses, Congress has avoided this temptation; in passing the 1976 [Copyright] Act, Congress made it clear that off-the-air videotaping was to be permitted only in very limited situations.

Original material has been traditionally protected by copyright on its creation, and the creator has exclusive rights to reproduce and distribute the work, as well as the right to create other works derived from the original. The courts, particularly in the early iterations of the Internet, failed to agree on how to apply rights in cyberspace. One court concluded that uploading a document, copying an Internet posting, and retransmitting the posting without permission all could violate copyright law (*Playboy Enters., Inc. v. Frena*, 1993). Even browsing through material on a computer—which causes a copy of the digital information to be held in temporary memory—could possibly constitute illegal copying. Alternately, another court found that this might be interpreted as the digital form of reading a book in a library or a magazine at a newsstand, and not a copyright infringement (*Religious Technology Ctr. v. Netcom On-Line Communications Servs.*, 1995).

The growth of Internet sites that link to or frame a small section of another site content raised a new question of copyright infringement. In 1997, the *Washington Post* and five other news organizations charged a group of Internet news sites with copyright infringement for republishing and repackaging the media web pages for profit (*Washington Post Co. v. Total News Inc.*, 1997). Media organizations charged that the news sites copied their trademarks for use as links and republished their copyrighted material with advertising that generated revenue for the pirates rather than the copyright holders. The suit was settled out of court when the news sites agreed to provide links but not to frame any content from the news organization sites. The *Los Angeles Times* won its case against FreeRepublic.com, a web site that reprinted full-text online news stories and invited caustic reader reaction (*Los Angeles Times v. Free Republic*, 1999).

Courts have considered questions concerning intellectual property in a variety of contexts. Each new technology has presented unique legal issues and reasoning, examples of which are next reviewed in turn.

CD-ROM

Although a live TV broadcast is not fixed in a tangible medium and is not protected unless, for example, it is simultaneously put on videotape, the same cannot be said for computer disks. A document saved on a floppy, hard disk, or CD-ROM is fixed (*Mai Systems Corp. v. Peak Computer, Inc.*, 1993). The question then is whether material created on a computer, sent over the Internet (as e-mail, for example), and then not saved on the computer is fixed in a tangible medium and thus subject to copyright protection. One court has held that material in a computer random access memory (RAM) is sufficiently fixed to be protected (*Triad System Corp. v. Southeastern Express Co.*, 1995).

Webcasts and Music Distribution

In 1995 and again in 1998, Congress updated copyright laws in part to deal with these issues. Congress enacted the Digital Millennium Copyright Act (DMCA) of

1998 to eliminate an apparent loophole in the Digital Performance Rights in Sound Recordings Act that had permitted webcasters to digitally transmit recordings without copyright licenses or fees. The new law grants webcasters a statutory license to transmit, with certain limitations, recordings at fees established voluntarily or by a Copyright Arbitration Royalty Panel (Title IV, 1998). The established fees put an end to many webcasts as cost-prohibitive. For example, some college radio stations ended webcasts when forced to pay as much as $20,000 in additional music licensing fees.

Peer-to-Peer Sharing

The rock music band Metallica sued the Napster.com site in 2000 for copyright infringement and racketeering over its free software that allowed users to trade MP3 copies of music on an unrestricted basis. The digital music format allowed compact disk quality songs to be converted into a personal computer format, and this made possible the easy distribution of pirated copies over the Internet. The software made it so easy that thousands of music pirates clogged university campus network systems. The case was an opening salvo in a legal and economic battle over the role newcomers would be allowed to play in the recording and publishing industries.

The international distribution via the Internet of illegal copies is a problem for Internet service providers (ISPs) and computer users. A German judge ruled that ISPs may be held liable for their role in distributing pirated copies. However, the Supreme Court has let stand a New York Court of Appeals ruling that said ISPs are not liable for transmissions, but are akin to telephone and telegraph companies accorded common law qualified privilege against liability. That ruling protected the Prodigy service from a defamation lawsuit. In the Metallica case, two of three universities, Yale and Indiana, blocked access to Napster.com in exchange for being dropped from the lawsuit. The band claimed large numbers of university students were using campus computer systems to access the site and trade pirated music. Metallica also identified 335,000 Napster screen names in 60,000 pages of documents. The Napster.com site posted warnings about copyright, but it had looked away as its software users trade music files. However, Napster ultimately yielded to legal pressures by removing 317,377 users, a move that did not prevent the traders of pirated music from returning under new screen names.

At the same time, the company MP3.com settled its lawsuit by halting access to major label songs after a federal district court ruled its database of more than 80,000 songs was copyright infringement against Time Warner, Sony, and others. The web site agreed to pay licensing fees to music companies. The Napster case, however, attracted much attention.

"Napster had built a business based on large-scale piracy," the complaint alleged. The Recording Industry Association of America (RIAA) also sued over the distribution of MP3 music. A district court held that record companies and artists owned the music and issued an injunction prohibiting Napster from "engaging in, or facilitating…copying, downloading, uploading, transmitting, or distributing…copyrighted recordings…." The Ninth U.S. Circuit Court of Appeals agreed, holding that Napster infringed on the music reproduction and distribution rights and that the sharing software went beyond a fair use (*A&M Records, Inc. v. Napster, Inc*, 2001). The court distinguished Internet file sharing from the Supreme Court ruling in *Sony Corp. v. Universal Studios, Inc.* that videocassette recorders constituted a fair use.

Following the decision, Napster was unable to reach an agreement with music companies to compensate them. Subsequently, federal District Court Judge Marilyn Patel in San Francisco ordered Napster to curb trading of copyrighted music. Napster went ahead with plans to develop a fee-based system, even though the company had no initial agreement with record companies. At the same time, alternative sites such as Gnutella and Aimster prompted new legal challenges in 2001, as record companies struggled to develop their own pay sites. Ultimately, the controversy appeared to be moving toward resolution via business deals: Vivendi Universal (teaming with Sony and Yahoo to form Duet) purchased MP3.com for about $372 million, and Bertelsmann aligned with Napster, EMI, AOL Time Warner, and RealNetworks to form MusicNet. In 2002, four of five record companies suspended legal action against Napster in order to avoid inquiries into their copyright actions; Napster continued to be threatened with bankruptcy; and Bertelsmann ultimately agreed to pay $8 million for Napster assets.

This raised the issue of whether such sites could survive economically by charging user fees to supplement limited advertising revenues. Napster was forced to show that it was eliminating illegal file swapping and creating new software to monitor the behavior of users around the world. Thus, the one-time legal adversaries became partners.

The copyright liability of online service providers remains unclear. If a subscriber uses one of these services to send material that clearly is protected under copyright, is the service as well as the user guilty of copyright violation? For example, a company could print a hard copy of a magazine and then put the contents online. A user could copy a story from the online magazine and send it to millions of other Internet users. A key factor is whether the provider had, or should have had, prior knowledge of the infringement. Online content providers who electronically screen or preview material may be held accountable for copyright infringements. By comparison, Internet service providers such as the telephone companies, cable carriers, or independent Internet connection service providers are, by law, considered common carriers, and thus not allowed to alter any online content that they help carry and deliver.

DVD copying is also a serious problem, one assuming global proportions now that intellectual copyright represents America's top export commodity (Lin & Atkin, 2002). Representatives of the Motion Picture Association (MPA)—formerly the MPAA—have testified to Congress that an estimated 350,000 be outlawed in a revision of the Digital Millennium Copyright Act. However, measures to protect intellectual property by gathering software user data clash with those wishing to increase privacy protection for online computer users. Despite regulatory efforts, file sharing remains a legal issue.

In *MGM v. Grokster* (2005), the U.S. Supreme Court addressed the problem of "billions of files" that were being "shared across peer-to-peer networks each month" (125 S.Ct., at 2764). The downloading of primarily copyrighted files was found to be illegal by the court because the software companies that operate P2P can profit by selling advertising while failing to protect property rights. These networks did not attempt to filter illegal downloads. The U.S. Supreme Court ruled that Grokster and other such services *contribute and promote* copyright infringement and may be held liable for damges. In ruling this way, the Court distinguished what happens in these networks from the earlier legal videotape copying equipment, such as Sony Betamax and popular VHS formats (*Sony v. Universal Studios*, 464 U.S. 417, 1984). Although videotape equipment allowed individuals to record and play TV programs on a single tape, the digital software networks are different because they create an environment for widespread copying and distribution electronically with ease.

The *MGM* court found that:

(a) secondary liability has emerged from the common law principles;
(b) the legal uses of peer-to-peer sharing software do not negate the liability posed by illegal downloads;
(c) promotion of illegal downloading of copyrighted material may be considered by the courts;
(d) services earned advertising revenues based upon the high volume of illegal downloads; and
(e) there was strong evidence of actual copyright infringement (MGM v. Grokster, 125 S.Ct., at 2769).

The Supreme Court reversed a lower court, which had ruled in favor of the peer-to-peer services. Thus, the case was returned to the lower courts for further action.

SOCIAL CHANGE IMPLICATIONS

Rogers (2003) suggested that the dispersion of ideas, including practices or objects, constitutes diffusion when these are "perceived as new" by individuals or others in the adoption process (p. 12). In the case of media technologies, perceived newness traditionally has been and continues to be an important motivation for social change. Presumably, diffusion is most likely to occur when legal constraints are least restrictive. In the case of copyright to own monopoly rights over content, a balance has been struck between freedom of distribution and protection of intellectual property rights.

Regulators have termed *new communication technologies* as those mass media developed following the analog age of broadcast and cable TV, but we also know that the printing press, telegraph, telephone, radio, and TV all were considered revolutionary when adopted. New media appear to borrow or are said to be a form of "remediation" from "old" media (J. D. Bolter and R. Grusin, 1999, p. 5). The convergence of printed text, photographs, audio, and video created challenges across the legal spectrum, from libel (*Zeran v. America Online*, 1997; *Blumenthal v. Drudge and America Online*, 1998; *Cubby v. CompuServe*, 1991) and privacy (*Jessup-Morgan v. American Online*, 1998) to obscenity (*ApolloMedia Corp. v. Reno*, 1998) and, of course, copyright issues (*Leibovitz v. Paramount*, 1998).

Yu (2004) has suggested that we are in the midst of an escalating copyright war—one in which individual users' abilities to download Internet files battle against industry and government efforts to stop it. He suggested that industry strategies include battling new technologies with others, legal battle strategies waged against Napster and others to restrict file-sharing behaviors, economic battles via creation of legal music services such as iTunes, and deployment of encryption and other means of protecting copies: "Notwithstanding these technologies, however, the entertainment industry remains vulnerable" (p. 920).

Lessig (2005) argued that copyright laws now construct a significant barrier to the free flow of ideas. Beginning in the 20th century, publication rights were extended to the area of copy protection, and changes in law extended the reach beyond a previous minority of works. In traditional copyright law, one could lend a book to a friend because "it creates no copy" (p. 50). However, in the digital conversion of content to computer bits, this changes. "It is the nature of digital technologies that every use produces a copy" (p. 50). In such a world, all uses "presumptively require permission" (p. 52). Alternative models, however, are emerging, such as the production of movies with Apple computers at very low cost:

> "It is enough to recognize that many more people (indeed millions more) could make good films. New digital technologies could enable an explosion of creative work" (p. 52).

It remains to be seen how this clash between huge global media corporations and independent producers will play out. Because the corporations benefit from integration of distribution services, they continue to have a huge advantage over everyone else hoping to defeat the barriers to entry into the media business.

Courts have struggled with the impact of illegal and legal copying because the diffusion of digital technologies makes it impossible to track the millions of songs out there:

> For the sake of round numbers, assume there are 500 million people using the internet today, and that much of the world's demand for popular music would be satisfied by the availability of something like 5 million individual songs (Apple's iTunes, by way of comparison, is a twentieth of that size). Because people outnumber songs,

> if every user had one MP3 each, there would be a average of a hundred copies of every song somewhere online. A more realistic accounting would assume that at least 10% of the online population had at least 10 MP3 files each, numbers that are both underestimates, given the popularity of both ripping and sharing music. (Shirky, 2003, para 21)

Although the music labels at first embraced the iTunes model of selling all songs for ninety-nine cents, more recently they have expressed the desire to exert pricing controls over Apple that would allow them to charge more for each song. Critics claim that this would simply drive music enthusiasts back underground into illegal networks. It is best not to engage in quantitative estimates because any attempt is likely to be an incomplete and out-of-date accounting. Further, it is even less empirical to guess about the negative impact on copyright holders. Thus, courts have tended to be offered little data and disregarded such analyses.

Coles, Harris, and Davis (2004) concluded that the international legal system of copyright has had difficulty keeping pace with the fast-changing landscape of peer-to-peer Internet file sharing behaviors with its movement from an illegal Napster service to iTunes and other legal purchasing and then back to back alleys of the Web:

> Closing Napster could have worked in favour of the large firms through controlling user access to music on the web and preventing the diffusion of music promoted by smaller labels.... Here there was a distinct clash of interests between the desire of lesser known bands to gain some online exposure to build a fan base and the need to control unlimited assess to the popular, profitable music. Continuous innovation in the peer-to-peer format, however, has exposed the limitations of slow and incumbent legal challenges in the digital environment. Fundamental to these events is the apparent lack of consumer loyalty to and trust of the music labels, and this issue poses another marketing challenge. The fragmentation of the net into genres could favour the smaller "grassroots" labels and could also put greater power into the hands of individual consumers and musicians to revolutionise the existing industry structure. (p. 29)

If one were hoping for social, legal, or technological solutions to the copyright–fair use quagmire, this would be a serious mistake. We seem destined for ongoing social conflict between artists (and their legal business representatives) and consumers of media.

We live within a set of legal rules that value rights of content producers—essentially to earn money by creating media content and protecting it with a simple copyright. Davis and Neacsu (2001), from a political economy perspective, consider that it is economics, which is the engine of social change: "That is, it seems, an adoption of the Marxist view that social change cannot be effected (or at least not solely) through legal change but only through materially real economic change because law simply reflects the material conditions of a society" (p. 736).

The social conflict between fair use and copyright is likely to continue for the foreseeable future. As U.S. lawmakers attempt to react to the lobbying pressure to protect property rights of large media production and distribution companies, innovative and relatively inexpensive technologies allow for global copying and sharing of material that challenge the limits of fair use.

REFERENCES

A&M Records, Inc. v. Napster, Inc., 239 F.3d 1004 (9th Cir. 2001).

Apollo Media Corp. v. Reno, 19 F. Supp. 2d 1081 (N.D. Cal. 1998).*Blumenthal v. Drudge and America Online*, 992 F. Supp. 44 (D.D.C. 1998).

Bolter, J. D., & Grusin. R. (1999). *Remediation: Understanding new media.*

Bowles, D. (2005). Intellectual property. In W. Hopkins (Ed.), *Communication and the law* (pp. 119–150). Northport, AL: Vision.

Coles, A. M., Harris L., & Davis, R. (2004). Is the party over? Innovation and music on the web. *Information, Communication & Ethics in Society, 2*, 21–29.

Copyrights, Chapter 1. Subject Matter And Scope Of Copyright, Definitions, 17 USCS, § 101, 103. (Oct. 19, 1976, P. L. 94-553, Title I, § 101, 90 Stat 2545.)

Cubby v. CompuServe, 776 F. Supp. 135 (S.D.N.Y. 1991).

Davis, M. H., & Neacsu, D. (2001). Legitimacy, globally: The incoherence of free trade practice, global economics and their governing principles of political economy. *UMKC Law Review, 69*, 733.

Duration of Copyright: Works created on or after January 1, 1978, 17 USCS §§ 302, 305.

Eldred v. Ashcroft, 537 U.S. 186 (2003).

Elkin-Koren, N. (1996). Cyberlaw and social change: A democratic approach to copyright law in cyberspace. *Cardozo Arts & Entertainment Law Journal, 14*, 215.

Emerson, T. I. (1970). *The system of freedom of expression*. New York: Random House.

Hilt, M. L., & Lipschultz, J. H. (2005). *Mass media, an aging population and the baby boomers.* Mahwah, NJ: Lawrence Erlbaum Associates.

Jessup-Morgan v. America Online, Inc., 20 F. Supp. 2d 1105 (E.D. Mich. 1998).

L.A. Times v. Free Republic, 2000 U.S. Dist. LEXIS 5669, 54 U.S.P.Q.2D (BNA) 1453, 28 Media L. Rep. 1705 (C.D. Cal. 2000).

Leibovitz v. Paramount, 137 F. 3d 109 (2d Cir. 1998).

Lessig, L. (2005). The people own ideas. *Technology Review, 108*(6), 46–53.

Levy, L.W. (1985). *Emergence of a free press*. New York: Oxford University Press.

Limitations on Exclusive Rights: Fair use, 17 USCS §§ 107, (Oct. 19, 1976, P.L.94-553, Title I, § 101, 90 Stat. 2546; Dec. 1, 1990, P.L. 101-650, Title VI, § 607, 104 Stat. 5132; Oct. 24, 1992, P.L. 102-492, 106 Stat. 3145.)

Lin, C. (1985). Copyright issues for home videotaping. *Telecommunications Policy, 9,* 334–350.

Lin, C., & Atkin, D. (2002). *Communication technology and society: Audience adoption and use*. Cresskill, NJ: Hampton.

Lipschultz, J. H. (2000). *Free expression in the age of the internet: Social and legal boundaries*. Boulder, CO: Westview.

Lloyd, F. W., & Mayeda, D. M. (1986). Copyright, fair use, the First Amendment and new communications technologies. *Federal Communications Law Journal, 38,* 59.

Mai Systems Corp. v. Peak Computer, Inc., 991 F.2d 511, 518 (9th Cir. 1993), *cert. dismissed,* 510 U.S. 1033 (1993).

The Media Audit. (2002, April 25). http://www.themediaaudit.com/ growth.htm. *MGM v. Grokster*, 125 S.Ct. 2764 (2005, June 27).

Playboy Enters., Inc. v. Frena, 839 F. Supp. 1552 (M.D. Fla. 1993).

Religious Technology Ctr. v. Netcom On-Line Communications Servs., 907 F. Supp. 1361, 1378 n.25 (N.D. Cal. 1995).

Rogers, E. M. (2003). *Diffusion of innovations* (5th ed.). New York, NY: Free Press.

Shih Ray Ku, R. (2002). The creative destruction of copyright: Napster and the new economics of digital technology. *University of Chicago Law Review, 69*, 263.

Shirky, C. (2003, October 12). File-sharing goes social. *Networks, Economics and Culture Mailing List*. http://www.shirky.com/writings/file-sharing_social.html

Sony Corp. v. Universal City Studios. (1984). 464 U.S. 417.

Stallman, R. M. (2002). *Free software free society: Selected essays of Richard M. Stallman.* Boston: GNU Press.

Title IV of the Digital Millennium Copyright Act, H.R. 2281 (105th Cong., 2d Sess. 1998).

Triad System Corp. v. Southeastern Express Co., 64 F.3d 1330 (9th Cir. 1995).

Tussey, D. (2001). From fan sites to filesharing: Personal use in cyberspace. *Georgia Law Review, 35*, 1129.

Washington Post Co. v. Total News Inc., No. 97 Civ. 1190 (S.D. N.Y., complaint filed Feb. 20, 1997).

Yu, P. K. (2004). The escalating copyright wars. *Hofstra Law Review, 32*, 907.

Zeran v. America Online, 129 F.3d 327 (4th Cir. 1997).

14

Digital Media Technology and Individual Privacy

Laurie Thomas Lee

New communications technologies offer extraordinary opportunities to individuals, businesses, and government to efficiently collect, compile, and disclose personal information, as well as identify, track, target, and reach people. These capabilities arguably serve to promote economic efficiency and productivity, national security, innovation and progress, as well as an open society. Yet these new technologies also carry the potential for unfettered use and abuse, presenting substantial privacy concerns. Increasingly common are disruptions or intrusions from cell-phone marketing messages, online instant messages, and ad Internet pop-ups.

Computer users are being confronted by such threats as employee e-mail monitoring, hacking and spying from computer key-loggers and spyware, and "phishing" for personal information by identity thieves. Government surveillance has also been stepped up, from citywide spycams and global positioning systems (GPS), to catch speeders and car thieves, to FBI searches of library, telephone, Internet, and cable TV subscriber records to investigate terrorism and other criminal activity. Personal consumer data are also being mined and shared by businesses, with advertisers and marketers taking advantage of such data-collection techniques as cookies, online agreement forms, and the use of radiofrequency identification (RFID) on products purchased.

How have new technologies and social change affected privacy perceptions today? What are the implications of privacy and social change? Multidimensional theories have been proposed to comprehend the concept of privacy from different disciplines, including communication, psychology, and law. A number of empirical studies have also been conducted to understand the multidimensional nature of privacy. As the following examination reveals, the concept of privacy is not simple, although its impacts can be far-reaching.

BACKGROUND

Before the development of complex, literate societies, the notion of privacy was relatively unknown. Even as more complex communities developed, the notion of privacy was not distinguished in any meaningful way that we would recognize today (Schoeman, 1992). In many ways, the desire for privacy throughout history has stemmed from some type of larger relationship power struggle. Westin (1996) has identified five types of social authorities that have provoked conflict and a desire to be free from controlling surveillance or disclosure: (a) the employer–landlord, (b) the church, (c) associations to which an individual belongs, (d) local governmental officials, and (e) national regimes. The physical separation of work and home later allowed people to have more control over the amount of information they chose to make available to others and limited the scrutiny of their neighbors (Katz, 1987). As a result, the ability to live separately, have anonymity, and enjoy the psychological pleasures of privacy grew to become important social values.

Samuel Warren and future U.S. Supreme Court Justice Louis Brandeis (Warren & Brandeis, 1890) wrote the most famous early discussion of privacy in an 1890 law review essay entitled "The Right to Privacy." They saw technological advances such as high-speed cameras and the mass production of newspapers as having "invaded the sacred precincts of private and domestic life" (p. 195), and suggested that the law should "protect the privacy of private life" (p. 215). Telecommunications development of the 19th and early 20th centuries, particularly the telephone, raised new questions about what types of communications should be considered private. The initial telephone system's use of party lines and switchboard operators meant that conversations with outsiders could easily be monitored by family members, neighbors, strangers, the telephone switchboard operators, and unexpected others (e.g., the police).

By the 1960s, the U.S. Supreme Court addressed the idea of a legal right to privacy when it tackled issues associated with search and seizure (*Katz v. United States*, 1967), contraceptive use (*Griswold v. Connecticut*, 1965), interracial marriage (*Loving v. Virginia*, 1967), access to pornography in the home (*Stanley v. Georgia*, 1969), and the issue of abortion (*Roe v. Wade*, 1972). During the 1970s, an individual's medical, educational (e.g., transcripts), and other personal information received statutory protection (e.g., the Privacy Act of 1974, the Family Educational Rights and Privacy Act of 1974, the Right to Financial Privacy Act of 1978) as high-speed computers and computer databanks were beginning to threaten individuals' informational privacy (Allen, 2001).

Later on, Congress revised and created laws to protect the privacy of communications over newer electronic communications technologies. These laws include the Electronic Communications Privacy Act of 1986, which addresses wiretapping, the Cable Communications Policy Act of 1984, which protects cable TV subscriber records, and the Video Privacy Protection Act of 1988, which protects videotape rental records. By the end of the century, Congress had expanded its privacy laws to include the Internet. For example, the Children's Online Privacy Protection Act of 1998 (15 U.S.C. 6501-06, 2000) restricts the use of online information gathered from children, and the 2001 USA PATRIOT Act (Uniting and Strengthening, 2001) expands government authority over the monitoring of Internet communications.

But even as privacy laws evolved over time, no explicit protection for privacy has existed in the U.S. Constitution. The U.S. Supreme Court instead refers to such provisions as the Fourth Amendment (prohibiting unwarranted government searches and seizures) and primarily weighs such factors as whether an individual has a "reasonable expectation of privacy" in the case of government infringements. As a result, legal disputes and public debates about all things related to digital information privacy continue to evolve in our society.

THEORETICAL EXPLANATIONS

The concept of privacy has taken on different meanings, functions, and dimensions over time. Our awareness of privacy and the manner in which we perceive, value, and demand it has also changed. This has led to changes in our motives and strategies for seeking, negotiating, and preserving privacy. Indeed, theorists cannot agree on what privacy is, calling it a behavior, attitude, process, goal, or phenomenal state (Margulis, 1977). Definitions also vary across disciplines. Psychologists and sociologists, for example, focus on such notions as intimacy and personal space (e.g., Derlega & Chaikin, 1977; Pedersen, 1979). Legal practitioners focus on privacy as a "right" against government interference (e.g., Gavison, 1980; Tribe, 1988; Warren & Brandies, 1890). Communications scholars consider privacy in the context of interpersonal communications and relative to the changing dynamics posed by communications technologies (e.g., Burgoon, 1982). Nonetheless, we can still see certain themes emerge from a multitude of theoretical conceptions. The following discussion explores each of these theoretical themes.

Boundaries and Balance

The simplest explanation for privacy is that it is an innate need, instinctive to animals. Studies have shown that all animals, including humans, have a tendency toward territoriality and will seek periods of seclusion (Altman, 1975). Animals, like humans, will also try to balance their privacy participation with others (Westin, 1970).

Altman (1975), for example, considers privacy as a bidirectional "boundary-regulation" process, whereby individuals use barriers to control inputs from others and outputs to others. According to Altman, the regulation process is driven by a desire to maintain, restore, or reach an optimal level of privacy. Shapiro (1998) also suggests that privacy might be thought of in terms of the confluence of various boundaries, both physical and virtual. Not only is there a boundary around the physical human body, but a boundary between the public and the private as well as a boundary between producers and consumers (e.g., sellers and buyers). Boundary permeability and directionality serve to determine the level of privacy balance. For example, the telephone renders the boundary of the home more permeable to outside contact, but also enhances privacy by permitting contact outside public view. Yet Shapiro states that the boundary between producers and consumers is more permeable to information about buying habits, and that this permeability is unidirectional, representing a significant imbalance.

Personal Growth and Intimacy

Privacy is also considered necessary for personal growth by allowing a person to get in touch with him or herself and to experience a sense of unity (Fischer, 1980). As such, privacy is required for an individual to maintain his or her personal dignity and self-esteem (Laufer & Wolfe, 1977). Privacy is also important for intimacy, love, trust, and friendship, and it is considered necessary for developing and maintaining meaningful interpersonal and social relationships (Fried, 1970; Rachels, 1975). Indeed, DeCew (1997), who proposed that privacy may be best understood as a cluster concept, described one area of privacy as pertaining to one's ability to make important decisions about family and lifestyle.

Westin (1970) proposed four privacy functions, stating that privacy provides for personal autonomy, self-evaluation, emotional release, and confidentiality. In this sense, privacy provides people with an opportunity to share confidences, assess their experiences, meditate, and plan strategies, as well as relax from social roles without worrying about how they look, dress, or behave. Pedersen (1999) later identified four similar privacy functions or needs as autonomy, contemplation, rejuvenation, and confiding, and added creativity as a fifth function. In terms of today's communications technologies, one can see how the Internet, for example, allows individuals to avoid physical contact with others, thereby enhancing privacy, while providing an opportunity to share confidences through blogs and e-mails.

Scholars have also created categories or dimensions of privacy. Westin (1970) proposed the first systematic analysis of the multidimensional nature of privacy, identifying several basic states of privacy. Other scholars expanded on Westin's dimensions with support from factor analytical research. Marshall (1972, 1974) and Pedersen (1979), for example, found dimensions similar to Westin's. Pedersen (1979), however, split the intimacy dimension into intimacy with family and intimacy with friends, recognizing that people may keep some private family matters away from friends and vice versa. Certainly, interests in intimacy and reserve are relevant in the context of modern communications as individuals choose when, what, and how to communicate with friends, family, and others via such means as instant messaging, e-mails, blogs, and cell phones.

Solitude and Accessibility

Some definitions of *privacy* describe it as a desire for control over outside interactions or stimuli. For example, Altman (1975) defines *privacy* as "selective control of access to the self or to one's group." Bonavia and Morton (1998) also define privacy in the context of personal accessibility.

Various scholars have identified a privacy dimension pertaining to personal access. Westin (1970) termed this category as "solitude," referring to the need for "physical privacy." Pedersen (1979), who also recognized solitude, added a dimension called seclusion or isolation, acknowledging a privacy desire to live away from traffic and neighbors, distinct from mere solitude. DeCew (1997) distinguished one privacy cluster concept as control over access to oneself, both physical and mental. Lee (2000), who conceptualized privacy involving online or Internet communications, described one dimension as "intrusion," referring to a solitude interest in controlling Internet access to one's self. Spam, pop-ups, "junk" (unsolicited) faxes, and viruses are examples of such intrusions that infringe on an Internet user's solitude (Lee, 2000). A ringing telephone is also an example of a violation of personal solitude (Lee & LaRose, 1994).

Burgoon (1982), who proposed and confirmed (Burgoon et al., 1989) several broadly defined dimensions of communication-related privacy, also identified a "physical privacy" dimension as freedom from unwanted intrusions and added a dimension she called "impersonal violations." Burgoon's dimension of impersonal violations resulted from a factor analysis that loaded on such statements as "demands your time" and "sends junk mail" (Burgoon et al., 1989); it focuses on privacy violations from unknown or unspecified others. This may be interpreted as primarily relating to unwanted intrusive communications from strangers, such as telemarketing calls and spam.

Along the same lines, Laufer and Wolfe (1977) suggested a category called *interaction management*, meaning the ability to manage interactions with others, constrained by such factors as physical environment and one's role in a group or society. Burgoon (1982) further distinguished verbal from nonverbal "interactional privacy" (control over interactions). Verbal and nonverbal interactional privacy might help distinguish today's communication privacy in cases involving, for example, text messages and instant messages, which convey little to no nonverbal cues, from cell phone calls, which carry voice inflections in addition to verbal messages.

Surveillance and Observability

Other definitions of privacy emphasize control over being observed or monitored. Allen (1988), for example, calls privacy "the degree of inaccessibility...to the senses and surveillance of others." Lee (2000) considered surveillance as a distinct category relative to Internet privacy concerns, given the capabilities of the

Internet to monitor and track users. Although both solitude and surveillance pertain to personal access and some scholars lump the interests together (Burgoon et al., 1989; DeCew, 1997), surveillance may be distinguished because it does not involve physical intrusion. Lee considered surveillance violations to include such actions as employee e-mail monitoring and the use of "cookies"—code sent to and stored on a person's hard disk with the intention of tracking the person's web-site visits. Today, surveillance threats may be similarly posed by spyware, radiofrequency identification (RFID), cell-phone location monitoring, hidden cameras, and city cams (cameras located at city intersections).

Autonomy and Information Control

Other definitions of privacy express more of an interest in controlling information about one's self. Westin (1970) defined *privacy* as "the claim of individuals, groups, or institutions to determine for themselves when, how, and to what extent information about them is communicated to others." DeCew (1997) and Bonavia and Morton (1998) also defined privacy in the context of control over information about oneself. Laufer and Wolfe (1977) categorized such interests as "information management," pertaining to the ability to manage the disclosure or nondisclosure of one's personal information. Lee (2000) identified such privacy interests on the Internet as an interest in autonomy.

Burgoon (1982; Burgoon et al., 1989) proposed the dimension of informational/psychological privacy (personal data control or control over with whom information is shared). This dimension holds strong relevance relative to today's communications technologies. For example, privacy invasive techniques such as data mining and "phishing" (i.e., tricking someone into providing confidential information online) can be seen as violating subdimensions of informational/psychological privacy. Many scholars today consider "information privacy" as the primary privacy concern because of the increased ability for computers to collect, assemble, and distribute personal information.

Anonymity

Privacy is also considered by some as an interest in shielding one's identity or to be anonymous. Jourard (1966) says privacy is the "desire to be an enigma to others;...self-concealment." Westin (1970), Marshall (1972, 1974), and Pedersen (1979) recognized anonymity as a distinct dimension of privacy. Lee (2000) referred to anonymity as an Internet privacy dimension relating to a basic privacy interest in surfing the Web and communicating online anonymously. Indeed, people have come to expect a degree of confidentiality in private correspondence and telephone calls, and this expectation of privacy has carried over to the Internet (Lee, 1996). Examples of the interest in anonymity include the use of screen names or "pseudonyms" as well as anonymous remailers—a free computer service through which users route e-mail messages with fake return addresses. The use of Caller ID blocking also constitutes an interest in communicating anonymously (Lee & LaRose, 1994).

Social Safety Valve and Control

Although not a personal privacy dimension, it is important to note that society also requires some level of privacy for its citizens in order to function properly. Effective social structure is dependent on some nonobservability of behavior (Burgoon, 1982). For example, democratic societies need some individual anonymity in order to encourage debate and run fair elections. Anonymity also serves as a safety valve, whereby citizens can vent their frustrations and opinions without retaliation and resorting to criminal action. Corporate and government whistle-blowers, for example, must be able to report abuse without fear of losing their jobs and without resorting to sabotage. Full surveillance would lead to social dysfunction through fear. When law enforcement has unrestricted access to private information, for example, public trust may diminish, leading to paranoia (Will, 1998).

For the same matter, societies may have a need to limit or regulate individual privacy. In less developed societies, little privacy is afforded because all persons depend on the open participation of others. Societies may also have a need to engage in surveillance of their members in order to guard against antisocial conduct. Certainly, the basic functioning of a society depends on individuals disclosing some information about themselves, such as for tax collection (Westin, 1970). Yet by manipulating levels of privacy availability, societies can also exert social control.

In turn, an individual's social experiences and surrounding environment are factors affecting an individual's need for privacy. Stress has been associated with a perceived lack of privacy (Webb, 1978), and crowdedness and "stimulation overload" are thought to increase the desire for more privacy. Social pressures of role responsibilities and having to adhere to expected behaviors can also motivate individuals to pursue privacy (Pastalan, 1970). When people are anonymous, social sanctions cannot be applied, and individuals can enjoy such freedoms as free expression (Burgoon, 1982).

This relationship between society and individual privacy has motivated leading scholars in the field of surveillance studies to seek new directions for conceptualizing the social consequences of modern technologies (Hillyard & Knight, 2004). Lyon (2003), for example, argues that privacy concerns are too often relegated to the individual realm, when today's techniques for collecting and analyzing personal data put privacy in the social and not just the individual realm. Hillyard and Knight (2004) contend that research neglects to observe privacy as a fundamental social process and has not fully explored privacy as the creation, transmission, and institutionalization of cultural practices. They argue that privacy should be reconceptualized as a kind of activity, rather than as a condition, in order to better understand the relationship among privacy, technology, and social change.

To summarize, many privacy definitions are relevant to the field of communication because of the ability of new telecommunications technologies to intrude, monitor, and collect personal information. An interest in solitude continues to be a factor as telecommunications users are intruded on by viruses, telemarketing calls, junk faxes, ad pop-ups, and spam. Anonymity has become a more prominent dimension because of the remarkable ability over the Internet to remain anonymous, adopt pseudonyms, or take on entirely new identities, as well as expose identities. Autonomy or control over one's personal information becomes especially salient as data mining, collection, sharing, hacking, and theft become major threats. A strong interest in controlling and protecting personal information has certainly developed. The proliferation of "spy-cams," "cookies," spyware, RFID, GPS, and other monitoring and tracking technologies renders the surveillance dimension of privacy a social phenomenon that is worthy of investigation. Further categorizing of the concept of privacy can be expected with the introduction of new technologies that create even more opportunities for infringement and protection. Corresponding changes in social norms and privacy preferences will also lead to changes in how privacy is conceptualized and categorized.

EMPIRICAL FINDINGS

A number of empirical studies have examined the nature of privacy. Some of these studies have sought to better understand or confirm the various privacy dimensions theorized. Most assess privacy perceptions, examining the variables and conditions affecting privacy concerns. Others focus on privacy protection behaviors and measures utilized. The empirical findings discussed next review studies that address privacy issues related to communication and information technology use.

Concerns Over Privacy

Most surveys show that an overwhelming number of people are increasingly concerned about their personal privacy. One online survey of 1,500 Internet users found nearly 100% were highly concerned about invasive promotional tactics and the collection of personal information on web sites. Three fourths were also highly concerned about the tracking habits of web sites (Kandra & Brandt, 2003). A 2005 study found most consumers try to resist exposure to marketing and advertising and have a more negative opinion about the marketing and advertising they encounter today than they did a few years ago (*Yankelovich*, 2005). Not surprisingly, two thirds of Americans are taking steps to protect their privacy, such as deciding not to shop at a store or requesting that a company remove personal information from a database (Privacy & American Business, 2004).

Several studies have been conducted to better understand how people respond to each type or dimension of privacy. Burgoon et al. (1989), for example, found that people feel more strongly about some dimensions than others. They found that informational/psychological violations, such as sharing personal files with others, were considered most invasive. Meanwhile, intrusions, such as from a ringing telephone, were not regarded as especially serious. From this, one might predict that intrusions (e.g., ad pop-ups and junk faxes) would be perceived as less serious privacy violations than data mining and phishing.

Nonetheless, how people perceive the different dimensions of privacy can change. According to a 2003 Harris Interactive survey, people are somewhat less concerned about being able to control the collection of information than they were in 1994. The survey also found that people were slightly less concerned about someone watching or listening to them without their permission ("Most People," 2003). Although concerns about autonomy and surveillance nonetheless remain high, this slight decline may be due to some level of awareness and acceptance, given the widespread use of computer cookies, telephone call center monitoring, and company policies stating the legal right to monitor calls and e-mails.

However, the same study found an increase in concerns associated with the dimensions of intrusion/solitude and anonymity ("Most People," 2003). For example, more respondents said it was extremely important to not be disturbed at home. This concern coincided with an increase in telemarketing calls and unsolicited commercial e-mails. More people also said it was extremely important to be able to go unidentified in public. This burgeoning interest comes at a time when newfound opportunities for anonymity are available while browsing and communicating over the Internet.

Various studies have characterized the levels of concern. According to Westin ("Most People," 2003), about one fourth of adults are "privacy fundamentalists" who tend to feel that they have lost a lot of their privacy and are strongly resistant to any further erosion of it. Nearly two thirds are considered "privacy pragmatists," in that they have strong feelings about privacy and the need to protect themselves, yet are often willing to allow some access and use of their personal information in certain situations. About 10% are considered to be "privacy unconcerned," with no real concerns or anxiety about privacy. Sheehan (2002) found even more consumers (one fourth) to be unconcerned.

Risk–Benefit Perceptions of Privacy

Other studies have assessed the types of conditions that affect privacy perceptions. For example, Culnan (1993) found that when information was collected in the context of an existing relationship, privacy was less of a concern. Thus, private information held by a company one does business with may be considered less of a privacy concern than if it were held by an unfamiliar company. Research also reveals some online privacy perception differences based on income, gender, and race. O'Neil (2001) found those with lower incomes seem to be more concerned about Internet privacy than those with higher incomes. Concern was also higher for women as well as for Latinos and Hispanics.

The amount of information and its purpose or judged benefit can also affect perceptions of privacy. Research shows that people are less concerned if the information collected is relevant to the transaction (Culnan, 1993). In terms of telecommunications today, having one's cell-phone location information forwarded in conjunction with a 9-1-1 call would likely be less of a privacy violation than if that location information were forwarded to marketers for the purpose of text messaging an advertisement.

Perceptions of risks and benefits have an effect on privacy concerns and resulting Internet use. Using an economic model of risky consumption decisions, Kim and Montalto (2002) found that a perceived risk of privacy invasion significantly reduces the probability that a person will use online technology. A study of teenagers also found that a higher level of risk perception of information disclosure led to less willingness to provide online information (Youn, 2005). Nonetheless, such teens were more willing to disclose information online when they perceived more benefits from online disclosure (Youn, 2005).

Yet the amount of perceived risk and awareness of information-collection processes seems to have an effect. One study found that people lacking in sufficient risk information were complacent about what happens to their personal data, while considering the perceived benefits gained through online transactions and the available strategies for minimizing whatever any potential existing dangers (Viseu, Clement, & Aspinall, 2004). Indeed, many users seldom consult web-page privacy policies and have inaccurate perceptions of their own knowledge about privacy technologies and vulnerabilities (Jensen, Potts, & Jensen, 2005).

Where there is awareness of risk, however, Olivero and Lunt (2004) found less trust and a tendency to increase the demand for control and rewards, complicating the relationship between online retailers and consumers. Tolchinsky (1981), for example, found that failure to ask permission to give out information was considered a greater invasion of privacy than where the information might be sent or the possible consequences. Culnan (1993) also found that individuals were not as concerned about their privacy if they believed that information collected would be used to draw reliable and valid information about them.

Privacy Protection Behavior

Some empirical work has assessed privacy protection behaviors and responses. In one study, more than two thirds of Internet users said privacy was their primary consideration when deciding to use a particular web site ("Survey Reveals,"

2004). Sheehan and Hoy (1999) found that as privacy concerns increased, online users were less likely to register for web sites requesting information and more likely to provide incomplete information, request removal from mailing lists, notify Internet service providers about spam, and retaliate by sending a "flame" to spammers. Another survey found nearly three fourths of respondents provided false data about their identity because of concern about their personal information becoming public (Survey Reveals, 2004). Youn (2005) found that as perceptions of risk increased, teenagers tended to engage in risk-reducing strategies such as falsifying information, providing incomplete information, or going to alternative web sites that do not ask for personal information. Yet LaRose and Rifon ("Consumers Should," 2004) found that the protection that people perceive from online privacy seals, such as TRUSTe and BBBOnline, encouraged disclosure from two vulnerable groups: those who worry about online privacy but do not know how to protect themselves, and those who know how to protect themselves but are careless about doing so.

Certain people are also more likely to purchase privacy protection products. A Harris Interactive survey ("U.S. Consumers," 2003) found that middle-age men with higher levels of education and income are most likely to buy privacy products. Certainly as privacy fears escalate, more people may conceal their identities and engage in privacy protection behaviors while the market for privacy-enhancing products grows.

Privacy and the Law

While academic scholars study the nature of privacy through empirical studies, legal scholars examine the nature of privacy under the law. Lawmakers, including legislators, the court system, and legal practitioners, continue to grapple with the question of what constitutes an invasion of privacy. They look to the relevant constitutional law, federal and state statutory law, case law precedents, and administrative law for guidance.

As with the academic studies of privacy, the definitions and contexts for privacy under the law are varied. One study categorizes the legal notion of privacy into four areas: bodily integrity (intrusive searches and seizures), decisional privacy (decisions about oneself and family), information privacy (controlling the flow of information about the self to others), and communications privacy (protecting the confidentiality of communications) (Kent & Millett, 2003). The following highlights some of the major areas of legal analysis and regulatory landmarks relative to these privacy themes under the law.

"Bodily Integrity"—Intrusive Searches/Seizures. There are a number of laws addressing "bodily integrity" in terms of intrusive searches and seizures and other intrusions, pertinent to privacy interests in solitude. Although bodily integrity is not explicitly protected by the Constitution, most of the U.S. Supreme

Court's jurisprudence concerning a right to privacy has, in fact, centered on the Fourth Amendment's prohibition of unreasonable searches and seizures. The Fourth Amendment provides that the "right of people to be secure in their persons, houses, papers, and effects against unreasonable searches and seizures, shall not be violated." Most states also have a similar constitutional provision that provides similar protection.

An example of recent legislation that may test the Fourth Amendment right to privacy is the USA PATRIOT Act (2001). The PATRIOT Act allows for greater law enforcement access to Internet service provider (ISP) and cable company business records and subscriber records, as well as voice-mail messages and personal communications property such as a home computer. Provisions allowing "sneak and peek" searches, roving wiretaps (without a location specified), the potential search of entire online message content, and wiretaps without probable cause that a crime has occurred arguably contravene core Fourth Amendment protections (Lee, 2003).

As a constitutional right, the Fourth Amendment right to privacy only applies when the government is the infringer and does not apply when third parties (such as an ISP or the media) provide the information. If private entities, such as employers, marketing firms, the media, and hackers, intrude on a person's solitude, then relief may be found in a common law cause of action known generally as "invasion of privacy." Courts tend to consider (a) whether there was an intentional intrusion, (b) the location and private nature of the activity involved, (c) whether the intrusion was "highly offensive to the reasonable person", and (d) whether the infringer had a legitimate purpose warranting the intrusion. Thus, in a private lawsuit, a court would likely find no invasion of privacy if a computer system administrator accidentally read a company employee's e-mail while fixing the computer network (Lee, 1994) or a news photographer took photos in a public place (Overbeck, 2005). But a court may find a privacy invasion if a TV camera operator barged uninvited into a private home or a photographer relentlessly photographed a person, causing harassment and endangering safety (*Galella v. Onassis*, 1973).

There are also statutory laws that address privacy concerns associated with solitude. There are restrictions on intrusions from unsolicited faxes (Telephone Consumer Protection Act of 1991) and unsolicited commercial e-mails (CAN-SPAM Act of 2003). The latter bans misleading header and subject lines, requires such e-mail to be identified as an advertisement, and requires an opt-out method be provided. The Federal Trade Commission (FTC) is authorized to enforce the CAN-SPAM Act, while the Federal Communications Commission (FCC) has adopted rules to protect consumers from receiving unsolicited commercial messages on their wireless devices (Rules and Regulations, 2004).

"Decisional Privacy"—Decisions About Self/Family. When it comes to personal decisions about oneself and one's family, the law has treated privacy as a fundamental right, based in part on the Fourteenth Amendment's concept of

personal liberty (e.g., *Roe v. Wade*, 1973). As noted earlier, the most controversial U.S. Supreme Court privacy cases have dealt with personal decision making associated with such matters as contraceptive use (*Griswold v. Connecticut*, 1965), interracial marriage (*Loving v. Virginia*, 1967), access to pornography in the home (*Stanley v. Georgia*, 1969), abortion (*Roe v. Wade*, 1973), and sexual conduct (*Lawrence v. Texas*, 2003). Government intervention into such intimate matters is only allowed if there is a compelling state interest and the law is narrowly drawn (*Roe v. Wade*, 1973).

Decisional privacy may concern telecommunications indirectly. For example, a decision to view obscenity off of the Internet while in the privacy of one's home would violate statutory law that forbids the electronic transmission of obscenity (18 U.S. Section 1464). Yet a decision to share one's intimate secrets over the Internet would likely be protected by the First Amendment (see e.g., Perritt, 2001).

"Information Privacy"—Controlling Information Flow. A strong interest in the right to control the flow of information about oneself has developed as new technologies make it easier for personal information to be gathered, compiled, and distributed. Laws are being proposed and created to address the privacy concerns that electronic monitoring or surveillance presents, while lawmakers are also addressing autonomy rights associated with the collection and sale of personal data.

The key federal statutory law that addresses electronic monitoring and interceptions is the Electronic Communications Privacy Act ([ECPA] 1986). The ECPA bars confiscation of computer files (*Steve Jackson Games v. U.S. Secret Service*, 1993) and the interception of e-mail, for example. Privacy protections under the ECPA are limited, however, because it only protects messages sent over public networks (interstate), permits one-party prior consent, and allows for business use exceptions (Lee, 1994). As a result, privacy may not be protected if e-mail, for example, is intercepted within intracompany networks. Employers and ISPs may also monitor and intercept communications if users are informed of company policies that state that their communications may be monitored in the normal course of business (Lee, 1994).

Recent legislation that deals with surveillance is the federal Video Voyeurism Prevention Act of 2004, which prohibits taking surreptitious videos, impacting the use of miniature cameras, camera phones, and video recorders in public places. Legislation to restrict the use of spyware has also been introduced (e.g., Internet Spyware Prevention Act of 2005), although there are no laws prohibiting the use of cookies, which place information on one's personal hard disk to track usage history and preferences largely without one's knowledge (Randall, 1997).

Infringements by private entities such as marketers have prompted the demand for federal statutory laws to address the privacy interest in controlling the collection, compilation, and transfer or sale of one's personal information to others.

This includes the desire to keep one's name and other tracking information off marketing lists, for example. Data integrity also comes into play, where personal data maintained by others is expected to be secure, with limited access to others. Notification and awareness are also key because individuals are often not aware that information is collected, how it is used, and how they can access and correct their records.

There are a number of statutory laws or policies that regulate the collection, use, and distribution of personal information. Specific to telecommunications, the ECPA (1986) prohibits the intentional disclosure of the contents of a personal electronic communication intercepted by anyone, government or private. Prior consent is often a distinguishing factor under the law. For example, the Telecommunications Act of 1996 requires telephone companies to obtain customers' approval before using information about users' calling patterns to market new services (Li, 2003). The Children's Online Privacy Protection Act of 1998 (COPPA, 2000) requires commercial web sites and other online services directed at children to obtain consent prior to collecting information from children. The Cable Communications Policy Act of 1984 places fairly strong protections on the privacy of cable subscribers, stating that cable operators may not collect or disclose consumers' personally identifiable information without written or electronic consent (Section 551[b]).

Sometimes other rights may come in conflict with privacy laws. For example, in *Bartnicki v. Vopper* (2001), where broadcasters aired cell-phone recordings obtained illegally by someone else, the U.S. Supreme Court held that prohibiting all disclosures of intercepted information in accordance with federal antiwiretapping law violates the First Amendment, particularly when the banned information is of significant public concern. Essentially, information of public concern, such as a political or social issue, may be lawfully distributed if illegally intercepted by a third party, as long as there is no encouragement or participation in the illegal interception.

"Communications Privacy"—Protecting Confidentiality. A related area of law focuses more on protecting the confidentiality of communications. The U.S. Supreme Court has recognized a First Amendment right to speak anonymously in a case involving the distribution of anonymous election-related literature (*McIntyre v. Ohio Elections Commission*, 1995). But the High Court said that a ban that is limited to fraudulent, false, or libelous speech may pass constitutional scrutiny. One state unsuccessfully tried to ban anonymous (and pseudonymous) online communication (*ACLU of Georgia v. Miller*, 1997). Commercial anonymity may be limited, however, with a number of states and the CAN-SPAM (2003) federal law forbidding false headers and misleading subject lines in unsolicited commercial e-mail. Federal law also requires broadcasters to identify the names of all sponsors on the air (Communications Act, 1934).

Some statutes protect individual identity along with other aspects of personal data collection (e.g., the Privacy Act of 1974, the Family Educational Rights and

Privacy Act of 1974, the Right to Financial Privacy Act of 1978). For example, the Video Privacy Protection Act of 1988 states that videotape service providers (such as Blockbuster and Hollywood Video) are liable if they knowingly disclose personally identifiable information about their customers. The Cable Communications Policy Act of 1984 allows cable operators to collect aggregate data about their customers but not to disclose personally identifiable information. Although the Internet affords tremendous opportunities for anonymous and pseudonymous communications, lawsuits aimed at identifying anonymous posters have increased (Traynor, 2005). ISPs are not liable for the actions of subscribers (Sadler, 2005) and may voluntarily disclose subscriber identities, but at the risk of a lawsuit. Instead, third-party subpoenas have been filed against ISPs in defamation, trade secret, and copyright cases with the primary purpose of identifying anonymous subscribers. These are known as John Doe cases, and in them courts will sometimes balance First Amendment rights to anonymous speech with the need to identify the defendant (Traynor, 2005).

SOCIAL CHANGE IMPLICATIONS

Today, privacy interests and rights are in the midst of considerable social change. As new technologies have multiplied into the new millennium, so too has public fear that privacy rights are eroding. The ability of high-speed computers and telecommunications networks to engage in data mining and electronic surveillance, for example, now means that personal information can be gathered, merged, and distributed with remarkable ease and little oversight (Green, 2001). The widespread adoption of instant messaging and cell phones also facilitates more around-the-clock intrusions on solitude. Privacy interests have shifted from concerns about small-town gossip to fears of bureaucratic and corporate surveillance and data collection (Allen, 2001). As a result, Americans are increasingly uneasy about potential threats to their privacy, and this fear is magnified as more software and new devices are introduced and regularly denounced as privacy-invasive.

These privacy fears may, in turn, have certain socioeconomic impacts. For example, some individuals are steering clear of unfamiliar web sites and not clicking on e-mails from unknown senders in order to avoid spyware, cookies, adware, hijackings, spam, and other privacy invasions (Gellman, 2002; Sullivan, 2005). Studies have found that many people cite personal privacy concerns as their single biggest reason for staying off the Internet, with some saying that they would use the Internet more if their privacy were protected (Dempsey, Anderson, & Schwartz, 2003). This carries with it the social repercussions of a divided society consisting of connected and engaged citizens and those who are unconnected and less involved. The development of full-scale e-commerce and other Internet growth opportunities may also be hindered when individuals opt out of engaging in an online marketplace.

Another implication of privacy and social change is the lack of control individuals have over their personal information. Personal release of and outsider access to private information and data are often imposed on individuals by various public, social and commercial institutions with or without prior negotiation or informed consent. For example, people have to give out personal information in order to obtain a driver's license or a bank account, get admitted to a school or a hospital, or subscribe to a magazine or a frequent flyer program. In the digital age, the range of scenarios where an individual may willingly or unwillingly give out personal information in order to complete a transaction or task is also widening.

As more information and products become available only online, personalized and targeted advertising and entertainment programming present incredible efficiencies for the consumer who must otherwise surrender some personal information (McLean, 1995; Posner, 1978). Government pressures and fears in the event of continued terrorism attacks, such as 9/11, may also result in citizens giving up more individual privacy needs in favor of national security needs (Black, 2001). Social pressures from friends, family, and others will continue to prompt people into adopting privacy-invasive technologies such as cell phones and instant messaging (Vos, Hofte, & Poot, 2004). Indeed, it will be increasingly difficult to enjoy privacy in the form of isolation or seclusion in a world where society expects people to be reachable by cell phone and computer 24/7.

Despite these concerns, however, there are many individuals today who are nonetheless willing to relinquish their privacy to intrusions and information sharing via cell phones, instant messaging, e-mail, online chat, web cams, and so forth. The ability to use pseudonyms, take on different personas, and communicate anonymously online has led to more self-exposure (Kiesler, Siegel, & McGuire, 1984). For example, online chatrooms and personal blogs have become a source for personal revelations, where individuals can essentially post their private thoughts and journal entries for strangers to read (Noguchi, 2005). People are willing to share their childbirth experience and have invasive medical procedures shown live on the Internet to thousands of complete strangers (Allen, 2001). Although people can choose to expose their private lives, others can choose to anonymously observe these private details from the confines of their homes. Television and the Internet have made dramatic changes in erasing the lines between private and public, with voyeuristic strangers now able to watch others detailing their sex lives, personal oddities, and problems (Horn, 1998).

There also continue to be those who are not overtly alarmed by or ignorant of the potential invasion of information privacy and are willing to surrender their privacy in exchange for something else (Viseu, Clement, & Aspinall, 2004). For instance, online magazine subscribers may get a free subscription in exchange for providing the publisher with survey information that can be used to secure advertisers. Internet users may get free e-mail service in exchange for enduring pop-up ads or being accessible to unsolicited commercial e-mails. Users wishing to get information from an online travel service may have to provide personal information to get a password and information. Internet users may reluctantly or unwittingly permit "cookies" as a means to navigate efficiently, as they expose themselves to online businesses intent on tracking information for marketing. Even if legislative and policy solutions to enhance digital information privacy continue to grow, individuals will be increasingly faced with decisions as to how much privacy they are willing to give up or trade.

A potential consequence is that less privacy may be enjoyed in the future. Rust (2002) asserts that the amount of privacy enjoyed by consumers will decline over time and that privacy will be increasingly expensive for consumers to maintain. Despite the emergence of a market for privacy that enables customers to purchase a certain degree of privacy, he asserts that without government intervention, the overall amount of privacy will erode. Essentially, if information is less expensive to obtain and process, then it will become more valuable to consumers who will be forced to pay more to protect it.

Yet privacy may also thrive and gain additional protection. One positive implication for individual privacy is the potential opportunity for more individual control. As people become more aware of privacy concerns, the marketplace will likely respond with more privacy protection products, and those who seek out and utilize such products will become more empowered. A more informed consumer will also likely engage in self-protection strategies. Already, people are taking advantage of the ability to ignore or delay their replies to instant messages and cell-phone text messages in order to protect solitude, manage their interactions, and help control the flow of personal information. The increase in awareness and control may also mean a greater ability to negotiate one's privacy.

Expectations of privacy will also evolve, which carries implications for how privacy is interpreted and protected under the law. For example, greater expectations of anonymity and accessibility may become customary as online users get used to communicating anonymously, whereas expectations to intrude on others via cell phones and instant messaging may increase (e.g., Wong, 2004). Public opinion polls seem to confirm that individuals are more aware of, expect, value, and demand more privacy (Center for Survey, 2002). Indeed, compared with the late 1990s, people today are more likely to report that the right to privacy is "essential" (Center for Survey, 2002).

As a result, with more people aware of privacy threats and expectations of privacy increasing, more people will demand appropriate legal protection. Although a number of statutory laws have been created to address privacy concerns, as previously outlined, there continues to be strong demand for more protection. The creation of a "Do Not Call" list ("Controlling the Assault," 2003) and a push for state and federal spam legislation, for example, have occurred in conjunction with technological advances and increases in solitude and anonymity interests ("Most People," 2003), while federal and legislative action and changes have been prompted by highly publicized privacy concerns associated with identity theft, spyware, and the USA PATRIOT Act. More bills aimed at protecting privacy are introduced each year (e.g., identity theft [Stross, 2005] and spam ["Security," 2005]), and additional laws—such as to protect consumer data from identity theft and to further control spam—may be expected in the future ("Regulatory, Legislative, Economic," 2005).

In summary, sparked by advances in new technologies and social change, privacy interests today present new conflicts and challenges. All users of communication technology have a need or desire for some level of information and access to others while also possessing a relatively strong need or desire for their own privacy. Individuals, businesses, and government can gain incredible efficiencies by taking advantage of technological capabilities that allow greater reach, speed, and capacity when it comes to collecting, compiling, and disseminating personal information. At the same time, individuals are increasingly concerned about the effect that the security of digital information collection, storage, retrieval, and access has on their privacy interests.

Scholars contend that there has been no "golden age" of privacy from which we have fallen (Sykes, 1999). Despite recent attention given to privacy concerns, the degree to which people have enjoyed privacy over the centuries has actually ebbed and flowed. In general, as gains in privacy within certain relationships have been made over time, they are typically replaced by an increase in surveillance and disclosure by other authorities. The same society that allows us to live anonymously in turn relies on surveillance to keep track of us in a world of strangers. Perhaps what citizens see more today is an increase in awareness, potential risks, technological opportunities, and choices available when it comes to managing and balancing their personal zones of privacy against the society's needs to intrude on that privacy. In this digital information age, finding and creating the right balance between these two opposing interests has become more challenging as more communication and information technologies come online with conduits that can easily spy and prey on the vast array of information databases that are continuously becoming available.

REFERENCES

ACLU of Georgia v. Miller, 977 F. Supp. 1228 (N.D. GA. 1997), affirmed, 194 F.3d 1149 (10th Cir. 1999).

Allen, A. L. (1988). *Uneasy access: Privacy for women in a free society*. Totowa, NJ: Rowman & Littlefield.

Allen, A. L. (2001). Is privacy now possible? A brief history of an obsession. *Social Research, 68*(1), 301–306.

Altman, I. (1975). *The environment and social behavior*. New York: Wadsworth.

Bartnicki v. Vopper, 53 U.S. 514 (2001).

Black, J. (2001, Nov. 30). Uncle Sam needs watching, too. *Business Week Online*. Retrieved March 22, 2002, from LEXIS, Nexis Library.

Bonavia, M., & Morton, L. (1998). Personal information privacy issues relating to consumption in the U.S. marketplace. *Consumer Interests Annual, 44*, 25–29. Burgoon, J. K. (1982). Privacy and communication. In M. Burgoon (Ed.), *Communication yearbook: 6* (pp. 206–249). Beverly Hills, CA: Sage.

Burgoon, J. K., Parrott, R., LePoire, B. A., Kelley, D. L., Walther, J. B., & Perry, D. (1989). Maintaining and restoring privacy through communication in different types of relationships. *Journal of Social and Personal Relationships, 6*, 131–158.

Cable Communications Policy Act of 1984, 47 U.S.C. Section 551 (2003).

CAN-SPAM Act of 2003 (Controlling the Assault of Non-Solicited Pornography and Marketing Act), 15 U.S.C. Sections 7701 et seq. (2003).

Center for Survey Research and Analysis. (2002, August). *State of the First Amendment 2002*. Retrieved October 9, 2005, from http://www.freedomforum.org/templates/document.asp?documentID=16840

Children's Online Privacy Protection Act ("COPPA"), 15 U.S.C. 6501-06 (2000).

Communications Act of 1934, 47 U.S.C. Section 309 (1934).

Consumers should not be lulled by Internet privacy seals. (2004). *Ascribe Newswire*. Retrieved June 10, 2004.

Controlling the assault of the non-solicited pornography and marketing act of 2003 (CAN-SPAM), Pub. L. No. 108-187, 117 Stat. 2699 (2003), codified at 15 U.S.C. 7701–7713, 18 U.S.C. 1037 and 28 U.S.C. 994.

Culnan, M. J. (1993). How did they get my name?: An exploratory investigation of consumer attitudes toward secondary information use. *MIS Quarterly, 17*(3), 341–361.

DeCew, J. W. (1997). *In pursuit of privacy: Law, ethics, and the rise of technology*. Ithaca, NY: Cornell University Press.

Dempsey, J., Anderson, P., & Schwartz, A. (2003). *Privacy and e-government: A report to the United Nations Department of Economic and Social Affairs as background for the world public sector report: E-Government*. Retrieved October 10, 2005, from www.internetpolicy.net/privacy/20030523cdt.pdf

Derlega, V. J., & Chaikin, A. L. (1977). Privacy and self-disclosure in social relationships. *Journal of Social Issues, 33*(3), 102–115.

Electronic Communication Privacy Act of 1986, 18 U.S.C. Sections 2510–2521, 2701–2710, 3117, 3121–3126.

Fischer, C. T. (1980). Privacy and human development. In W. C. Bier (Ed.), *Privacy: A vanishing value*? (pp. 37–45). New York: Fordham University Press.

Fried, C. (1970). *An anatomy of values*. Cambridge, MA: Harvard University Press.

Galella v. Onassis, 487 F.2d 986 (2d Cir.1973).

Gavison, R. (1980). Privacy and the limits of law. *Yale Law Journal, 89*, 421–461.

Gellman, R. (2002, March). *Privacy, consumers, and cost: How the lack of privacy costs consumers and why business studies of privacy costs are biased and incomplete*. Retrieved October 9, 2005, from the Electronic Privacy Information Center (EPIC) web site http://www.epic.org/reports/dmfprivacy.html

Green, H. (2001, November 6). A defining moment for info privacy. *Business Week Online*.

Griswold v. Connecticut, 381 U.S. 479 (1965).

Hillyard, D. P., & Knight, S. M. (2004). Privacy, technology, and social change. *Knowledge, Technology & Policy, 17*(1), 81–101.

Horn, M. (1998). Shifting lines of privacy. *U.S. News & World Report, 125*(16), 57.

Internet Spyware Prevention Act of 2005 (I-Spy Act), H.R. 744, 109th Cong. (2005).

Jensen, C., Potts, C., & Jensen, C. (2005). Privacy practices of Internet users: Self-reports versus observed behavior. *International Journal of Human-Computer Studies, 63*(1–2), 203–227.

Jourard, S. M. (1966). Some psychological aspects of privacy. *Law and Contemporary Problems, 31*, 307–318.

Kandra, A., & Brandt, A. (2003, November). The great American privacy makeover. *PC World*. Retrieved July 31, 2005, from http://www.pcworld.com/howto/article/0,aid, 112468,00.asp

Katz, J. E. (1987). Telecommunications and computers: Whither privacy policy? *Society, 25*(1), 81–86.

Katz v. U.S., 389 U.S. 347 (1967).

Kent, S., & Millett, L. (Eds.). (2003). *Who goes there? Authentication through the lens of privacy, committee on authentication technologies and their privacy implications.* Washington, DC: National Research Council, National Academics Press. Retrieved July 28, 2006, from http://www.nap.edv/catalog/10656.html?onpi—newsdoc032503

Kiesler, S., Siegel, J., & McGuire, T. W. (1984). Social psychological aspects of computer-mediated communication. *American Psychologist, 39*(10), 1123–1134.

Kim, S., & Montalto, C. P. (2002). Perceived risk of privacy invasion and the use of online technology by consumers. *Consumer Interests Annual, 48*, 1–9.

Kolsti, N. (2005, March 23). *Study says students dependent on cell phones, e-mail.* University of North Texas News Service. Retrieved October 10, 2005, from web2.unt.edu/news/story.cfm?story=9143

Laufer, R. S., & Wolfe, M. (1977). Privacy as a concept and a social issue: A multidimensional development theory. *Journal of Social Issues, 33*(3), 22–42.

Lawrence v. Texas, 539 U.S. 558 (2003).

Lee, L. T. (1994). Watch your e-mail! Privacy law in the age of the "electronic sweatshop." *John Marshall Law Review, 28*(1), 139–177.

Lee, L. T. (1996). On-line anonymity: A new privacy battle in cyberspace. *The New Jersey Journal of Communication, 4*(2), 127–146.

Lee, L. T. (2000). Privacy, security, and intellectual property. In A. B. Albarran & D. H. Goff (Eds.), *Understanding the web: Social, political, and economic dimensions of the Internet* (pp. 135–164). Ames: Iowa State University Press.

Lee, L. T. (2003). The USA Patriot Act and telecommunications: Privacy under attack. *Rutgers Computer & Technology Law Journal, 29*(2), 371–403.

Lee, L. T., & LaRose, R. (1994). Caller ID and the meaning of privacy. *The Information Society: An International Journal, 10*(4), 247–265. *Loving v. Virgima*, 388 U.S.I (1967).

Lyon, D. (2003). *Surveillance as social sorting: Privacy, risk, and digital discrimination.* New York: Routledge.

McIntyre v. Ohio Elections Commission, 514 U.S. 334 (1995).

Margulis, S. T. (1977). Conceptions of privacy: Current status and next steps. *Journal of Social Issues*, *33*(3), 5–21.

Marshall, N. J. (1972). Privacy and environment. *Human Ecology, 1*(2), 93–110.

Marshall, N. J. (1974). Dimensions of privacy preferences. *Multivariate Behavioral Research, 9*, 255–272.

McLean, D. (1995). *Privacy and its invasion*. Westport, CT: Praeger.

Most people are "privacy pragmatists" who, while concerned about privacy, will sometimes trade it off for other benefits, says Harris Interactive survey. (2003, March 19). *PR Newswire.N.C. State professor victim of online theft*. (2005, October 11). Retrieved June 10, 2004, from www.hackinthebox.org/modules.php?op=modload&name=News&file=index&catid=&topic=9

Noguchi, Y. (2005, October 12). Cyber-catharsis: Bloggers use web sites as therapy. *The Washington Post*. Retrieved October 24, 2005, from http://www.washingtonpost.com/wp-dyn/content/article/2005/10/11/AR2005101 101781.html

Olivero, N., & Lunt, P. (2004). Privacy versus willingness to disclose in e-commerce exchanges: The effect of risk awareness or relative role of trust and control. *Journal of Economic Psychology, 25*(2), 243–262.

O'Neil, D. (2001). Analysis of Internet users' level of online privacy concern. *Social Science Computer Review, 19*(1), 17–31.

Overbeck, W. (2005). *Major principles of media law* (2006 ed.). Belmont, CA: Thomson Wadsworth.

Pastalan, L. A. (1970). Privacy as a behavioral concept. *Social Science, 45*(2), 93–97.

Pedersen, D. M. (1979). Dimensions of privacy. *Perceptual and Motor Skills, 48*, 1291–1297.

Pedersen, D. M. (1999). Model for types of privacy by privacy functions. *Journal of Environmental Psychology, 19*(4), 397–405.

Perritt, H. (2001). *Law and the information superhighway* (2nd ed.). New York: Aspen Law & Business.

Posner, R. (1978). The right of privacy. *Georgia Law Review, 12*, 393.

Privacy & American Business. (2004, June 10). *New national survey on consumer privacy attitudes to be released at Privacy & American Business landmark conference*. Privacy and American Business Press Release. Retrieved from http://www.marketwire.com/mw/release_html_b1?release_id=68484

Rachels, J. (1975). Why privacy is important. *Philosophy and Public Affairs, 4*, 323–333.

Randall, N. (1997, September 9). Cookie managers. *PC Magazine*, pp. 159–162.

Regulatory, legislative, economic. (2005, January 15). Retrieved October 1, 2005, from LEXIS, Nexis Library.

Roe v. Wade, 410 U.S. 113 (1973).

Rules and Regulations Implementing the Controlling the Assault of Non-Solicited Pornography and Marketing Act of 2003 (CAN-SPAM Act), Order, 19 FCC Rcd 15927 (2004).

Rust, R. T. (2002). The customer economics of Internet privacy. *Journal of the Academy of Marketing Science, 30*(4), 455–464.

Sadler, R. L. (2005). *Electronic media law*. Thousand Oaks, CA: Sage.

Sanders v. ABC, 20 C. 4th 907 (1999).

Schoeman, F. D. (1992). *Privacy and social freedom*. New York: Cambridge University Press.

Security. (2005, April 21). *Washington Internet Daily, 6*(77). Retrieved November 20, 2005, from LEXIS, Nexis Library.

Shapiro, S. (1998). Places and spaces: The historical interaction of technology, home, and privacy. *The Information Society, 14*(4), 275–284.

Sheehan, K. B. (2002). Toward a typology of Internet users and online privacy concerns. *The Information Society, 18*(1), 21–32.

Sheehan, K. B., & Hoy, M. G. (1999). Flaming, complaining, abstaining: How online users respond to privacy concerns. *Journal of Advertising, 28*(3), 37–51.

Steve Jackson Games v. U.S. Secret Service (W.D. Tex. 1993).

Stanley v. Georgia, 394 U.S. 557 (1969).

Stross, R. (2005, February 13). How to stop junk e-mail: Charge for the stamp. *The New York Times*. Retrieved November 20, 2005, from http://coba.usf.edu/abhatt/eb/NYI-Spam.htm

Sullivan, B. (2005, June 15). *Data leaks stunt e-commerce, survey suggests: Half say they avoid making purchases online*. Retrieved October 13, 2005, from http://msnbc.msn.com/id/8219161.

Survey reveals increased privacy concerns cause consumers to provide false identity data. (2004, 19). *PR Newswire*. Retrieved June 10, 2004, from LEXIS, Nexis Library, Financial News File.

Sykes, C. J. (1999). *The end of privacy*. New York: St. Martin's Press.

Telecommunications Act of 1996, Pub. L. 104–104, 47 U.S.C. Section 222 (1996).

Telephone Consumer Protection Act. 1991. 47 U.S.C. Section 227.

Tolchinsky, P. D., McCuddy, M. K., Adams, J., Ganster, D. C., Woodman, R. W., & Fromkin, H. L. (1981). Employee perceptions of invasion of privacy: A field simulation experiment. *Journal of Applied Psychology, 66*(3), 308–313.

Traynor, M. (2005). Anonymity and the Internet. *The Computer & Internet Lawyer, 22*(2), 1–16.

Tribe, L. (1988). *American Constitutional law* (2nd ed.).

Uniting and Strengthening America by Providing Appropriate Tools Required to Intercept and Obstruct Terrorism (USA PATRIOT) Act of 2001, Pub. L. No. 107–56, 115 Stat. 272 (2001) (amending scattered sections of 18, 47, 50 U.S.C.).

U.S. consumers flock to products to protect privacy, study says. (2003, July 7). *Electronic Commerce News, 8*(14).

Viseu, A., Clement, A., & Aspinall, J. (2004). Situating privacy online. *Information, Communication & Society, 7*(1), 92–114.

Vos, H., Hofte, H., & Poot, H. (2004). IM [@Work] adoption of instant messaging in a knowledge worker organization. *Proceedings of the 37th Annual Hawaii International Conference on System Sciences (HICSS'04)*, Track 1, 10019a. Retrieved October 20, 2005, from http://csdl2.computer.org/comp/proceedings/hicss/2004/2056/01/2056 10019a.pdf

Warren, S., & Brandeis, L. (1890). The right to privacy. *Harvard Law Review, 4*, 193–220.

Webb, S. D. (1978). Privacy and psychosomatic stress: An empirical analysis. *Social Behavior and Personality* 6(2), 227–234.

Westin, A. (1967). *Privacy and freedom.* New York: Atheneum.

Westin, A. (1996). Privacy in America: An historical and socio-political analysis. In Connecticut Freedom of Information Commission and the Connecticut State Library (Eds.), *Report of the National Privacy and Public Policy Symposium* (pp. 155–179). Hartford, CT: The Connecticut Foundation for Open Government, Inc.

Will, R. (1998, October 5). Proper vigilance or paranoia? *Inter@ctive Week, 5*(38).

Wong, M. (2004, December 7). New social medium fosters impatience. *Azstarnet.com* (Arizona Daily Star). Retrieved July 28, 2006, from http:/www.azstarnet.com/sn/specialreports/51394

Yankelovich Monitor. (2005, April 18). *2005 Marketing Receptivity Survey*. Retrieved July 27, 2006, from http://www.yankelovich.com/thought/TL2005MarketingReceptivityStudy.pdf

Youn, S. (2005). Teenagers' perceptions of online privacy and coping behaviors: A risk-benefit appraisal approach. *Journal of Broadcasting & Electronic Media, 49*(1), 86–110.

VIII

SUMMARY

15

An Integrated Communication Technology and Social Change Typology

Carolyn A. Lin

This volume adopted a unique approach, one that addresses various social settings—namely, home, workplace, surveillance, entertainment, consumer, and legal settings—to examine the relations between technology and social change. By presenting scholarly explications in a range of theoretical and empirical venues, we thus integrated these perspectives using conceptual themes in socially relevant contexts—rather than technology modalities—as the unit of analysis. Drawing from our various contributions, it is clear that the influence of these mediated communication technologies has been evolutionary. Newton's Third Law states that for every action there is an equal and opposite reaction. Based on this "dual effects" conception, it would appear that for all of the benefits accruing from a new technology, there are likely to be countervailing negative effects. Naturally, not all predictions concerning the social consequences of mediated communication technology adoption have come to pass. Orwell's (1949) dystopian prophecies involving digital "Big Brother" governments—where average citizens endure a fish bowl existence in the eye of government surveillance—provides a case in point. Instead, the rise of telematic "technologies of freedom" (Poole, 1983) has actually helped undermine totalitarianism in some instances and encourage a freer flow of information across borders.

At the global level, spurred by the deregulation of state-supported media monopolies and protectionist policies (Anowkwa, Lin, & Salwen, 2003), many nations now are opening up their media systems to competition, hoping to hasten cross-media competition and convergence (see e.g., Noam, Groebel, & Gerbarg, 2004; Pelton, 2003; Straubhaar, Burch, Guilherne Duarte, & Sheffer, 2002). As a result, Ogan (chap. 2, this volume) notes that interactive media technologies such as the Internet are altering the one-way media domination or information flow from the industrialized to the developing nations of the yesteryear (see Dizard, 2000). Hence, there is a shift from the traditional media/cultural imperialism paradigm to the new glocalization paradigm (Robertson, 1995)—where cultural regionalism and linguistic commonality dominate how data, information, news, media, and communication content are shared, exchanged, and marketed. Moreover, just as new media technologies represent a threat to totalitarianism, they also present channels for oppositionary discourse aimed at the world's remaining superpower.

As a case in point, Yahoo!, Google, and MSN have all consented to comply with the censorship rules issued by the communist Chinese government, which include stripping all references to the word *democracy*. Even so, Chinese bloggers have been able to circumvent the government's surveillance and system control to convey their free thoughts with the outside world. In a similar vein, the Dubai-based Arab language satellite TV network Al Jazeera regularly reports stories that overshadow the ideological hegemony and propaganda diffused by American and Arabic national media outlets on the issues of democracy, religion, and war.

Although technology innovations have the potential to become the equalizer for social, political, and economic development inequities among societies, a vast digital divide that exists between the digital *haves* and *have-nots* has, nevertheless, created unprecedented information and knowledge gaps. Surprisingly, these information and knowledge gaps also exist within the borders of an information society like ours, where concerns over the digital divide are considered minimal. These information and knowledge gaps, along with other social changes, are a result of how individuals, groups, and institutions adopt and use communication and information technologies to function as part of the social system in different social settings. The section to follow outlines these avenues of social influence in the context of our various book chapters, each of which is outlined in turn.

INDIVIDUAL AND GROUP SETTING

In an information society like ours, where information is readily available and communication technology is easily within reach, the social implications of technology adoption and uses are drastically different from the discussion presented earlier. Nonetheless, as pointed out by Perse and Dunn (1998), a focus on social impacts may be premature until we fully understand how and why people use computer technology. Caplan, Perse, and Recchiuti's discussion of individual user-level influences suggests that, due to reduced sensory channels involved and queues delivered in computer-mediated communication, this hyperpersonal communication (Walther, 1996) nonetheless provides additional communicative advantages over traditional face-to-face interaction.

A hyperpersonal message sender is hence given the capacity to plan, develop, review, and edit the information and communication content to send a self-selective and/or optimized presentation of oneself to others (Walther, 1996; Walther & Burgoon,

1992). This process of communication tailoring, in effect, can provide a purposeful and goal-oriented communication experience that may reflect a more socially categorical interaction instead of the traditional personal versus impersonal interaction experience based on the social identification model of deindividuation effects (Lea & Spears, 1992; Reicher, Spears, & Postmes, 1995; Spears & Lea, 1992, 1994; Spears, Postmes, & Lea, 2002). This suggests that individual users are afforded the freedom to compose their communication and messages when there is a time delay involved in their social interaction. In real-time interaction, the dynamics may be different in that less contemplated, more thoughtless, and even impulsive reactionary communication may be exchanged instead.

Whether socially categorized and tailored communication takes place in real or nonreal time—with either a more or less calculated exchange—this individual computer-mediated communication scenario can play out in either a weak- or strong-tie network. Therapeutic relational communication online conducted through a weak-tie network—such as an online discussion—was evidenced to be a productive means of providing social support by some (Walther & Parks, 2002; Wright, 1999, 2000, 2002; Wright & Bell, 2003), although the comparative effectiveness of this therapeutic communication venue has not been established relative to that of face-to-face social support (Finfgeld, 2000; Owen, Yarbrough, Varga, & Tucker, 2003; Walther & Boyd, 2002).

As computer-mediated communication can serve as a tool for facilitating personal expressions in social interaction, it can also help serve as a window to the world for the younger members of our society and play a role in their socialization process. According to Krcmar and Strizhakova (chap. 4, this volume), social effects conceptions of the relations between children and interactive communication technologies are still evolving. Although the long-standing concerns over the potential media effects of violent and sexual content on children persist, our understanding of the role that these potential effects play in children's socialization process remains inconclusive.

Specifically, in applications of social cognitive theory (Bandura, 1994)—which emphasizes social modeling behavior—the empirical evidence tends to indicate limited short-term effects of violent media content on children's aggressive behavior. For instance, players of violent videogames, based on a number of studies (Anderson, 2004; Anderson & Bushman, 2001), demonstrated a stronger tendency to express aggression induced by a vicarious violent act. But lacking solid empirical evidence showing that such tendencies are activated within relevant contextual cues (e.g., interpersonal conflicts), it is difficult to know whether the effects of these violent videogames could take root in the real world.

At the other end of the spectrum, children's use of interactive media technologies also purportedly enhances their cognitive learning in formal as well as informal settings. For instance, children utilize these technologies to do research on various curricular topics, converse with their teachers and classmates to collaborate on projects, and complete homework assignments, both at school and in the home. Children were also found to be more motivated to learn, as technology use helped stimulate their interest by offering nontraditional modes of instruction, including learning through playing (Strasburger & Wilson, 2002), aside from having gained improved spatial skills through playing videogames (e.g., Greenfield, De Winstanley, & Kilpatrick, 1994; Subbrahmanyam & Greenfield, 1994). However, lacking sustained evidence indicating that computer-mediated learning actually can help improve the academic performance and intellectual growth of children, in either formal or informal learning environments, interactive technologies may simply be a fun tool that could distract children from reaching desired learning objectives. Stoll (1995) decries the opportunity costs of investments in academic technology, which he terms "Silicon Valley snake oil," noting that the costs of acquiring and maintaining computers in the classroom displace more effective human resources (i.e., teachers).

As children spend more time with interactive media technologies and engage in more frequent computer-mediated communication, they also tend to spend less time with their family and become more intertwined with their own social network (Kraut, Patterson, Lundmark, Keisler, Mukopadhyay, & Scherlis, 1998; Kundanis, 2003). This social network usually revolves around Internet technology—which includes a web of instant messaging, e- mailing, blogging, podcasting, and chatting channels—as well as wireless communication devices such as cellular phones and other smart-phone devices. Staying connected to the peer group culture with the help of interactive communication technologies suggests neither a positive nor negative social outcome, as the social outcomes are instead a result of *how* these technologies are used, not whether they are used. A number of real-life incidents, which profiled how popular teen web sites (e.g., Myspace.com) provide a neutral outlet for teens to express themselves, demonstrate how they could also be abused by willing teen victims attracted by sex offenders (Bucy & Newhagen, 2004).

WORK AND ORGANIZATIONAL SETTING

Similar to the scenario described earlier, where children use these technologies for engaging in informal learning and homework, the term *home office* has also become synonymous with the notion of where the computers designated for home work are located in the household (Bracken & Lombard, 2004; Choie & Cheon, 2005). As the diffusion of computer-mediated and wireless communication technologies reached a critical mass and peripheral devices—including document/image-generating software programs (e.g., MS-Word and Adobe pdf)—also became commonplace in homes, these technologies have made home work a viable option for many workers in both formal (i.e., working from home) and informal (i.e., working at home) settings.

The social influence of the home workplace, as explored in Atkin and Lau's chapter (chap. 5, this volume) on telecommuting (or telework), remains a phenomenon that is still evolving. From an organizational perspective, the decision and policy of operating a home work or telecommuting network have to do with whether an organization is capable of perceiving the relative advantage or cost-efficiency of this type of operation. This particular organizational behavior can be explained by the concept of absorptive capacity (Boynton, Zmud, & Jacobs, 1994; Cohen & Levinthal, 1990), which states that an organization's ability to recognize and successfully incorporate innovative ideas into productive commercial applications helps determine whether certain innovations will be adopted for organizational use (Cohen & Levinthal, 1990). In the case of telework, corporations that are willing to either accommodate the lifestyle choice of their valued workers or impose this type of operational structures on certain workers often do so after considering a set of social as well as economic factors.

Social factors such as the need to attract and retain highly valued workers are often the primary reasons for allowing telework to selective workers, be they those who prefer a more casual work setting, those who reside outside of a reasonable commuting distance, or those whose work responsibilities tend to be independent, whose work routines are rather transparent, or whose work performance requires little supervision. Economic factors for operating a telework unit often encompass such reasons as financial savings on physical space, travel costs, supervisory resources, or even reduced fringe benefits (for non-full-time employees). These organizational dynamics, according to Kraut, Galegher, Fish, and Chalfonte (1992), can be best explained by contingency theory, which suggests that organizational task performance is contingent on successfully matching the technological designs, social factors (i.e., worker perception and preference of certain technologies), and organizational structure, which in turn dictates technology adoption and task management—to optimize the completion of the designated tasks.

Hence, to successfully implement a telework structure into an organization, technology choice by the organization and the workers is crucial. According to Trevino, Daft, and Lengel (1990), communication unequivocality is an important aspect for workers when they adopt and use computer-mediated communication technologies in an organizational setting. This observation is relevant to the concept of social presence (Lombard & Ditton, 1997; Wellman, Salaff, & Dimitrova, 1996), which explores such social factors as the lack of social intimacy and realism (e.g., working together) either due to asynchronous communication exchanges or the lack of physical proximity between workers and their relations with the organization.

To date, telework remains a small, single-digit percentage of the workforce at large. It is entirely likely that telework, despite the optimistic forecasts made by those who abide by the paradigm of technology determinism, will always remain a fringe or peripheral workplace choice, one that is adopted by organizations and workers that can find common ground in matching the organization's task requirement to the worker's personal needs. After all, there is something to be said about "going to work" at a physical location that symbolizes a public work space that is also a social space, instead of a private personal space. For most individuals in society, such a distinction remains a significant and personal preference.

Telework (or telecommuting) thus aims at creating an alternative venue, one that enables virtual workers to reduce a number of time-consuming, in-person communication transactions and physical processing of paperwork. A concurrent trend in accomplishing the same objectives has been seen in organizations—both large and small—that have not been at the forefront of technology-savvy hyperbole. Rice and Schneider (chap. 6, this volume) help us to better understand the office information workers—namely, what they do with adopting and adapting to new information and communication technology and the impact of such technologies on their productivity as workers in an increasingly informated workplace.

In chronicling the rise of the paperless office, the concept of informating (i.e., processing and making the information "enactable"), as opposed to automating the workplace, is an important one. In that vein, the concept of how workers incorporate different artifacts in their workspace (Davis, 1984) as part of their information processing and communication routine can be explored within the confine of symbolic interaction (between worker and medium)—and social construction of the medium—as both a delivery channel of messages and a symbol of messages (Feldman & March, 1981; Sitkin, Sutcliffe, & Barrios-Choplin, 1992; Rice, 1987, 1999). For instance, e-mail is a channel for sending messages; as a channel of message delivery, it may also take on a unique symbolic meaning. The symbolic meaning can differ depending on whether the e-mailing is prefaced as formal communication (e.g., an official policy announcement) or informal communication (e.g., an informal discussion of a policy).

This social construction of a medium (e.g., an e-mail or an intranet system) may or may not be meaningful to workers from an organizational perspective if workers do not cognitively perceive and physically utilize "affordances" (or the available technical or functional actions associated with the use) of the technology or medium to maximize them to their fullest (Gibson, 1979; Hartson, 2003). Specifically, as revealed by Rice and Schneider (see chap. 6, this volume), during and after the period of transitioning from a paper-based to a digital network-based information-processing system, workers could often bypass, alter, or reinvent different aspects of the new informating system; this might be done at the expense of failing to maximize the affordances of the costly technologies that have been put in place to facilitate such action (Goodman, Griffith, & Fenner, 1990; Johnson & Rice, 1987; Majchrzak, Rice, Malhotra, King, & Ba, 2000; Rice & Gattiker, 2000).

Hence, the mere act of transferring workflow management from a primarily paper-based physical environment to a network-based virtual environment requires a lot more than a simple replacement of these information workers' informating tools. Instead, this transfer or transition requires a shift in organizational culture, in addition to cultivating in workers a sense of efficacy in adapting a technology system design that, more often than not, has neglected to take into consideration the human-engineering factors. These human-engineering factors often hold the key to success in the implementation of a new information and communication technology system, as they reflect the ways that humans perceive, comprehend, contemplate, and activate their interface actions with a medium (Lievrouw & Livingstone, 2002; Scheneiderman, 2005). Information or communication technology systems that were not designed to prevent the human–technology interface errors will then not be able to fully achieve the intended affordances expected by the organizations that deploy them.

SURVEILLANCE SETTING

Departing from the organizational (work) setting, the discussion of the next social phenomenon at the group level involves the community setting, which is the building block of civic life in a democratic society. The role that communication technologies play in civic life is often not readily visible, although the social functions these technologies help to fulfill can be relatively significant. As suggested by Jeffres (chap. 7, this volume), these functions can include social surveillance, service, and integration. In particular, the concept of social capital (e.g., Putnam, 1995) reflects how communication technologies can help enhance the delivery of services that aim to serve and support a community economic and cultural life for the purposes of maximizing the community's existing social capital.

For instance, local cable TV systems allot a number of channels to help deliver an array of community-service oriented programs generated by various local community organizations or individual residents, via both government and public access channels. These community organizations can include the local public library, senior center, high school, community college or university, fire/police station, municipal government, city council, state legislature, and so on, which aim to inform, educate, and entertain the various segments of a community's residents. Hence, when the local Lions Club or Girl Scouts group holds an auction to raise funds to stock the local food bank, or when the local community college is offering a free computer summer camp to disadvantaged children, a number of these public service-oriented cable TV channels can help promote and recruit volunteers.

The arrival of the community-oriented Internet services has offered an even more effective and expedient channel for carrying out the same social service functions. In fact, due to the nature of virtual channels that can deliver a nearly limitless amount of information, the Internet has become the new frontier for renewing the community identity that has increasingly dwindled due to urban decay and sprawl (Jeffres, 2003; Matei & Ball-Rokeach, 2003). This is evidenced by how all municipalities and most community-based organizations, whether they be not for profit (nonprofit) or for profit, now operate their own web sites to communicate with their constituents within and outside of the community. The Internet has thus provided an opportunity to help reenergize a sense of community identity and common purposes. Even more important, the Internet has been found to help strengthen civic involvement, economic activities, and a sense of civic duty/pride (Baker & Ward, 2002; Dutta-Bergman, 2005; Pigg, 2001; Shah et al., 2000).

When using interactive media technologies to communicate with, inform, and serve the local communities, economically disadvantaged residents often lag behind in their access to both cable TV and the Internet to reap the full benefits of this information abundance. This phenomenon then helps result in a social outcome that has long been dubbed the *knowledge gap* (Tichenor, Donohue, & Olien, 1980), which highlights the classic problem between the haves and the have-nots in terms of the knowledge gains that keep them apart in seeking social mobility. In an information age, an information gap and a digital gap can help accelerate and widen such knowledge gaps even further (Matei & Ball-Rokeach, 2003).

This knowledge gap, as facilitated by an information and digital gap, is not limited to any particular communities. From a macrosocial perspective, this knowledge gap is also found in the larger society when one considers the amount and the speed of information and communication traveled and exchanged in the 24-hour cycle. Bucy, Gantz, and Wang (chap. 8, this volume) explored this 24-hour cycle of news and information diffusion. In this vastly fragmented media environment and 24-hour news cycle, the news media's agenda-setting role (McCombs & Shaw, 1972) remains unchanged. As the news and information dissemination function is still dominated by the mainstream electronic media's offline and online channels, this diffusion process now takes place at a rapid pace with an infinite volume of information interlinked from countless diversified sources.

Aside from the concern about a knowledge gap that exists between the information haves and have-nots, a different kind of gap—one that is defined by information-access choice instead of information-access capability—is rising to widen the gap between what different segments of the public may or may not know. This is because, as the amount of news and information becomes available to the public 24-7 across a vast number of channels, the users' available time that can be devoted to accessing these channels remains fixed. To limit their information and entertainment media consumption at a manageable level, users often develop both an off- and online "channel repertoire" (Heeter, 1985; Perse & Courtright, 2004), which contains a number of channels or web sites that are regularly accessed. In doing so, the function of media displacement may come into play.

The concept of displacement functions can be traced back to Himmelweit, Oppenheim, and Vince's (1958) three displacement principles—displacement due to marginal activities, functional equivalence, or transformed activities. Based on these principles, users may, for instance, enact time displacement by reducing the amount of time spent with one medium (e.g., broadcast TV news) and favoring another medium that offers more time flexibility (e.g., cable TV news); they may also engage in functional displacement by replacing the function performed by one medium (e.g., newspapers) by another medium that can provide additional value-added functions (e.g., online news sites; Kayany & Yelsma, 2000).

These types of media displacement activities in turn help generate intense competition among various 24-hour news and information channels that are vying for user ratings and advertiser support. Intense market competition has thus prompted these channels to present hasty coverage or even irresponsible reporting to attract users, in addition to filling the spectrum with divisive or radical political talks. In other words, the information abundance that is the fruit of channel abundance also can help create the phenomenon of knowledge fragmentation. According to that dynamic, a web-based one-man gossip-news operation such as *The Drudge Report* or political editorialists on a 24-hour cable TV news channel such as Bill O'Reilly could become a key or even the sole news source for a large segment of the public. As the Bucy et al. chapter notes, the rise of this "personal me" media focus may soon erode some of the common cultural reference points, and resultant social capital (Putnam, 1995), that enable Americans to cohere as a nation due to open and civil social and political discourse.

ENTERTAINMENT SETTING

The growth of 24-hour national news and community news/event channels is not an isolated phenomenon, as these channels are usually being delivered as part of the rich-media offerings that are embedded in a cable TV system, satellite TV service, and/or the Internet environment. The latest (and perhaps last) major entrant to the digital media world is digital TV broadcasting, which joins digital compression technology applications in nearly every other type of electronic media (including cable, satellite, and terrestrial radio broadcast channels) and mediated-communication modality (including wired/wireless telephony, two-way radio communication, and the Internet). The Federal Communications Commission (FCC) has slated the launch of this much-delayed digital media service for early 2009. When this technology service becomes available, local terrestrial broadcast TV stations—similar to their radio counterparts—will each be able to deliver up to six or eight channels of digital signals.

Grant (chap. 9, this volume) surveys the new and emerging entertainment media and the roles these entertainment media play in our cultural and social lives. As available leisure pursuits and entertainment media spending are derived from a fixed amount of "free" time and disposable income, respectively, any acquisition of new media-use activity or technology units will then "take" from this fixed pool of resources based on the principle of relative constancy (McCombs & Nolan, 1992). This suggests that certain media displacement or substitution dynamics will need to take place to balance the media-use "portfolio," so to speak. The process of determining whether to increase or reduce the amount of time spent with each media content outlet—or whether to acquire a new technology product—is often related to user perceptions of the functions of each of these media content outlets and technology units.

In essence, users will make a conscious judgment with regard to the functional similarities and dissimilarities—as well as the relative advantage and technical complexity—of displacing, supplementing, or complementing existing media-content outlets and technology choices with alternative options. This selection process is also related to an adoption process that reflects the concept of technology clusters (Rogers, 2003), whereby technology units that share similar technical functions tend to be adopted before those that do not, to ease the transition for displacement, supplementation, or complementation. Hence, old media technologies do not die out unless they have truly outlived their technical advantages. For instance, relative to the DVD players that have displaced a portion of the prerecorded video market, VCRs are still widely used for time-shifting purposes due to their recording and playback function—something that the DVD players cannot easily duplicate. By comparison, a satellite TV service shares nearly identical technical capabilities in delivering the same type of media content and services to users as cable TV, so these media are thus highly substitutable as well.

All of these new media content and delivery modalities, including the impending arrival of digital TV broadcasting, compete against each other for the users' adoption and patronage. The best predictor for keeping users interested in continued adoption and repeat use remains the degree to which users' media and technology use needs are satisfied. This explanation parallels the theoretical assumption of the uses and gratifications perspective (Katz, Blumler, & Gurevitch, 1974), which asserts that users are motivated to select, use, and reuse specific media content and modalities to satisfy a set of cognitive and affective needs. These motivational factors typically include (a) cognitive needs such as news surveillance, information seeking, and knowledge learning; (b) affective needs such as escape, passing time, and stimulation; and (c) socializing needs such as companionship, parasocial communication, or social identity seeking.

Although hundreds of TV channels on cable and satellite TV systems—as well as numerous prerecorded video entertainment choices—are available to users, users tend to sample only a small percentage of these content offerings (Jeffres, Atkin, Neuendorf, & Lin, 2004). Hence, the concept of channel repertoire already discussed in the news/surveillance setting is also applicable here, in that the media abundance and content diversity also can further audience and knowledge fragmentation. For instance, aside from the Superbowl and the season- or show-ending finale of selected TV programs such as *The American Idol* or *Seinfeld*, no one media outlet is capable of capturing a sizable audience in today's media market.

The entry of online media has helped hasten this audience fragmentation trend, as it brought about a seemingly limitless number of information and entertainment sources from around the world. Although the Internet represents a medium of information diversity, one that facilitates user access to every imaginable subject matter and topic area, it is first and foremost a medium of technology convergence (e.g., Atkin, 2002; Klopfenstein, 2002). The Internet allows two-way real-time transmission and reception of text, data, graphics, photo images, animation, and audio and video streams from any locations that have a simple telephone connection or a sophisticated dedicated wireless communication network. According to the theory of technology fluidity (Lin, 2004), the most striking characteristic of the Internet is its fluid nature, which enables its users to have a dynamic multifunctional, multiplatform, and multitasking interface and interactive communication experience.

As a medium of information diversity and communication that is not bound by time, distance, and spectrum space, the Internet has the capacity to deliver information and entertainment content that can meet the widest possible range of user interests, in addition to linking any Internet user who wishes to engage in a one-to-one, one-to-many, many-to-one, or many-to-many mode of communication. Although the information and entertainment content made available on the Internet tends to emulate its offline counterparts, social institutions ranging from governments and businesses to local schools as well as individual members of society have all become an integral part of the Internet universe through their participation in contributing information to and communicating with each other through the Internet.

After a little over a decade of opening up the Internet for public consumption, scholars have only just begun to understand the cultural and social influences that the Internet can bring about to our society. Considering the number of e-mail messages exchanged, the amount of data transferred, the number of newsgroup/chatroom messages posted, the volume of blogs published, and the streams of audiocasts and videos uploaded/download via the Internet on a daily basis, it is easy to see how the Internet has been become a medium that is intimately woven into our work as well as daily life routine. Research that attempted to explain how and why we use the Internet, what we do with the Internet, and what we get from using the Internet has been largely successful in adapting the dominant media diffusion and uses theories in existence.

For instance, research applying the uses and gratifications theory was able to establish that the Internet-use motivations are similar to those of traditional offline media use, but motivations for accessing online media are not predictive of TV use and vice versa (Lin, 1999). This suggests that users perceive a difference between what gratifications they may expect to receive between online and offline media. As such, the Internet news and entertainment sources have been found to facilitate varying degrees of media substitution functions across functionally similar technologies or contents (e.g., Lin, 2001; Waldfogel, 2002). Similar to the phenomenon where users develop a "channel repertoire" containing a limited number of media consumption options, Internet users also select a finite repertoire of web sites for consumption (Perse & Courtwright, 2004), drawing from an infinitely large number of content options. By comparison, the Internet's role in shaping the aforementioned knowledge fragmentation is relatively stronger than that of offline video entertainment media, because the users can customize or create the Internet content for their own individual as well as like-minded individuals' consumption via chatrooms, newsgroups, blogs, podcasts, and so on.

Given the private and easy nature of accessing the Internet and its plentiful supply of stimulation to occupy one's time and mind, some users have developed a certain dependency on the Internet to help moderate their dysphoric moods (e.g., depression) or satisfy their specific obsession (e.g., gambling). Over time, when users suffer from deficient self-regulation of obsessive Internet-use patterns, their Internet-use activity can evolve into a compulsion or even an addiction behavior (LaRose, Lin, & Eastin, 2004), one that is similar in scope to compulsive or addictive teleshopping behavior. Hence, although the Internet as—an entertainment source—provides numerous choices for recreation, escape, companionship, passing time, and parasocial interaction, it also presents potential downsides, like unregulated and addictive use of the Internet.

CONSUMER SETTING

Although media consumption (whether it be off- or online) is usually tied to advertising and marketing message delivery, consumer behavior theory suggests that consumers perceive the value of their shopping behavior as being utilitarian (or more task-oriented) and/or hedonistic (or more play-oriented; Babin, Darden, & Griffin, 1994). As a consumer's shopping experience may contain both utilitarian and hedonistic values, teleshopping transactions (or shopping directly via an electronic channel)—which may or may not involve interpersonal interactions—may encompass even a broader set of cognitive and affective responses. For instance, in applications of uses and gratifications theory (Katz et al., 1974), motivational factors such as information, conversation, and social escapism (Korangaonkar & Wolin, 1999), as well as diversion, role playing, social interaction, and recreational functions (Salomon & Koppelman, 1992), were all said to help motivate online shopping. Parasocial relations (between the shopper and the salesperson/shopping show host) represent another factor that can help motivate repeat purchases from teleshopping consumers (Skumanich & Kintsfather, 1998).

As teleshopping continues to grow with its online shopping sector, the social change implications are also on the rise. On the one hand, teleshopping allows consumers to make purchases in a private, in-home, and instantaneous environment. On the other hand, this type of shopping activity also raises concerns about the potential for consumers to develop impulsive, followed by compulsive and then, in the worst-case scenario, addictive shopping behavior. This downward spiral of obsession often starts with consumer dependency on the source of their purchase activity to help fulfill certain psychological needs such as parasocial interaction (Grant, Guthrie, & Ball-Rokeach, 1991). When this dependency parallels a lack of will power to resist the shopping stimuli, due to one's desire to experience instant gratification, then impulsive and compulsive shopping behavior can ensue (Hoch & Loewenstein, 1991). Once this behavior starts to harm the consumer's psychological well-being, relations with others, and financial health, and if the consumer is incapable of controlling or changing this behavior, it can grow into a clinically defined shopping addiction (Faber, O'Guinn, & Krych, 1987).

Interestingly, although the two-way interactive communication technologies utilized in teleshopping activity can provide a venue for gratifying consumer psychological needs and a source for creating damaging psychological addiction, they can also provide the tools for treating such psychological disorders and other health conditions. The practice of telemedicine, which includes the e-health (online pharmacies and online clinics) and telehealth market segments, is a growing industry. This market growth is a result of escalating health care costs, rising demands for universal health care access, and the need to serve a population that is rapidly aging and vastly underserved, groups that are often economically disadvantaged and/or have special needs (e.g., needing home care, impaired mobility, mentally depressed, lacking transportation means, residing in a remote area, etc.).

Using communication technologies to help deliver disease prevention messages, disease intervention programs, and health care services is reflective of a larger digitization trend in social and health marketing. For instance, the U.S. government is currently pushing the conversion of patient records from paper to electronic modalities to facilitate dynamic data retrieval and portability, thereby improving patient care, billing efficiency, and insurance-program transfer. Although a typical telemedicine delivery involves regular medical personnel communicating with the patients at a remote location via two-way audiovisual channels, the communication dynamics between the patients and the health care professionals and user acceptance/use of technologies remain the most important aspects of this transaction.

Whitten (chap. 12, this volume) examined the phenomenon of telemedicine communication. According to Whitten, structuration theory (Giddens, 1981), when applied to the telemedicine setting, treats the telemedicine system as an independent social system and the health care providers/patients as a separate social entity. The integration between these two separate social units occurs when the human players utilize the rules, resources, and social relations governing the social system to generate adequate social-action dynamics, particularly to produce and reproduce the social functions intended for the social system (or telemedicine service).

Within the telemedicine system, technology acceptance and usage are two important factors that enable successful functioning and marketing of a telemedicine service. Research has examined these technology adoption factors via the technology acceptance model (TAM; Davis, 1989), which focuses on examining users' (e.g., health-care providers and patients) perceptions about the technology system's usefulness and ease of use as the basis for predicting users' attitude and intention to adopt the technology system. By implication, delivering health messages, programs, products, and services through technology channels requires the health care providers, the health care systems (including the insurance programs), and the patients (or end users) to recognize and believe in the relative advantage and ease of using technologies to make the telemedicine system function effectively and efficiently.

Telemedicine got its start by providing telecare to the underserved population—who usually reside in rural areas or are not physically mobile—by removing the barriers of time, costs, and distance required for travel. The practice of telehealth has thus come a long way. For instance, there is a growing elderly or chronically ill population that utilizes the online clinics to obtain medical prescriptions and then order their prescription drugs through online pharmacies operated outside the United States (e.g., Canada) due to economic reasons. Although satellite and telephony technologies have long been the staples for connecting providers and patients in the delivery of telehealth care, computer conferencing involving live video streams, and wireless technologies involving cellular phones have also become the technologies of choice.

Cellular phones, in particular, have been outfitted with different types of medical devices to record, store, and transmit such patient data as blood pressure or glucose level to patients' physicians' network servers through either a wireless or a wired Internet connection. Handheld UPC code scanners, integrated with cellular phone devices, have also been utilized to record, monitor, and store patients' food consumption and safety data to help provide dietetic education, implement dietary interventions, and treat diet-induced illness. As more innovative use of new communication technologies is integrated into delivering telemedicine services at a large scale, the potential for providing cost-effective care to all segments of our population—including the underserved and uninsured, to improve health—will also be pragmatic and reachable. Desirable as it may be, the development of telemedicine has not kept in step with the progress of technology innovations. The primary reason for this application gap is related to electronic and digital information privacy and security concerns. In fact, the information privacy and security concerns are neither new nor recent. These concerns carry social implications that go beyond telemedicine delivery and affect other aspects of our lives in an information society, as discussed in the next section.

LEGAL AND REGULATORY SETTING

The issue of who should have access to use various information contents, and at what cost, has long generated social controversy. This debate has been exacerbated by the introduction of new technologies that can copy, store, retrieve, and transmit information content. Lipschultz (chap. 13, this volume) focuses, in particular, on digital access issues concerning music, film, and other intellectual electronic property rights. The Copyright Act of 1976, the statutory basis for protecting copyright holders' intellectual rights, contains a provision (Section 107) commonly known as the "fair use doctrine," which allows an exception to an otherwise comprehensive set of safeguards. The principle behind the fair use doctrine is that the public has the right to reproduce copyrighted work—as long as the use of such work is of noncommercial nature or educational purposes and such use has no impact on the commercial value of such work.

In the first landmark case addressing the fair use and copyright infringement of an electronic medium, the Sony Betamax case (*Sony Corp of America v. Universal City Studios*, 1984), the U.S. Supreme court ruled that individual user actions to tape copyrighted content via a videocassette recorder for private, noncommercial were was considered fair use and thus protected by the 1976 Copyright Act (see e.g., Lin, 1985). Fast forward to the late 1990s, where advanced development in digital technology has enabled the copying and sharing of media content much easier and more prevalent. Individual users now can easily download copyrighted materials from the Internet—whether they be text, graphics, images, music, audiocasts, or videocasts—and place them on a digital storage device (e.g., a disk or an MP-3 device). From there, users can share copied materials electronically via the Internet with a limitless number of Internet users. Vigilant policing of the online file-sharing activities of a vast number of individual users, either by the copyright holders or the government, can be difficult. In particular, because the Internet universe has no centralized authority for exercising such surveillance functions, legislative solutions have emerged and are still evolving.

Recognizing that digital copyright protection is an issue that knows no national boundaries, the Digital Millennium Copyright Act of 1998 represents the first U.S. legislative attempt to help implement the World Intellectual Property Organization's copyright protection treaties. Although the fair use doctrine remains intact, the two best-known test cases for this act were indicative of the new limitations for this long taken-for-granted doctrine. In *A&M Records, Inc. v. Napster, Inc.* (2001) and *MGM v. Grokster* (2005), the courts ruled that the use of peer-to-peer sharing software does not incur copyright infringement liability, but brokering the online digital file sharing to help facilitate widespread distribution of illegally copied materials does.

It appears that as information in all forms is increasingly becoming digitized, the transmission of that digital output has been subject to ever-more comprehensive copyright restrictions. As a result, user access to information may, ironically, become more restricted as we enter a more advanced stage of the information age. All told, the fundamental issues at stake here are twofold, encompassing (a) the commercial value of copyrighted materials and (b) under what circumstances noncommercial and private peer-to-peer file sharing can be considered fair use.

With regard to the commercial value of copyrighted materials such as music recording, major record companies are usually a part of a media conglomerate that can set the market price for digital music or video sales without any real concern for market competition. Hence, allowing the market to set the price for such music or video file sales does not necessarily protect the public from undue price burdens. In terms of the circumstances under which peer-to-peer file sharing should be considered legal, if using the file sharing software is not a violation of copyright law, then individual users should be allowed to acquire this software and set up their private noncommercial limited peer-file sharing between friends and families. This is because electronic sharing of digital files is being done on a daily basis with text, graphic, and other types of files for noncommercial purposes.

As pointed out by Lessig (2005), even the electronic equivalent of loaning a book may run afoul of the law, given that every digital technology use creates a copy. The legal issues that challenged the fair use doctrine with cases involving Napster and Grokster signal the emergence and evolution of additional digital copyright debates, as current legal thinking seems utterly incapable of articulating a set of principles that could balance the public's right to know versus the copyright holder's right to profit. It remains to be seen whether this legal pendulum may swing towards protecting the public's fair use right by not overrewarding the copyright holder's commercial interests, which seems to favor mega-media companies that strive to profit from a self-serving, noncompetitive pricing structure.

As the intellectual property piracy issue is rooted in its market value, so too is the information privacy issue, where an individual citizen's private information often becomes a commercial commodity. Lee (chap. 14, this volume) introduces the issue of information privacy by reviewing the evolution of our cultural and social perceptions of what privacy means in today's digital information age. According to Lee's summary, the concept of privacy is complex and multidimensional (Gavison, 1980; Derlega & Chaikin, 1975; Margulis, 1977; Pedersen, 1979; Tribe, 1988; Warren & Brandeis, 1890). Depending on the tradition of the theoretical perspective, the concept of privacy can be seen as an individual's right to preserve the freedom of personal movement, communication, or information within the confines of a specific zone of physical space—or virtual space—without allowing external parties to reap benefits or profits from intruding on (or interfering with) such freedom.

In the digital age, intrusion on an individual's information and communication privacy has become an alarming social problem, due to the far-reaching financial harms that it could inflict on the massive number of individuals in society. For instance, although most companies monitor employee e-mail and computer use at the workplace, other commonly known surveillance phenomena could involve using hidden spycams or a global positioning system to monitor individuals' movement or whereabouts in public places (e.g., street corners, stores, moving vehicles, etc.). Personal purchase activity tracking or data mining via spyware and other logging software (e.g., cookies or radiofrequency identification tags) for building marketing databases, along with clogging individual e-mail boxes with unsolicited spam mail, are prime examples of commercial prying. Digital data hacking for the purpose of phishing for individuals' personal identification information to accomplish identity theft is the ultimate act of privacy invasion at anothers' expense.

Unfortunately, aside from the Fourth Amendment—which provides a broad principle of prohibiting unwarranted government searches and seizures on private citizens—there remains a lack of explicit constitutional protection of individual privacy rights. Hence, even with a set of federal laws in place—ranging from shielding individual citizens from being illegally wiretapped to having their video rental records tracked—many have suffered large financial losses, and endured layers of difficulty in reestablishing personal identities stolen by others. The implementation and the renewal of the 2001 USA PATRIOT Act ("Uniting and Strengthening," 2001), which greatly expands government authority to electronically gather individual citizens' personal data and monitor their communication activities, have continued to irk the sensitivities of civil libertarians as well as citizens hoping to protect individual rights to privacy.

As with copyright piracy, the current federal laws do not adequately address the already devastating statistics of privacy violations on a daily basis in society. Such inadequacy in legal protections of the citizens' rights to privacy then will continue to stir up additional debate on how best to deter the overwhelming tide of privacy thefts. The task of striking a balance between opposing interests in the case of personal digital privacy will become another important challenge in the digital age frontier. The essential question remains over whether a dystopian prophecy, which foretells a dark future where the new technologies facilitate oppressive governmental surveillance and thought control (e.g., Orwell, 1949), will materialize alongside the continuing growth of institutionally supported spy technologies.

SYNTHESIS: A SOCIAL CHANGE TYPOLOGY

As Lievrouw and Livingstone (2002) note, "The social contexts and uses of the new media are as important as the technologies themselves" (p. 11). By integrating the social change implications synthesized from the different social settings described earlier, a social change typology is proposed here to broadly capture how communication and information technologies function to help shape the pros and cons of social changes in society.

A. Surveillance Function

- Pros: Enable individuals to meet their self-preservation needs and to exercise control over their environment via 24/7 surveillance tools.
- Cons: Widen the information gap and challenge information-privacy boundaries between those who are surveillance-minded and those who are not.

B. Knowledge Function

- Pros: Enable individuals to meet their cognitive stimulation and intellectual growth needs with abundant and diversified knowledge resources, which help develop information workers for a knowledge-based economy.
- Cons: Broaden the knowledge gap and increase knowledge fragmentation between individuals who have greater substantive breadth and depth in their channel repertoire and those who do not.

C. Communication Function

- Pros: Enable individuals, groups, and institutions to meet their social interaction, social support, and/or organizational communication needs through a web of borderless open-architecture networks.
- Cons: Disinhibit individuals from engaging in precarious social interaction and create a venue for individuals to become obsessed with or addicted to pseudosocial and parasocial interactions.

D. Entertainment Function

- Pros: Enable individuals to meet their affective release and stimulation needs by pursuing an abundant and diversified set of leisure modalities in a private and home setting.
- Cons: Further the audience, information and knowledge fragmentation as well as create venues for individuals to become obsessed with or addicted to such entertainment fare as videogaming, gambling, and pornography.

E. Commerce Function

- Pros: Provide outlets for individuals to meet their utilitarian and hedonistic consumption needs as well as create offline and online marketing synergy.
- Cons: Challenge the intellectual-property piracy boundaries and create venues for individuals to become obsessed with or addicted to compulsive consumption.

CONCLUSION

Echoing the work of Bell (1976, 1980), Porat (1977), and others, Toffler and Toffler (2006) note that information and communication—the capital goods of the knowledge economy—carry the potential to fundamentally alter our political, social, and economic relations in an age marked as the Third Wave. In the Tofflers' lexicon, the Third Wave is characterized by a postindustrial emphasis on networks rather than hierarchies (see Rice & Schneider, chap. 6, this volume) along with de-massification. The latter is observed in the media by several of our contributors (e.g., Bucy et al., chap. 8. this volume), but the Tofflers note that domains ranging from education to production will be tailored to specific users. They note, in particular, that information is a nondiminishing resource; that is, although industrial "rival goods" like steel are used up when consumed, information is a nonrival good that can remain with senders and the receiver:

> You and a million other people can use the same chunk of knowledge without diminishing it. In fact, the greater the number of people who use it, the greater the likelihood that someone will generate more knowledge with it. (Juskalian, 2006, p. 5)

This digital revolution will continue and cannot be muted by any individuals, capitalists, or governments. Technology-future pessimists present the dystopian visions of the future that foretell dehumanization via technology, power inequalities, devaluation of human labor, losses in privacy (via omnipresent surveillance cameras), and even the rise of totalitarian government (e.g., Gandy, 1993; Orwell, 1949; Schiller, 1981; Stoll, 1995; Rosen, 2000). Technology optimists, on the other hand (e.g., Gates, 1995; Rogers, 2002) are more sanguine about the prospects for social change in the information age, one that is defined by several "megatrends" (Naisbett, & Aurburdene, 2001) that generally presage peace, higher standards of living, and more empowerment of the individual via communication and information technology.

Unlike other man-made resources that we can modify or maneuver, communication and information technologies are at our disposal to help us achieve the social development goals that we desire. As with any innovations, the implementation of innovation adoption may result in the following phases: adaptation, acceptance, routinization, and reinvention (Lin, 2002). Within each of these phases, individuals, groups, and institutions may initiate a different adaptation approach—one that embraces a different acceptance capacity, set of implementation routines, and set of inventive applications—to best accommodate and satisfy their technology-use aims. Hence, neither the utopian nor the dystopian predictions about what social changes lie ahead should blind or misguide our perceptions, beliefs, and attitudes about the relations between technology and society. As the fluid nature of technology applications helps define the hallmark technological development, the social changes that communication and information technologies portend will also be fluid in nature.

REFERENCES

A&M Records, Inc. v. Napster, Inc., 239 F.3d 1004 (9th Cir. 2001).

Anderson, C. A. (2004). An update on the effects of playing violent video games. *Journal of Adolescence, 27*(1), 113–122.

Anderson, C. A., & Bushman, B. J. (2001). Effects of violent video games on aggressive behavior, aggressive cognition, aggressive affect, physiological arousal, and prosocial behavior: A meta-analytic review of the scientific literature. *Psychological Science, 12*(5), 353–359.

Anokwa, K., Lin, C., & Salwen, M. (2003). *Mass media around the globe*. New York: Wadsworth.

Atkin, D. (2002). Convergence across media. In C. A. Lin & D. Atkin (Eds.), *Communication technology and society: Audience adoption and uses* (pp. 23–42). Cresskill, NJ: Hampton.

Atkin, D. (2003). The Americanization of global film. In K. Anowkwa, C. Lin, & M. B. Salwen (Eds.), *International communication: Concepts and cases* (pp. 175–189). Belmont, CA: Wadsworth.

Babin, B. J., Darden, W. R., & Griffin, M. (1994). Work and/or fun: Measuring hedonic and utilitarian shopping value. *Journal of Consumer Research, 20*, 644–656.

Baker, P. M. A., & Ward, A. C. (2002). Bridging temporal and spatial "gaps": The role of information and communication technologies in defining communities. *Information, Communication & Society, 5*, 207–224.

Bandura, A. (1994). Social cognitive theory of mass communication. In J. Bryant & D.Zillmann (Eds.), *Media effects: Advances in theory and research* (pp. 61–90). Hillsdale, NJ: Lawrence Erlbaum Associates.

Bell, D. (1976). *The coming post-industrial society*. New York: Basic Books.

Bell, D. (1980). *The winding passage: Essays and sociological journeys, 1960–1980*. Cambridge, MA: Abt Books.

Boynton, A., Zmud, R., & Jacobs, G. (1994). The influence of IT management practice in IT use in large organizations. *Management Information Systems Quarterly, 18*(3), 229–318.

Bracken, C. C., & Lombard, M. (2004). Social presence and children: Praise, intrinsic motivation, and learning with computers. *Journal of Communication, 54,* 22–37.

Bucy, E. P., & Newhagen, J. E. (2004). *Media access*. Hillsdale, NJ: Lawrence Erlbaum Associates.

Choie, H., & Cheon, H. J. (2005). Children's exposure to negative Internet content: Effects of family context. *Journal of Broadcasting and Electronic Media, 49*, 488–510.

Cohen, W. M., & Levinthal, D. A. (1990). Absorptive capacity: A new perspective on learning and innovation. *Administrative Science Quarterly, 35*, 128–152.

Davis, F. D. (1989). Perceived usefulness, perceived ease of use, and user acceptance of information technology. *MIS Quarterly, 13*(3), 319–340.

Davis, T. R. V. (1984). The influence of the physical environment in offices. *Academy of Management Review, 9*(2), 271–283.

Derlega, V. J., & Chaikin, A. L. (1977). Privacy and self-disclosure in social relationships. *Journal of Social Issues, 33*(3), 102–115.

Deriega, V.J., & Chaikin, A.L. (1977). *Sharing intimacy: What we reveal to others and why*. New York: Prentice Hall.

Digital Millennium Copyright Act, H.R. 2281 (105th Cong., 2d Sess. 1998).

Dizard, W. (2000). *Old media, new media*. New York: Longman.

Dutta-Bergman, M. J. (2005). Access to the Internet in the context of community participation and community satisfaction. *New Media & Society*, *7*, 89–109.

Faber, R. J., O'Guinn, T. C., & Krych, R. (1987). Compulsive consumption. In M. Wallendorf & P. Anderson (Eds.), *Advances in consumer research* (Vol. 14, pp. 132–135). Provo, UT: Association for Consumer Research.

Feldman, M. S., & March, J. G. (1981). Information in organizations as signal and symbol. *Administrative Science Quarterly, 26*, 171–186.

Finfgeld, D. L. (2000). Therapeutic groups online: The good, the bad, and the unknown.*Issues in Mental Health Nursing, 21*(3), 241–255.

Gavison, R. (1980). Privacy and the limits of law. *Yale Law Journal, 89*, 421–461.

Gandy, O.H. (1993). *The panoptic sort: A political economy of personal information*. New York: Perseus.

Gates, W. (1995). *The road ahead*. New York: Viking Press.

Gibson, J. (1979). *The ecological approach to visual perception*. Boston: Houghton-Mifflin.

Giddens, C. A. (1981). *A contemporary critique of historical materialism*. Berkeley: University of California Press.

Goodman, P., Griffith, T., & Fenner, D. (1990). Understanding technology and the individual in an organizational context. In P. Goodman, L. Sproull, & Associates (Eds.), *Technology and organizations* (p. 45–86). San Francisco: Jossey-Bass.

Grant, A., & Meadows, J. (2004). *Communication technology update*. Boston: Focal.

Grant, A. E., Guthrie, K., & Ball-Rokeach, S. J. (1991). Television shopping: A media system dependency perspective. *Communication Research, 18*(6), 773–798.

Greenfield, P. M., De Winstanley, P., & Kilpatrick, H. (1994). Action video games and informational education: Effects on strategies for dividing visual attention: Special Issue. Effects of interactive entertainment technologies on development. *Journal of Applied Developmental Psychology, 15*, 105–123.

Hartson, R. (2003). Cognitive, physical, sensory, and functional affordances in interaction design. *Behaviour & Information Technology, 22*(5), 315–338.

Heeter, C. (1985). Program selection with abundance of choice: A process model. *Human* Communication Research, 12, 126–152.

Hiebert, R. E. (2003). Public relations and propaganda in framing the Iraq war: A preliminary review. *Public Relations Review, 29*, 243–255.

Himmelweit, H. T., Oppenheim, A. N., & Vince, P. (1958). *Television and the child: An empirical study of the effects of television on the young*. London: Oxford University Press.

Hoch, S. J., & Lowenstein, G. F. (1991). Time-inconsistent preferences and consumer self-control. *Journal of Consumer Research, 17*, 492–507.

Jeffres, L. (2003). *Urban communication systems*. Cresskill, NJ: Hampton.

Jeffres, L. W. (2002). *Urban communication systems: Neighborhoods and the search for* community. Cresskill, NJ: Hampton.

Jeffres, L. W., Atkin, D. J., Neuendorf, K. A., & Lin, C. A. (2004). The influence of expanding media menus on audience content selection. *Telematics & Informatics, 21*(4), 317–334.

Johnson, B. M., & Rice, R. E. (1987). *Managing organizational innovation*. New York: Columbia University Press.

Juskalian, R. (2006, May 15). Knowledge drives future, creates wealth, authors say. *USA* Today, p. 5b.

Katz, E., Blumler, J. B., & Gurevitch, M. (1974). Utilization of mass communication by the individual. In J. G. Blumler & E. Katz (Eds.), *The uses of mass communication* (pp. 19–32).

Kayany, J. M., & Yelsma, P. (2000). Displacement effects of online media in the socio-technical contexts of households. *Journal of Broadcasting & Electronic Media, 44*, 215–229.

Kiesler, S. (Ed.). (1997). *Culture of the Internet*. Hillsdale, NJ: Lawrence Erlbaum Associates.

Klopfenstein, B. (2002). The Internet and web as communication media. In C. A. Lin & D. Atkin (Eds.), *Communication technology and society: Audience adoption and uses* (pp. pp. 353–378). Cresskill, NJ: Hampton.

Korgaonkar, P. K., & Wolin, L. D. (1999). A multivariate analysis of web usage. *Journal of Advertising Research, 39*(2), 53–68.

Kraut, R., Galegher, J., Fish, R., & Chalfonte, B. (1992). Task requirements and media choice in collaborative writing. *Human-Computer Interaction, 7*, 375–407.

Kraut, R., Lundmark, V., Patterson, M., Mukopadhyay, T., et al. (1998). Internet paradox: A social technology that reduces social involvement and psychological well-being? *American Psychologist, 53*(9), 1017–1031.

Kraut, R., Patterson, M., Lundmark, V., Kiesler, S., Mukopadhyay, T., & Scherlis, W. (1998). Internet paradox: A social technology that reduces social involvement and psychological well-being? *American Psychologist, 53*(9), 1017–1031.

Kundanis, R. M. (2003). *Children, teens, families, and mass media: The millennial generation.* Mahwah, NJ: Lawrence Erlbaum Associates.

LaRose, R., Lin, C. A., & Eastin, M. S. (2003). Unregulated Internet usage: Addiction, habit or deficient self-regulation? *Media Psychology, 5*, 225–253.

Lea, M., & Spears, R. (1992). Paralanguage and social perception in computer-mediated communication. *Journal of Organizational Computing, 2*, 321–341.

Lessig, L. (2005, June). The people own ideas. *Technology Review, 108*(6), 46–53.

Lievrouw, L. A., & Livingstone, S. (Eds.). (2002). *Handbook of new media: Social shaping and* consequences of ICTs. London: Sage.

Lin, C. (1985). Copyright issues for home videotaping. *Telecommunications Policy, 9*, 334–350.

Lin, C. A. (1998). Exploring personal computer adoption dynamics. *Journal of Broadcasting & Electronic Media, 42*, 95–112.

Lin, C.A. (1999). Online-service adoption likelihood. *Journal of Advertising Research, 39*, 79–89.

Lin, C. A. (2001). Audience attributes, media supplementation, and likely online service adoption. *Mass Communication & Society, 4*, 19–38.

Lin, C. A. (2002). A paradigm for communication and information technology. In C. A. Lin & D. Atkin (Eds.), *Communication technology and society: Audience adoption and uses* (pp. 447–476). Cresskill, NJ: Hampton.

Lin, C. A. (2004). Webcasting adoption: Technology fluidity, user innovativeness, and media substitution. *Journal of Broadcasting & Electronic Media, 48*(3), 446–465.

Lin, C. A., & Atkin, D. (2002). *Communication technology and social change: Audience* adoption and uses. Cresskill, NJ: Hampton.

Lombard, M., & Ditton, T. V. (1997). At the heart of it all: The concept of presence. *Journal of Computer Mediated Communication*, 3(2). Accessed September 15, 2005, from http:jcmc.huji.ac.il/vol3/issue2/lombard.html.

Majchrzak, A., Rice, R. E., Malhotra, A., King, N., & B. S. (2000). Technology adaptation: The case of a computer-supported inter-organizational virtual team. *MIS Quarterly, 24*(4), 569–600.

Margulis, S. T. (1977). Conceptions of privacy: Current status and next steps. *Journal of Social Issues, 33*(3), 5–21.

Matei, S., & Ball-Rokeach, S. (2003). The Internet in the communication infrastructure of urban residential communities: Macro- or mesolinkage? *Journal of Communication, 53*, 642–657.

McChesney, R.W. (2004). *The problem of the media*. New York: Monthly Review Press.

McCombs, M., & Nolan, J. (1992). The relative constancy approach to consumer spending for media. *Journal of Media Economics, 5*(2), 43–52.

McCombs, M., & Shaw, M. (1972). The agenda-setting function of mass media. *Public* Opinion Quarterly, 36, 169–174.

MGM v. Grokster (2005, June 27), 125 S.Ct. 2764.

Naisbett, J., & Auburdene, P. (2000). *Megatrends 2000*. New York: William Morrov and Sons.

Noam, E., Groebel, J., & Gerberg, D. (2004). *Internet television*. Boston: Allyn & Bacon. Orwell, G. (1949). *1984*. New York: Harper & Row.

Owen, J. E., Yarbrough, E. J., Varga, A., & Tucker, D. (2003). Investigation of the effects of gender and preparation on quality of communication in Internet support groups. *Computers in Human Behavior, 19*(3), 259–275.

Pedersen, D. M. (1979). Dimensions of privacy. *Perceptual and Motor Skills, 48*, 1291–1297.

Pelton, J. (1996). *Wireless and satellite telecommunication*. Upper Saddle River, NJ: Prentice-Hall.

Pelton, J. (2003). The changing shape of global telecommunications. In K. Anowkwa, C. Lin, & M. B. Salwen (Eds.), *International communication: Concepts and cases* (pp.267–284), Wadsworth.

Perse, E., & Courtright, J. (2004, November). *Functional images of communication channels: Mass and interpersonal alternatives in a fragmented media environment*. Paper presented at the National Communication Association annual conference, Chicago.

Perse, E., & Dunn, D. (1998). The utility of home computers and media use: Implications of multimedia and connectivity. *Journal of Broadcasting & Electronic Media, 42*, 435–456.

Pigg, K. E. (2001). Applications of community informants for building community and enhancing civic society. *Information, Communication & Society, 4*, 507–527.

Poole, I. (1983). *Technologies of freedom: On free speech in an electronic age*. Cambridge, MA: Harvard Press.

Porat, M. (1977). *The information economy: Definition and measurement* (Vol. 1, pp. 15–21). Washington, DC: Office of Special Publications.

Postman, N. (1985). *Amusing ourselves to death: Public discourse in the age of show business*. New York: Viking.

Putnam, R. (1995). *Bowling alone*. New York: Routledge.

Reicher, S. D., Spears, R., & Postmes, T. (1995). Effects of public and private self-awareness on deindividuation and aggression. *Journal of Personality and Social Psychology, 43*, 503–513.

Rice, R. E. (1987). Computer-mediated communication and organizational innovation. *Journal of Communication, 37*(4), 65–94.

Rice, R. E. (1999). What's new about new media? Artifacts and paradoxes. *New Media and Society, 1*(1), 24–32.

Rice, R. E., & Gattiker, U. (2000). New media and organizational structuring. In F. Jablin & L. Putnam (Eds.), *New handbook of organizational communication* (pp. 544–581). Newbury Park, CA: Sage.

Robertson, R. (1995). Glocalization: Time-space homogeneity-heterogeneity. In M. Featherstone, S. Lash, & R. Robertson (Eds.), *Global modernities* (pp. 27–77). London: Sage.

Rogers, E. M. (2002). The information society in the new millennium: Captain's log, 2001. In C. A. Lin & D. Atkin (Eds.), *Communication technology and society: Audience adoption and uses* (pp. 43–64). Cresskill, NJ: Hampton.

Rogers, E. M. (2003). *The diffusion of innovations.* New York: The Free Press.

Rosen, J. (2000). *The destruction of privacy in America.* New York: Vintage Books.

Salomon, I., & Koppleman, F. S. (1992). Teleshopping or going shopping? An information acquisition perspective. *Behavior & Information Technology, 11*(4), 189–198.

Salvaggio, J., & Bryant, J. (1989). *Media in the information age.* Hillsdale, NJ: Lawrence Erlbaum Associates.

Schiller, H. I. (1981). *Who knows: Information in the age of the Fortune 500.* Norwood, NJ: Ablex Publishing.

Shah, D., Schmierbach, M., Hawkins, J., Espino, R., Ericson, M., Donavan, J., & Chung, S. (2000, November). *Untangling the ties that bind: The relationship between Internet use and engagement in public life.* Paper presented at the annual conference of the Midwest Association for Public Opinion Research, Chicago.

Sitkin, S., Sutcliffe, K., & Barrios-Choplin, J. (1992). A dual-capacity model of communication media choice in organizations. *Human Communication Research, 18*(4), 563–598.

Skumanich, S., & Kintsfather, D. P. (1998). Individual media dependency relations within television shopping programming: A causal model reviewed and revised. *Communication Research, 25*(2), 200–220.

Sony Corp. v. Universal City Studios. (1984). 464 U.S. 417.

Spears, R., & Lea, M. (1992). Social influence and the influence of the "social" in computer- mediated communication. In M. Lea (Ed.), *Contexts of computer-mediated* communication (pp. 30–65). London: Harvester-Wheatsheaf.

Spears, R., & Lea, M. (1994). Panacea or panopticon?: The hidden power in computer-mediated communication. *Communication Research, 21*, 427–459.

Spears, R., Postmes, T., & Lea, M. (2002). The power of influence and the influence of power in virtual groups: A SIDE look at CMC and the Internet. *Journal of Social Issues, 58*, 91–108.

Stewart, T. A. (1994, April 4). The information age in charts. *Fortune*, pp. 70–74.

Stoll, C. (1995). *Silcon valley snake-oil: Second thoughts on the information highway.* New York: Doubleday.

Straubhaar, J. D., Burch, E., Guilherme Duarte, L., & Sheffer, L. (2002). International satellite television networks: Gazing at the global village or looking for "home" video? In C. Lin & D. Atkin (Eds.), *Communication technology and society* (pp. 307–328). Cresskill, NJ: Hampton.

Strasburger, B., & Wilson, B. (2002). *Children, adolescents and media.* Newbury Park, CA: Sage.

Subrahmanyam, K., & Greenfield, F. M. (1994). Effect of video game practice on spatial skills in girls and boys. *Journal of Applied Developmental Psychology, 15*, 13–32.

Telecommunication Act of 1996. Pub. L. No. 104-104, 110 State. 56, 11 (1996). (Codified as amended in 47 C.R.R. S. 73.3555).

Tichenor, P., Donohue, G. A., & Olien, C. N. (1980). *Community conflict and the press.* Beverly Hills, CA: Sage.

Tocqueville, A. (1969). *Democracy in America* (12th ed.; G. Lawrence, Trans.). New York: Harper & Row. (Original work published 1848)

Trevino, L., Daft, R., & Lengel, R. (1990). Understanding manager media choices: A symbolic interactionist perspective. In J. Fulk & C. Steinfield (Eds.), *Organizations and communications technology* (pp. 71–94). Newbury Park, CA: Sage.

Tribe, L. (1988). *American constitutional law* (2nd ed.). New York: Foundation Press.

Toffler, A., & Toffler, H. (2006). *Revolutionary wealth.* New York: Knopf.

Uniting and Strengthening America by Providing Appropriate Tools Required to Intercept and Obstruct Terrorism (USA PATRIOT) Act of 2001, Pub. L. No. 107-56, 115 Stat. 272(2001) (amending scattered sections of 18, 47, 50 U.S.C.).

Waldfogel, J. (2002, September). *Consumer substitution among media.* Philadelphia: Federal Communications Commission Media Ownership Working Group.

Walther, J. (1996). Computer-mediated communication: Impersonal interpersonal, and hyperpersonal interaction. *Communication Research, 23*, 3–43.

Walther, J. B., & Boyd, S. (2002). Attraction to computer-mediated social support. In C. A. Lin & D. Atkin (Eds.), *Communication technology and society: Audience adoption and uses* (pp. 153–188). Cresskill, NJ: Hampton.

Walther, J. B., & Burgoon, J. K. (1992). Relational communication in computer-mediated interaction. *Human Communication Research, 19*, 50–88.

Walther, J. B., & Parks, M. R. (2002). Cues filtered out, cues filtered in: Computer-mediated communication and relationships. In M. L. Knapp & J. A. Daly (Eds.), *Handbook of interpersonal communication* (3rd ed., pp. 529–563). Thousand Oaks, CA: Sage.

Warren, S., & Brandeis, L. (1890). The right to privacy. *Harvard Law Review, 4*, 193–220.

Wellman, B., Salaff, J., & Dimitrova, D. (1996). Computer networks as social networks: Collaborative work, telework, and virtual community. *Annual Review of Sociology, 22*, 213–218.

Williams, F. (1988). *Measuring the information society.* Newbury Park, CA: Sage.

Wood, A. F., & Smith, M. (2001). *Online communication: Linking technology, identity,* culture. Mahwah, NJ: Lawrence Erlbaum Associates.

Wright, K. (1999). Computer-mediated support groups: An examination of relationships among social support, perceived stress, and coping strategies. *Communication Quarterly, 47*(4), 402–414.

Wright, K. (2000). Perceptions of on-line support providers: An examination of perceived homophily, source credibility, communication and social support within on-line support groups. *Communication Quarterly, 48*(1), 44–59.

Wright, K. (2002). Social support within an on-line cancer community: An assessment of emotional support, perceptions of advantages and disadvantages, and motives for using the community from a communication perspective. *Journal of Applied Communication Research, 30*(3), 195–209.

Wright, K., & Bell, S. B. (2003). Health-related support groups on the Internet: Linking empirical findings to social support and computer-mediated communication theory. *Journal of Health Psychology, 8*(1), 39–54.

Author Index

A

Abclman, R., 171, 178, *179*
Åborg, C., 102, *118*
Action, H., 224, *235*
Adams, J., 266, *278*
Affleck, 223, *239*
Agostino, D., 103, *137*
Aguillo, I. F., 25, *32*
Aidman, A. J., 130, *141*
Alavi, M., 188, *197*
Albarran, A., 147, *157*
Alexander, A., 178, *181*
Allen, A., 225, 231, *234, 237*
Allen, A. L., 258, 261, 271, 272, *274*
Allen, I. R., 147, *157*
Alreck, P. L., 211, *222*
Alterman, E., 145, *157*
Althaus, S. L., 28, *32,* 145, 146, 152, *157,* 188, *179,* 188, *194*
Altman, I., 259, *274*
Alvcrson, D., 226, *236*
Amachai-Hamburger, Y., 192, *194*
Anantho, S., 25, *35*
Anderson, C. A., 62, 63, 67, *72,* 285, *301, 302*
Anderson, J. F., 44, *51, 57*
Anderson, P., 271, *275*
Andreason, A. R., 22, *32*
Andreason, M., 178, *179*
Andrews, P., 150, *158*
Anokwa, K., 4, *14,* 21, *32*, 283
Applebaum, S., 20, *32*
Archea, J., 104, *118*
Arlin, J., 63, 67, *72*
Armstrong, L., 189, 191, 192, *194*
Arnaldo, C. A., 69, *72*
Arnold, S. J., 206, *219*
Arroyo, N., 25, *32*
Aspden, P., 184, *197*
Aspenall, J., 266, 272, 279
Asya, I., 210, *220*
Atkin, D. J., 12, 28, *32,* 79, 89, 92, *95,* 125, 127, 129, 132, 136, *137, 138, 139, 140,* 153, 156, *159*, 168, 169, 171, 173, 175, 176, 177, 178, *179, 180, 181*, 185, 186, 187, 189, 193, *194, 195, 197,* 292, *302, 303, 304*
Atkins, E., 67, *73*
Atton, C., 30, *32*
Auberdene, P., 283, 301, *304*
Aufderheide, P., 130, *137*
Auter, P. J., 208, *218*
Aycsh, M. I., 27, *32*
Aylor, B., 41, *52*

B

Ba, S., 105, *120*, 288, *304*
Babin, B. J., 206, *218,* 294, *302*
Bachman, K., 126, *137*
Bae, H. S., 151, *158*
Bagdikian, B. H., 144, *158*
Bailey, D. E., 81, 82, 83, 87, 88, *95, 97*
Bain, H., 224, *235*
Bair, J. H., 47, *55*
Baker, P. M. A., 126, *137*
Baldwin, T. F., 130, *137*, 151, *158,* 185, *195*
Ball-Rokeach, S. J., 129, 132, 133, *139*, *187,* 196, 206, 208, *218, 220*, 289, 290, 294, *303, 304*

Baltes, S., 81, *95*
Bandura, A., 62, 63, *72,* 215*, 218,* 285, *302*
Baradagunta, S., 232, *237*
Bargh, J. A., 42, 48, 49, *54*, 193*, 195*
Barker, P. M. A., 290, *302*
Barker, R. G., 104, *119*
Barnard, J., 93t, 94t, *96*
Barr, R. F., 59, *73*
Barrera, M., 49, *51*
Barrett, M., 151, *158*
Barrios-Choplin, J., 104, *120*, 288, *306*
Barth, K., 148, 153, *162*
Basil, M. D., 153, *158*
Bateman, J., 184, *197*
Bates, B., 151, *158*
Baym, N. K., 40, *51*
Beard, K. W., 49, *51*
Beauclair, R., 40, *52*
Belier, M., 188, *195*
Bell, D., 285, 300, *302, 307*
Bellah, R. N., 125, *137*
Belson, W. A., 148, *158*
Benjamin, J., 63, 67, *72*
Bensley, L., 67, *72*
Berg, M., 226, *237*
Bergauer, N., 229, *237*
Berkovitch, E., 230, 232, *236*
Berkowitz, L., 62, 63, *72*, 211, *218*
Besser, A., 49, *52*
Bhavagnagri, N., 69, *74*
Bickford, T. R., 231, *237*
Billing, A., 102, *118*
Biocca, F., 187, *195*
Bird, K. T., 225, *237*
Black, J., 272, *274*
Blackwell, R. D., 206, *219*
Blanchett, G. T., 191, *196*
Blanton, W. E., 62, *73*
Block, 130, *137*
Blood, R., 145, *158*
Bloomer, S. E., 228, *237*
Blumler, J. G., 42, *53*, 149, 154, *160*, 173, *180*, 207, *220*, 292, 294, *303*
Boberg, E., 48, *52*
Bogart, L., 148, *158*
Boles, S. M., 49, *51*
Bolter, J. D., 253, *255*
Bonavia, M., 261, *274*
Boneva, B., 184, *197*
Bonfadelli, H., 156, *158,* 188, *195*
Bovill, M., 65, *74*
Bowering, R., 231, *235*
Bowles, D., 153, *158*, 243, 244, 245, *255*
Bowles, K. H., 232, *235*
Boyd, D., 26, *32*, 189, *197*
Boyd, S., 45, 48, 49, *57,* 189, *197,* 285, *306*
Boynton, A., 287, *302*
Bracken, C. C., 68, *73*, 156, *159*, 177, *179,* 286, *302*
Braima, M. A. M., 5, *14*
Braithwaite, D. O., 48, 49, *51*
Brandeis, L., 258, 259, 279, 298, *307*
Brandt, A., 264, *276*
Brants, K., 190, 191, *195*
Brenner, V., 49, *51*
Brody, E. W., 130, *137*
Bromley, R. V., 153, *158*
Brown, D., 175, *179*
Brown, M., 154, 157, *158*
Brownstein, C. N., 130, *137*
Bryant, J., 178, *179*
Bryant, J. A., 178, *179*
Bryant, K., 188, *195*
Buckingham, D., 65, 67, *73*
Buckingham, P., 225, *235*
Bucy, E. P., 7, 9, *14,* 144, 147, 156, 157, *158,* 286, *302*
Burch, E., 21, *35,* 295, *306*
Burgoon, J. K., 44, 45, *55*, *57*, 259, 261, 262, 263, 265, *275,* 284, *307*
Burleson, B. R., 48, *51*
Burnett, J. W., 231, *237*
Burns, R., 130, *137*
Burton, D. C., 230, 231, *236*
Bushman, B. J., 62, 67, *72*, 285*, 302*
Butler, B., 7, *15*

C

Cagiltay, K., 29, *34*
Cai, X., 153, *158*
Callas, P. W., 232, *238*
Calvert, S. L., 59, *73*

Camaioni, L., 68, *73*
Campbell, J., 188, *195*
Campbell, T., 232, *235*
Capaldo, G., 105, *119*
Caplan, S. E., 42, 44, 45, 47, 48, 49, 50, *51*
Cardwell, B., 231, *234*
Carnagey, N. L., 63, 67, *72*
Carpentier, F. D., 185, *197*
Carr, C. L., 215, *218*
Carrero, V., 47, *54*
Carson, S., 215, *218*
Carter, K., 211, *220*
Castells, M., 24, 31, *32*
Chachage, B. L., 29, *32*
Chae, Y. M., 231, *235*
Chaikin, A. L., 271, *275*, 298, *302*
Chaing, K. P., 211, *218*
Chalaby, J. K., 27, *32*
Chalfonte, B., 86, *97*, 287, *304*
Chan, A., 7, *15*, 48, *52*
Chan, W. M., 231, *235*
Chandler, A., Jr., 118, *119*
Chang, T. Z., 211, *218*
Chapman, S., 189, *195*
Charlton, J. P., 192, *195*
Charney, T., 28, *33*, 154, *158*, 184, 186, *195*
Chatterjee, R., 203, *220*
Chau, P. Y. K., 226, *235*
Chen, S. J., 211, *218*
Cheon, H. J., 286, *302*
Childers, T., 175, *179*, 215, *218*
Chinn, D. E., 130, *141*
Chiu, S. L., 192, *195*
Choie, H., 286, *302*
Chung, S., 129, 132, *141*, 290, *306*
Chung, T., 189, 191, *197*
Clark, G., 178, *181*
Clarke, P., 130, *137*
Clarke, P. H. J., 231, *235*
Clarkson, T., 231, *237*
Clayton, W., 230, 231, *236*
Clemens, C., 231, *235*
Clement, A., 105, 116, *119*, 266, 272, *279*
Cloutier, A., 225, *236*
Coffin, T. E., 148, *158*
Cohen, B. C., 145, 151, *158*, *159*
Cohen, E., 151, *159*
Cohen, G. E., 44, *51*
Cohen, J., 130, *137*
Cohen, W. M., 287, *302*
Coles, A. M., 254, *255*
Colfax, J. D., 147, *157*
Collins, B., 227, 232, *238*
Conrath, D. W., 224, 231, *235*, *236*
Cooke, D. J., 228, *239*
Cooke, M. C., 66, *74*
Coonan, C., 19, *33*
Cooper, J., 67, 71, *73*, 95, *96*
Corbett, R., 228, *237*
Cornwell, B., 48, *51*
Cortada, J., 118, *119*
Courtright, J. A., 47, *54*, 149, 154, *161*, 186, *195*, 209, 293, *305*
Couturier, P., 231, *235*
Cox, D. F., 209, *218*
Cox, R., 231, *234*
Crawford, A., 184, *197*
Creeber, G., 24, *33*
Crichton, C., 231, *236*
Crow, K., 135, *137*
Cullity, J., 27, *33*
Culnan, M. J., 42, 44, 45, *52*, 265, 266
Cummings, J., 184, *197*
Cunningham, N., 231, 232, *235*
Cunningham, T., 207, 209, 212, 214, *220*
Currell, R., 229, *235*
Czaja, S., 87, *99*

D

Daft, R. L., 43, *52*
Dainton, M., 41, *52*
D'Allesandro, D. M., 181, *195*
Dalton, R. J., Jr., 131, *137*
Dansky, K. H., 232, *235*
Darden, W. R., 206, *218*, 294, *302*
Darian, J. C., 207, 209, 212, 215, *218*
Davidow, W., 104, *119*
Davis, F. D., 226, *235*, 295, *302*
Davis, M. H., 254, *255*
Davis, R., 254, *255*
Davis, R. A., 49, 50, *52*, 179, *179*

Davis, T. R. V., 107, 111, *119, 120,* 288, *302*
Dawson, R., 178, *181*
Day, S. X., 45, *52*
DeCew, J. W., 260, 261, 262, *275*
DeFleur, M., 187, *195,* 206, 208, *218*
Dempsey, J., 271, *275*
Denis, J. L., 226, *237*
Derlega, V. J., 259, *275,* 298, *302*
De Sola Pool, I., 23, 30, *33*
De Winstanley, P., 68, *73,* 294, *303*
Dholakia, N., 6, *14*, 190, *195*
Dholakia, R., 6, *14*, 190, *195*, 206, *218*
Dick, S. J., 174, 175, *179*
Diehl, J., 22, *33*
Dill, J. C., 67, *72, 73*
Dill, K. E., 67, *73*
DilMaggio, P. E., 24, *33*
Dimitrova, D., 86, *96, 100*, 287, *307*
Dimmick, J., 40, 41, 43, 46, *52, 56,* 169, 176, 178, *179*
Dizard, W., 3, 13, *14*, 283, *302*
Dobrow, J., 10, *14*, 169, 173, 177, *179*
Doheny-Farina, S., 125, *138*
Dominick, J. R., 67, *73*
Dongier, M., 225, 231, *235*
Donohew, L., 192, *196*
Donohue, G. A., 127, 128, *141, 156, 163,* 186, 188, *196, 197,* 290, *306*
Donohue, T. R., 148, *159*, 173, 175, *180*
Donovan, J., 129, 132, *141*, 290, *306*
Donthu, N., 206, 207, 209, 211, 212, 214, 215, *219*
Doolittle, G. C., 229, 231, *235, 239*
Dordick, H., 129, *137*
Dorman, S. M., 48, 49, *57*
Dowling, G. R., 207, *221*
Driscoll, P. D., 147, *161*
Drori, G. S., 3, 25
Drotner, K., 62, *73*
Drucker, S. J., 214, *220*
D'Souza, M., 228, *235*
Duarte, L. G., 21, *35*
DuBrin, A., 93t, 94t, *96*
Duhachek, A., 211, 215, *221*
Dumais, S., 105, *119*
Dunham, P. J., 48, *52*
Dunn, D., 148, *162,* 284, *305*
Dunn, E. V., 224, 225, 231, *235, 236*
Dupagne, M., 168, 174, 176, *180*
Dutta-Bergman, M. J., 129, 132, *138,* 290, *302*
Duxbury, L., 84, 89, *96*

E

Eastin, M. S., 10, *15*, 184, 185, 188, 191, 193, *197,* 207, 215, *220,* 294, *304*
Eastlick, M. A., 212, 213, *219*
Edelstein, A. S., 125, *138*
Eighmey, J., 154, *158*
Elford, D. R., 231, *235*
Elkin-Koren, N., 246, *255*
Ellison, N. B., 82, 87, *96*
Elton, L., 130, *137*
Emerson, T. I., 245, *255*
Endean, E. D., 232, *236*
Engel, J. F., 206, *219*
Epstein, N., 22, *33*
Erbring, 152, *160*
Ercolani, P., 68, *73*
Ericson, M., 129, 132, *141*, 290, *306*
Espino, R., 129, 132, *141*, 290, *306*
Estrada, G., 146, *161*
Eubanks, J., 63, 67, *72*
Evans, P. E., 203, 205, *219*
Evans, S., 130, *137*
Eyal, K., 18, *35*, 169, 173, 174, 177, 178, *182,* 186, *199*

F

Faber, R. J., 206, 209, *219,* 295, *302*
Facer, K., 65, 69, *73*
Fairweather, N. B., 83, 85, 90, 93t, 94t, *96*
Fallows, J., 131, *138*, 217, *219*
Fathi, A., 147, *158*
Feil, E. G., 49, *51*
Feldman, M. S., 104, 105, *119*, 288, *303*
Fenner, D., 105, *119*, 288, *303*
Ferguson, D. A., 130, *140,* 154, 155, *158,* 173, 176, 178, *180*
Ferris, M., 88, *97*
Fine, A., 192, *194*
Fine, G. A., 147, *158*

Finfgeld, D. L., 45, 49, *52*, 285, *303*
Finholt, T., 148, *159*
Finn, J., 48, 49, *51, 52*
Finnegan, J. R., 130, *141*
Fischer, C. T., 260, *275*
Fischer, E., 206, *219*
Fish, R., 86, *97,* 287, *304*
Fisher, E., 232, *237*
Fisher, R. J., 214, *221*
Fisher, S., 64, *74*
Fishman, B. J., 188, *196*
Flaherty, L. M., 41, *52*
Flanagan, J., 229, *238*
Flanagan, M., 63, 67, *72*
Flanagin, A. J., 28, *33,* 44, *52,* 147, 154, *159*
Flett, G. L., 49, *52*
Fling, S., 67, *73*
Floyd, K., 44, *54*
Foehr, U. G., 41, 46, *55*
Foreman, D., 231, *237*
Forrest, E., 177, *182*
Fortin, J. P., 225, *236*
Fossi, C., 131, *138*
Fountain, K., 131, *138*
Fox, 93t, *96*
Francese, P., 87, *96*
Franco, A., 231, *235*
Freeman-Longo, R. E., 191, *196*
Fried, C., 260, *275*
Fritz, M. W., 89, 90, *96*
Fromkin, H. L., 266, *278*
Fry, D. L., 187, *196*
Fuchs, M., 225, 233, *236*
Fulk, J., 7, 8, *15*
Funk, J. B., 61, *73*
Furu, T., 148, *159*

G

Gable, L. L., 189, *196*
Gagnon, M. P., 225, *236*
Galegher, J., 86, *97*
Galloway, J. J., 127, *138*
Gandy, O. H., 12, *14*, 301, *303*
Ganley, G. D., 26, *33*
Ganley, O. H., 12, *14*, 26, *33,* 301, *303*
Ganster, D. C., 266, *278*
Gantz, W., 147, *159*
Garcia, A., 207, 209, 210, 211, 212, 215, *219*
Garcia, I., 25, *32*
Garrison, B., 147, *161*, 191, *199*
Gasser, L., 105, 116, *119*
Gates, W., 301, *303*
Gattiker, U., 288, *305*
Gavison, R., 259, *275*, 298, *303*
Geen, R. G., 63, *73*
Gelegher, J., 287, *304*
Gellman, R., 271, *275*
Gerberg, D., 283, *304*
Ghandi, M., 231, *235*
Gibson, J., 105, *119,* 288, *303*
Giddens, C. A., 295, *303*
Gilliland, D., 206, 209, 212, 214, *219*
Gilmor, D., 190, *196*
Gimeno, M. A., 47, *54*
Gitlin, T., 156, *159*
Glasgow, R. E., 49, *51*
Glazer, E., 231, 232, *235*
Gleason, M. E. J., 41, *54*, 189, *197*
Godbey, G., 148, *162*
Godin, G., 225, *236*
Goff, G., 214, *221*
Golan, G., 146, *163*
Golden, P. A., 40, *52*
Goldscheider, E., 125, *138*
Goldsmith, D. J., 48, *51*
Goldsmith, T. D., 189, 191, *197*
Goldstein, A., 192, *194*
Gomez, L., 188, *196*
Gonzalez, E., 232, *237*
Goodman, P., 105, *119,* 288, *303*
Gordon, D. N., 188, *196*
Graber, D., 156, *159*
Granka, L., 45, *57*
Granovetter, M., 227, *236*
Grant, A., 9, 11, *14*, 167, 169, 171, 172, 175, 177, *180, 182*, 186, 187, *196*, 208, 209, 211, 212, 213, 214, 215, *220*, 294, *303*
Gray, P., 80, *98*
Graziano, R. E., 48, *52*
Grebb, M., 190, *196*
Green, A. S., 189, *197*

Green, H., 271, *276*
Greenberg, B. S., 28, *33,* 130, *137,* 147, 151, 152, 154, *158, 159,* 178, *180*, 184, 186, *195, 196*
Greene, A. S., 41, *54*
Greenfield, F. M., 286, *306*
Greenfield, P. M., 68, *73, 75*, 286, *303*
Gregg, J. L., 225, *239*
Griffin, M., 206, *219*, 294, *302*
Griffith, T., 105, *119*, 288, *303*
Griffiths, M., 49, *52*, 186, 191, 192, *196*
Groebel, J., 283, *304*
Groleau, C., 112, *119*
Gronhaug, K., 148, *163*
Gross, A. M., 67, *73*
Gross, E. F., 189, *196*
Grusin, R., 253, *255*
Guilherme Duarte, L., 295, *306*
Gunn, D., 191, *198*
Gurevitch, M., 42, *53*, 149, 154, *160,* 173, *180*, 207, *220*, 292, 294, *303*
Gustafson, D. H., 48, *52*
Guthrie, K., 186, *196*, 208, 216, *220,* 294, *303*
Gwynne, G., 188, *195*

H

Hackforth, J., 26, *35*
Hafner, K., 20, *33*
Hailey, D., 228, 229, *236, 238*
Hall, A. S., 191, *196*
Hallin, D. C., 146, *159*
Hamelink, C. J., 127, *138*
Hampton, C., 232, *237*
Hampton, K. N., 44, *52*
Hancock, J. T., 48, *52*
Hardaf Segerstad, Y., 46, *53*
Hargittai, E., 24, *33*
Hargreaves, W. A., 228, *236*
Hargrove, T., 148, 153, *163*
Harpaz, I., 93t, 94t, *97*
Harper, R., 102, 105, 116, 117, *120*
Harrington, S. J., 83, *99*
Harris, C., 178, *181*
Harris, L., 83, 84, 88, 89, *97*, 254, *255*
Harrison, D. A., 116, *120*
Harrison, M., 24, *35*
Harrison, R., 230, 231, *236*
Hartson, R., 105, *119*, 288, *303*
Harvey, M., 148, *159*
Harwood, P., 127, *138*
Hashimoto, S., 228, *236*
Hassay, D. N., 215, 220, *220*
Hauber, R., 228, *238*
Hawkins, J., 129, 132, *141*, 290, *306*
Hawkins, R., 48, *52*
Haycox, A., 230, *239*
Hayes, B. A., 62, *73*
Hayes, M. P. A., 231, *234*
He, Z., 193, *197*
Heath, D., 226, *236*
Hedges, M., 151, *159*
Hee, H. S., 231, *235*
Heeter, C., 176, *180,* 186, *196,* 290, *303*
Hegelson, V., 184, *197*
Heintz, K. E., 130, *141*
Helfand, M., 229, *236*
Helitzer, D., 226, *236*
Heliö, S., 42, *53*
Henke, L., 148, *159*, 175, *180,* 186, *196*
Henry, M., 231, *237*
Heon, L. J., 231, *235*
Hernandea, M., 87, *99*
Herring, S. C., 39, 40, 41, *53*, *57*
Hersch, W. R., 229, *236*
Hetsroni, A., 210, *220*
Higgins, C. A., 231, *236*
Higgins, J. W., 125, *138*
Hill, A., 215, *221*
Hill, D., 231, *237*
Hill, E. J., 87, 88, *96, 97*
Hilt, M. L., 244, *255*
Hilty, D. M., 231, *237*
Hilyard, D. P., 263, *276*
Himmelweit, H. T., 148, *159,* 290, *303*
Hindman, E. B., 125, *138*
Hirschman, E. C., 207, *220*
Hjelm, N. M., 231, *235*
Hoag, A., 7, *15*
Hoch, S. J., 208, 209, *220, 221*, 295, *303*
Hoffman, D. L., 187, *196,* 203, 211, 215, *220, 221*
Hoffman, N., 231, *237*
Hofshire, L., 152, *159*

Hofte, H., 272, *279*
Holand, U., 231, *238*
Holdsworth, L., 82, 88, *98*
Hong Jun, K., 231, *235*
Hooper, F. J., 231, *237*
Horn, M., 272, *276*
Horrigan, J. B., 40, *53*
Horton, D., 186, *196*
Horwood, K., 229, *236*
Howard, G., 83, 84, 85, 90, *99*
Hoy, M. G., 265, 267, *278*
Hu, J., 148, *159*
Hu, P. J., 226, *235*
Huang, D. H., 192, *195*
Hudson, N., 80, *97*
Huesmann, L. R., 62, 63, *73*
Hui, E., 231, *235, 236*
Hunkeler, E. M., 228, *236*
Huston, J. L., 230, 231, *236*

I

Iacono, C. S., 227, *238*
Ibanez, I., 47, *54*
Irwin, A. R., 67, *73*
Itzak, B., 230, 232, *236*

J

Jackson, K. B., 232, *238*
Jacobs, G., 287, *302*
Jacobs, R., 125, 130, *138*
Jacques, D., 229, *237*
Ja Kim, S., 231, *235*
James, E. L., 207, 209, 212, 214, *220*
Jang, Y. S., 25, *33*
Jankowski, N., 132, *139*, *141*
Jansen, B. J., 184, *197*
Jarnowitz, M., 125, *138*
Järvinen, A., 42, *53*
Jeffres, L. W., 8, *14,* 28, *32,* 125, 127, 132, 133, *137, 138, 139,* 148, 153, 156, *159,* 168, 173, 175, 176, 178, *179, 180, 181*, 185, 186, 187, 189, 190, 193, *195, 197, 198*, 289, 292, *303*
Jensen, C., 266, *276*
Jensen, K., 188, *197*
Jessup, L., 216, *222*
Johnson, B. M., 1-5, 117, 118, *119,* 288, *303*
Johnson, K. F., 176, *181*
Johnson, P., 144, *160*
Johnson, T. J., 5, 9, *14*, 147, 155, *160*, 190, *197*
Johnson, V., 223, *239*
Johnston, R. E., 231, *237*
Joinson, A. N., 48, *53*
Jones, D. H., 231, *236*
Jones, M. J., 217, *220*
Jones, S., 41, *53*, 152, *160*
Jong, M. K. K., 229, *236*
Joslyn, M. R., 5, *14*
Jung, J.-Y., 133, *139*
Juskalian, R., 292, 301, *303*
Jutra, A., 224, *236*
Juvonen, J., 189, *196*

K

Kahn, B., 211, *211*
Kampshoff, J., 226, *238*
Kandra, A., 264, *276*
Kaneko, K., 228, *236*
Kaplan, S. J., 149, *160*
Kaplowitz, L., 232, *237*
Karlsen, K. O., 232, *237*
Karp, G. G., 188, *197*
Katz, E., 12, 42, *53*, *54*, *55*, 149, 152, *160*, 173, *180*, 207, *220*, 292, 294, *303*
Katz, J., 184, *197*, 207, *220*
Katz, J. E., 258, *276*
Kauffman, C. L., 231, *237*
Kaufman-Scarborough, C., 213, *220*
Kayany, J. M., 148, 149, 152, *160*, 290, *303*
Kaye, B. K., 147, 155, *160*, 190, *197*
Keck, P. E., 189, 191, *197*
Kellner, D., 65, *74*
Kelly, D. L., 259, 261, 262, 263, 265, *275*
Kendall, J., 103, 107, *119*
Kendall, K. E., 103, 107, *119*
Kendall, L., 41, *53*
Kerr, B. A., 44, *51*
Kerr, D., 188, *195*
Kessler, I. I., 231, *237*

Kestnbaum, M., 148, *162*
Khoslab, V. M., 189, 191, *197*
Kidd, A., 117, *119*
Kienzle, M., 231, *239*
Kiesler, S., 6, 42, *53*, 149, *160*, 184, 189, 192, *197*, 272, *276*
Kiley, K. C., 232, *238*
Kilpatrick, H., 68, *73*, 294, *303*
Kim, J., 215, *220*
Kim, S., 266, *276*
Kim, W., 184, 193, *197*
Kim, Y.-C., 133, *139*
King, N., 105, *120*, 288, *304*
Kintsfather, D., 208, 214, *222*, 294, *306*
Kinzer, S. L., 184, 189, 192, *195*, *197*
Kjaer, N. K., 232, *237*
Kline, F. G., 130, *137*
Kline, S., 231, *237*
Kline, S. L., 40, 41, *52, 56*
Kling, R., 30, *33*
Klopfenstein, B., 40, *53*, 173, 178, *180*, 292, *303*
Knight, S., 131, *139*
Knight, S. M., 263, *276*
Knobloch, S., 185, *197*
Kobb, R., 231, *237*
Koch, S., 105, *119*
Kohut, A., 148, 153, *162*
Komiya, M., 172, *180*
Koppelman, F. S., 206, 207, 210, 211, *222*, 294, *305*
Korengel, K., 135, *139*
Korgoankar, P. K., 154, *160*, 207, 211, 213, *220*, 294, *303*
Kosman, K., 232, *238*
Kotler, P., 22, *34*
Kowalski, K. B., 88, *97*
Kraut, R., 6, 7, *14, 15,* 42, *53*, 69, *73,* 82, 86, *97, 98, 99,* 105, *119,* 149, *160*, 286, 287, *304*
Krcmar, M., 66, *74*
Kreiter, C. D., 189, *195*
Krendl, K., 147, *159*, 178, *181*
Krick, S., 232, *237*
Krim, J., 131, *139*
Kristula, D., 183, *197*
Krugman, D. M., 170, 175, 176, *179, 181*
Krych, R., 206, 209, *219,* 295, *302*
Kubicek, H., 129, *139*
Kuchinskas, S., 67, *74*
Kuenneth, C. A., 231, *237*
Kundanis, R. M., 4, 6, *15*, 286, *304*
Kunkler, I. H., 231, *237*
Kuo, E. C. Y., 152, *160*
Kurland, N., 81, 82, 83, 87, 88, 93t, 94t, *95, 97*
Kurtz, H., 151, *160*
Kvammen, B., 231, *237*
Kvedar, J. C., 232, *237*
Kwolek, C. J., 232, *236*

L

Lachlan, K., 152, *159*
LaCroix, A., 226, *237*
Lalinec-Michaud, M., 225, 231, *235*
Lamothe, L., 225, *236*
Landmark, V., 189, 192, *197*
Lange, R., 189, 193, *195*
Lange, T., 226, *238*
LaRose, R., 10, 184, 185, 187, 191, 193, *197,* 207, 215, *220*, 261, 264, 266, *276*, 294, *303*
Larsen, O. N., 125, *138*
Lasica, J. D., 157, *160*
Lasswell, H. D., 128, *139*, 148, *160*, 173, *181*, 186, *187*
Lau, T. Y., 82, *95*
Lauber, B. A., 68, *73*
Laufer, R. S., 260, 261, 262, *276*
Lavin, M., 192, *197*
Lazarsfeld, P., 148, *160*
Lea, M., 45, *53*, *56,* 285, 286, *304, 306*
Leckliter, I. N., 130, *141*
Lee, C., 146, *163*
Lee, J. K., 231, *237*
Lee, J. W., 132, *138*
Lee, L. T., 261, 262, 263, 268, 269, *276*
Lee, M., 173, *182*
Lee, M. J., 182, 192, *197*
Lee, W., 152, *160*
Lehoux, P., 226, *237*
Lehr, J., 116, *119*
Lengel, R. H., 43, *52*
Lenhart, A., 25, *34,* 41, 46, *53*, *56*

Lent, J. A., 26, *32*
LePoire, B. A., 259, 261, 262, 263, 265, *275*
Lepper, M. R., 67, *74*
Lessig, L., 253, *255*, 298, *304*
Levinthal, D. A., 287, *302*
Levy, M. R., 10, *15*, 26, *34*, 169, 173, *181*, 245, *255*
Lewis, J., 87, *99*
Lewis, O., 46, *53*
Li, H., 151, *159*, 270, *276*
Li, S., 189, 191, *197*, 212, *220*
Liebermann, P. B., 233, *237*
Lievrouw, L. A., 8, *15*, 62, *74*, 289, 299, *304*
Lim, V., 84, *98*
Lin, C. A., 10, *14*, 18, 19, 22, *34*, 47, 50, *53*, *54*, *55*, *57*, 82, 92, *95*, 147, 148, 153, 154, *161*, 169, 173, 175, 176, 178, *180*, *181*, 184, 185, 186, 187, 191, 193, *197*, *198*, *199*, 207, 212, 213, *220*, *221*, 251, *255*, 292, 293, 294, 297, 299, 301, *303*, *304*
Lin, S., 67, *74*
Lin, S. H., 184, 192, *198*
Lindner, K., 189, *197*
Lindquist, J. D., 213, *220*
Lipschultz, J. H., 244, *255*
Litman, B., 172, *180*
Liu, M., 213, *219*
Livingstone, S., 8, *15*, *62*, 65, *74*, 289, 299, *304*
Ljungstrand, P., 46, *53*
Lloyd, F. W., 244, *255*
Lo, T. W., 192, *197*
Loane, M. A., 228, *237*
Lochen, M. L., 231, *237*
Lodge, R., 231, *237*
Loeffelholz, P., 228, *239*
Loh, T., 45, *57*
Lombard, M., 68, *73*, 286, 287, *302*, *304*
Lorch, E. P., 192, *196*
Lotz, S., 212, *219*
Lowenstein, G. F., 208, 209, *220*, *295*, *303*
Lowitt, M. H., 231, *237*
Lucas, A. G., 104, *119*
Lucas, W. A., 130, *139*
Lundgren, D., 48, *51*
Lundmark, V., 42, *53*, 69, *73*, *149*, *160*, 286, *304*
Lunt, P., 266, *277*
Lyle, J., 129, *137*
Lyon, M., 20, *33*, 263, *276*

M

Maccoby, E., 148, *161*
Macdonald, A., 231, *236*
Machlup, F., 300, *304*
Mackert, M., 231, *239*
Mackie, D., 67, *71*
Madden, M., 25, *34*, 40, 42, *54*
Maddex, B. D., 157, *163*
Maddigan, B., 231, *235*
Madsen, S., 88, *98*,125, *137*
Maheu, M., 225, *237*
Mahoney, M., 212, 213, *222*
Mair, F., 230, 231, *238*, *239*
Majchrzak, A., 105, *120*, 288, *304*
Malcolm, C. S., 215, *220*
Malcolm, T., 231, *237*
Malhotra, A., 105, *120*
Malhorta, N. K., 211, *221*, 288, *304*
Malina, A., 132, *139*
Mallon, L. I., 232, *236*
Malone, M., 104, 107, 111, *119*, *120*
Mangalindan, M., 203, *221*
Mann, S., 82, 88, 93t, 94t, *98*
Manning, J., 6, *15*
March, J. G., 104, 105, *119*, *120*, *121*, 288, *303*
Margolis, M., 3, *15*
Margulis, S. T., 259, *276*
Mark, R., 168, *181*
Mark, R. G., 223, 231, *237*, *239*
Marks, T., 232, *237*
Markus, M. L., 42, 44, 45, *52*
Marshall, C., 231, 232, *235*, 260, 262, *277*
Marshall, N. J., 260, 262, *272*
Martel, R. F., 232, *235*
Martinson, V., 88, *97*
Marvin, K., 192, *197*
Maslow, A. H., 207, *221*

Master, R. J., 231, *237*
Mastro, D., 185, *197*
Matei, S., 129, 132, 133, *139*, 289, 290, *304*
Mathwick, C., 211, *221*
Matrud, K., 226, *236*
Matthews, D. A., 233, *237*
Mäyrä, F., 42, *53*
Mayeda, D. M., 244, *255*
Mazdon, L., 24, *34*
Mazmanian, P., 232, *237*
Mazzarella, S. R., 130, *141*
McCain, T. A., 187, *196*
McCall, K., 93t, 94t, *98*
McCall, W., 131, *139*
McChesney, R. W., 9, 10, *15*
McCombs, M. E., 145, 148, *161*, 172, *181*, 290, 291, *304*
McCord, L., 154, *158*
McCorkle, D. E., 211, *222*
McCuddy, M. K., 266, *278*
McCue, M., 233, *237*
McDonald, D. G., 187, *197*
McElroy, S. L., 189, 191, *197*
McGuire, T. W., 272, *276*
McIlwraith, R. D., 191, *197*
McIntosh, S., 156, *161*
McKay, H. G., 49, *51*
McKenna, K. Y. A., 41, 42, 44, 48, 49, *54*, 189, *197*
McKinlay, J., 231, *236*
McLarney, A., 192, *197*
McLean, D., 272, *277*
McLeod, J. M., 130, *139*, *140*
McVeigh, T., 215, *221*
McVoy, D. S., 185, *195*
Meadows, J., 9, 11, *14*, 186, *196*, 211, *220*
Melkote, S. R., 22, *34*
Menn, E. R., 232, *237*
Menon, S., 211, *221*
Meresman, J. F., 228, *236*
Metzger, M. J., 28, *33*, 44, *52*, 147, 154, *159*
Meunier, D., 225, 231, *235*
Meyer, M., 231, *237*
Meyerowitz, J., 50, *54*
Midgley, D. F., 207, *221*
Miles, H., 27, *34*
Miniard, P. W., 206, *219*
Minion, D. J., 232, *236*
Mitchell, T., 130, *139*
Mohr, M., 226, *238*
Mokhtarian, P. L., 6, *15*, 81, 85, 87, *98*, *99*
Monge, P., 7, 8, *15*
Montalo, C. P., 266, *276*
Montgomery, K. C., 62, *74*, 178, *181*
Montgomery, W. L., 232, *238*
Moody, J., 227, *237*
Moorman, G. B., 62, *73*
Morahan-Martin, J. M., 49, 50, *54*, *192*, 197
Morgan, M., 178, *181*
Morris, M., 42, *54*, 154, *161*
Morrison, J., 229, *237*
Morrison, S., 29, *34*
Morten, L., 261, 262, *274*
Moseng, D., 231, *238*
Moss, M. L., 130, *139*
Mukhopadhyay, T., 6, 42, *53*, 69, *74*, 149, *160*, 189, 192, *197*, 286, *304*
Mundorf, N., 6, *14*, 190, *195*
Munushain, J., 80, *98*
Murgulis, S. T., 298, *304*
Murphy, C. R., 67, *72*
Murphy, L. H., 225, *237*

N

Naisbett, J., 301, *304*
Natajarian, R., 214, *221*
Neascu, D., 254, *255*
Needham, R. D., 132, *139*
Negroponte, N., 102, *120*
Neijens, P., 190, 191, *195*
Nelson, D., 105, *120*
Nelson, E. L., 228, *239*
Nerlich, M., 226, *238*
Nesbitt, T. S., 231, *237*
Neuendorf, K. A., 28, *32*, 125, 132, *137*, *138*, *139*, 156, *159*, 173, 175, 178, *179*, *180*, *181*, 186, 188, 189, 192, *195*, *197*, 292, *303*
Neufeld, D., 84, 89, *96*, *98*

Neuman, W. R., 24, *33*
Newcomb, H., 9, 10, *15*
Newhagen, J. E., 7, *14*, 149, *161*, 286, *302*
Nie, N. H., 152, *161*
Nielsen, P. E., 232, *238*
Nilles, J. M., 80, 82, 83, 88, *98*
Nixon, K., 67, *73*
Noah, L., 68, *74*
Noam, E., 12, *15*, 283, *304*
Noguchi, Y., 272, *277*
Noh, G. Y., 169, 172, 175, *182*
Nola, V., 192, *197*
Nolan, J., 172, *181*, 291, *304*
Nordal, E. J., 231, *238*
Nordenstreng, K., 21, *35*
Nossek, H., 140, *141*
Novak, T. P., 188, *196*, 203, 211, 215, *220*, *221*
Nowak, G. J., 206, 210, *221*
Nuttall, C., 29, *34*

O

O'Bryant, 132, *140*
Ogan, C., 18, 23, 26, 29, *34*, *42*, *54*, 154, *161*
O'Guinn, T. C., 206, 209, *219*, 295, *302*
Ohinmaa, A., 228, *238*
Ohmae, K., 23, *34*
Olien, C. N., 127, 128, *141*, 156, *163*, 188, 196, *197*, 290, *306*
Olivero, N., 266, *277*
O'Neil, 266, *277*
Oppenheim, A. N., 148, *159*, 290, *303*
Or, E., 188, *195*
Orlikowski, W., 103, 116, *120*, *121*
Ortet, G., 47, *54*
Osava, M., 130, *140*
Otto, C., 190, *197*
Otto, F., 231, *235*
Overbeck, W., 268, *277*
Owen, J. E., 45, *54*, 285, *305*

P

Palmgreen, P., 43, *54*, 192, *196*
Palser, B., 152, *161*
Papacharissi, Z., 41, 46, *54*, 150, 154, *161*, 184, *197*
Papastergiou, M., 29, *34*
Park, B., 223, *238*
Parker, B. J., 154, *161*
Parks, M. R., 39, 41, 44, 45, *54*, *57*, 285, *307*
Parpart, C. F., 232, *237*
Parrott, R., 259, 261, 262, 263, 265, *275*
Parsons, D., 116, *119*, 191, *196*
Pastalan, L. A., 263, *277*
Pastore, M., 205, *221*
Patrickson, M., 106, *120*
Patterson, M., 42, *53*, 69, *74*, 149, *160*, 189, 192, *197*, 286, *304*
Patterson, T. E., 144, *161*
Patwardhan, P., 28, *34*
Pava, C. H. P., 106, *120*
Pavlik, J. V., 156, *161*
Pea, R. D., 188, *196*
Pearce, K. J., 41, *52*
Pearlson, K. E., 81, 84, 85, 89, 90, 91, *98*
Peck, J., 215, *218*
Pedersen, D. M., 259, 260, 261, 262, *277*, 298, *305*
Pedersen, S., 231, *238*
Pelton, J. N., 3, 5, *15*, 283, *305*
Pennebaker, J. W., 47, *56*
Perdoma, B., 87, *99*
Peris, R., 47, *54*
Perls, T. T., 189, *197*
Perritt, H., 269, *277*
Perry, D., 259, 261, 262, 263, 265, *275*
Perse, E. M., 47, *54*, 130, *140*, 148, 149, 154, *161*, *162*, 176, 178, *180*, 186, *199*, 284, 290, 293, *305*
Perucchini, P., 68, *73*
Peters, P., 89, 90, *99*
Peterson, M. W., 189, 195
Petrie, H., 191, *198*
Phelps, J. E., 206, *221*
Phillips, M. A., 126, *140*
Phillips, V. L., 228, *238*
Piaget, J., 65, *74*
Pigg, K. E., 129, *140*, 290, *305*
Pinazo, D., 47, *54*
Pingree, S., 48, *52*

Pinkett, R., 132, *140*
Plank, R. E., 154, *161*
Pliskin, N., 40, *55*
Poole, I., 4, *15*, 283, *305*
Poor, H., 272, *279*
Posner, R., 272, *277*
Postman, N., 10, *15*, 145, *161*
Postmes, T., 45, *55, 56,* 285, *305*
Potter, E., 93t, 94t, *99*
Potter, J., 130, *141*
Potts, C., 266, *276*
Potts, S., 231, *236*
Powell, R. A., 186, *199*
Pringle-Specht, J. K., 229, *238*
Pulos, S., 64, *74*
Putnam, R. D., 129, *140*, 289, *305*

Q

Qui, J. L., 133, *139*

R

Rabby, M. K., 44, 48, *55*
Rachels, J., 260, *277*
Rafaeli, S., 149, *161*
Raffa, M., 105, *119*
Rafferty, P., 231, *237*
Rainie, L., 40, 42, 46, *53*, *54, 55*, 127, *140*, 152, *160*
Ramirez, A., 44, *55*
Ramirez, J. R. A., 44, *55*
Randall, N., 269, *277*
Ranerup, A., 133, *140*
Ratnasingham, P., 212, *221*
Reagan, J., 148, *162*, 172, 173, 175, *182*, 186, *197*
Recchuiti, J. K., 43, 45, 47, *55*
Reeves, B., 125, *141*
Reicher, S. D., 45, *55*, 285, *305*
Reinharz, D., 225, *236*
Reinsch, Jr., N. L., 82, 87, 89, *99*
Reis, S., 230, 232, *236*
Renner, J. B., 231, *237*
Rennie, E., 128, *140*
Resnick, D., 3, *15*
Rhee, M. C., 69, *74, 89, 90, 96*, 193, *197*
Rheingold, H., 32, *34*
Rhoads, C., 71, 89, 92, *99*
Rhul, C., 231, *235*
Riback, R., 71, *74*
Rice, R. E., 12, 40, 46, 47, *55*, 104, 105, 112, 116, 117, 118, *119, 120, 287,* 288, 292, *303*, *304*, *305*
Ricci, M. A., 232, *238*
Rich, S. U., 209, *219*
Richard, T., 192, *196*
Rideout, V. J., 41, 46, *55,* 59, *73*
Rieber, L. P., 68, *74*
Rifon, N., 266, *276*
Rigdon, E., 211, *221*
Riva, G., 44, *55*
Robbins, C. W., 229, *236*
Roberts, D. F., 41, 46, *55*
Roberts, L. D., 39, 41, *54*
Robertson, R., 23, *34*, 284, *305*
Robinson, J. P., 24, *33,* 148, 153, *162*
Rochlen, A. B., 45, *55*
Rodriguez, T., 67, *73*
Roeckelien, W., 226, *238*
Rogers, E. M., 3, 7, *15,* 17, 21, *34,* 102, 105, 116, 117, *120,* 127, 135, *140,* 172, *182*, 185, 193, 206, 207, 213, *221,* 226, *238*, 252, *256*, 292, 301, *305*
Rogers, R. C., 42, 43, 47, 49, *52, 55, 57*
Roine, R., 228, *238*
Roman, L., 231, *234*
Roman, R., 31, *35*
Romano, C., 204, 211, *221*
Romm, C., 40, *55*
Rook, D. W., 206, 208, 214, *221*
Rooney, B., 130, *141*
Rose, K., 88, *99*
Rosen, J., 301, *305*
Rosengren, K. E., 147, *162*
Rothe, J., 148, *159*
Roycroft, T. R., 25, *35*
Rubin, A. M., 18, *35,* 41, 42, 46, *52, 54, 55*, 150, 154, *161, 162*, 169, 173, 174, 177, 178, *182,* 184, 186, *199*
Rubin, R. B., 41, 42, *54*
Ruggiero, T. E., 150, *162*
Ruppel, C. P., 83, 84, 85, 90, *99*
Rust, R. T., 170, 175, *181*, 273, *278*
Ryan, S. D., 104, *120*

S

Sadasivan, R., 231, *234*
Sadler, R. L., 271, *278*
Salaf, J., 86, *100,* 287, *307*
Saling, L. L., 189, 191, 192, *194, 197*
Salomon, I., 206, 207, 210, 211, *222*, 294, *305*
Salwen, M. B., 10, *14,* 147, *161*, 191, *199*
Sanchiz, M., 47, *54*
Sapolsky, B. S., 177, *182*
Saracevic, T., 184, *197*
Saunders, B. F., 231, *237*
Saunders, C. S., 81, 84, 85, 90, 91, *98*
Sawhney, 193, *199*
Scanlon, M., 67, *73*
Schall, T., 226, *238*
Scherer, C. W., 148, *162*
Scherlis, W., 6, 42, *53*, 149, *160*
Schiller, H., 21, *35*, 301, *305*
Schkade, L. L., 154, 155, *163*
Schlachta-Fairchild, L., 232, *238*
Schleris, W., 189, 192, *197*
Schlosser, A. E., 188, *196*
Schmierbach, M., 129, 132, *141*, 290, *306*
Schneider, P. L., 45, *52*
Schoeman, F. D., 258, *278*
Schoenbach, K., 26, *35*
Schramm, W., 127, *140*
Schulman, M., 125, *140*
Schumacher, H., 130, *137*
Schumacher, P., 49, 50, *54*, 192, *197*
Schwarez, T. H., 232, *236*
Schwartz, A., 271, *275*
Schwartz, B., 271, *275, 278*
Schweitzer, J. C., 148, *162*
Scott, L., 192, *197*
Sebille, S., 228, *239*
Secunda, E., 169, *182*
Sefton-Green, J., 67, *73*
Sellon, A., 116, *120*
Serlin, R. E., 48, *52*
Settle, R. P., 211, *222*
Severin, W. J., 155, *162*
Shah, D., 129, 132, *141,* 290, *306*
Shanahan, J., 178, *181*
Shapira, N. A., 189, 191, *197*
Shapiro, S., 259, 260, *278*
Sharf, B., 49, *56*
Sharit, J., 87, *99*
Shaw, D. L., 145, 146, 148, *161, 290, 304*
Shaw, M. J., 217, *222*
Sheehan, K. B., 265, 267, *278*
Sheffer, L., 21, *35*, 295, *306*
Sherry, J. F., 206, *222*
Shih Ray Ku, R., 245, 246, *256*
Shim, S., 212, 213, *222*
Shimon, J. M., 188, *197*
Shirato, H., 228, *236*
Shirky, C., 254, *256*
Shiu, E., 41, 46, *56*
Short, J., 43, *56*
Shotton, M. A., 189, *197*
Sicotte, C., 226, *237*
Sidener, J., 131, *141*
Siefkin, A., 231, *237*
Siegel, E., 231, *237*
Siegel, J., 272, *276*
Silver, S., 71, 89, 92, *99*
Sime, D., 231, *236*
Sinclair, J., 24, *35*
Singer, J. B., 145, 150, *162*
Sitkin, S., 104, *120*, 288, *306*
Skalski, P., 168, 173, 175, 176, *179*, 189, 191, *193*
Skumanich, S., 208, 214, *222*, 294, *306*
Slater, M. D., 192, *197*
Sledge, W. H., 233, *237*
Sless, D., 115, *120*
Smith, A. E., 211, 213, *220*
Smith, L., 68, *74*
Smith, M., 5, 8, 10, *15*
Smith, R. L., 129, *141*
Smolkin, R., 144, *162*
Smulders-Meyer, O., 232, *237*
Snider, M., 171, *182*
Solomonidou, C., 29, *34*
Sommer, R., 104, *120*
Soroka, S. N., 146, *162*
Sothirajah, J., 5, *14*
South, J., 144, *162*
Sparkes, V. M., 148, *163*

Spears, R., 45, *53*, *55*, *56*, 285, *304*, *305*, *306*
Spears, S. C., 173, *180*
Speight, K., 188, *197*
Speyer, C., 45, *55*
Spink, A., 184, *197*
Spooner, T., 127, *141*
Sprague, R., 102, *120*
Sproul, L., 148, *159*
Sridhar, N., 89, 90, *96*
Stafford, L., 40, 41, 46, *52*, *56*
Stafford, M. R., 154, 155, *163*
Stafford, T. F., 154, 155, *163*
Stahl, B., 130, *141*
Stallman, R. M., 243, *256*
Stamford, P. P., 231, *237*
Stamm, K. R., 125, *141*
Stan, F., 26, *35*
Stanek, D. M., 6, *15*
Stark, S. D., 212, *222*
Steeves, H. L., 22, *34*
Stein, L., 128, *141*
Steinfield, C., 7, *15*, 46, *55*, 185, *195*
Stempel, G., 148, 153, *163*
St. John, K., 231, *235*
Stoll, C., 286, *306*
Stone, L. D., 47, *56*
Stoyanoff, N., 130, *137*
Strader, T. J., 217, *222*
Strasburger, B., 286, *306*
Straubhaar, J., 21, 22, 26, *32*, *35*, 295, *306*
Stross, R., 274, *278*
Strouse, G. A., 59, *73*
Subrahmanyam, K., 286, *306*
Suchman, L., 103, 104, 105, *120*, *121*
Suler, J., 41, 47, *56*
Sullivan, B., 271, *278*
Sullivan, E., 226, *236*
Sullivan, W. M., 125, *137*
Sundstrom, E., 104, *121*
Sundstrom, M. G., 104, *121*
Sunnefrank, M., 44, *55*
Surette, R., 190, *197*
Sussman, L., 40, *52*
Sutcliffe, K., 104, *120*, 288, *306*
Swanson, J. A., 88, *97*
Swanson, J. N., 225, *235*
Swidler, A., 125, *137*
Sykes, C. J., 274, *278*

T

Tan, B. W., 192, *197*
Tankard, J. W., 155, *162*
Tanriverdi, H., 227, *238*
Taub, E. A., 136, *141*
Taylor, C., 93n, *99*
Tempier, R., 225, 231, *235*
Tenner, E., 103, 118, *121*
Teo, T. S. H., 84, *98*
Tewksbury, D., 28, *32*, 157, *163*, 188, *194*
Thompson, S. H., 84, *98*
Thomson, B. A., 232, *238*
Thornton, D., 67, *73*
Thurlow, C., 39, 40, *56*
Tichenor, P., 127, 128, *141*, 156, *163*, 188, *196*, *197*, 290, *306*
Tidhar, C. E., 130, *141*
Tidwell, L. C., 44, 48, 49, *56*
Tietze, S., 91, *99*
Tijdens, K., 89, *99*
Tipson, S. M., 125, *137*
Toffler, A., 89, 91, *99*, 300, 301, *306*
Toffler, H., 300, 301, *306*
Tolchinsky, P. D., 266, *278*
Tomlinson, J., 21, *35*
Toms, J., 231, *236*
Tonetti, J., 231, *235*
Traynor, M., 271, *278*
Tribe, L., 259, *278*, 298, *306*
Trigg, R., 105, *120*
Troiano, C., 173, 178, *181*
Tucker, D., 45, *121*, 285, *305*
Tung, C., 191, *197*
Turkle, S., 39, *56*
Turner, J., 42, 45, 48, *51*
Turow, J., 156, *163*
Tussey, D., 245, *256*
Tyler, T. R., 24, *35*, 193, *197*
Tyrrell, J., 231, *235*

U

Uk Won, J., 231, *235*
Uotinen, J., 132, *141*

Urbanczewski, A., 216, *222*
Urquhart, C., 229, *235*
Uusitalo, O., 206, 211

V

Valacich, J. S., 188, *197*
Valentine, J. C., 63, 67, *72*
Valkenburg, P., 66, *74*
Van Eenwyk, J., 67, *72*
Van Maanen, 118, *121*
Varga, A., 45, *54,* 285, *305*
Varis, T., 21, *35*
Venkatesh, A., 148, *163*
Vesmarovich, S., 228, *238*
Vijayasarathy, L. R., 217, *220*
Vince, P., 148, *159,* 290, *303*
Viseu, A., 266, 272, *279*
Viswanath, K., 130, *141*
Vitalari, N. P., 148, *163*
Vivien, K. G., 84, *98*
Vogel, D., 188, *197*
Vos, H., 272, *279*

W

Wagner, R. M., 129, *139*
Wainwright, P., 229, *235*
Wakefield, B., 229, *238*
Waldfogel, J., 10, *15,* 152, *163,* 293, *306*
Waldron, V. R., 48, *51*
Walker, J., 226, *238*
Walker, O. C., 211, *218*
Wallace, J., 229, *236*
Wallace, P., 230, 231, *236*
Wallace, P. M., 39, *56*
Walt, V., 29, *35*
Walther, J. B., 44, 45, 48, 49, *55, 56, 57,* 189, *197,* 259, 261, 262, 263, 265, *275,* 284, 285, 289, *306, 307*
Walton, J. R., 211, *218*
Wanta, W., 146, *163*
Ward, A. C., 129, *137,* 290, *302*
Warren, S., 258, 259, *279,* 298, *307*
Wartella, E., 125, 130, *141*
Watad, M., 84, *100*
Waterman, D., 171, 174, *182*
Weaver, A. J., 157, *163*
Webb, S. D., 263, *279*
Webster, D., 23, *35*
Webster, J., 40, *55*
Weick, K., 103, 107, 112, *121*
Weinberger, T., 230, 232, *236*
Weiss, L., 68, *73*
Welch, M., 150, *163*
Wellman, B., 39, 50, *52, 57,* 86, *100,* 287, *307*
Werry, C., 41, *57*
Wesson, C., 214, *222*
Westfall, R., 93t, *100*
Weston, A., 260, 261, 262, 263, 265, *279*
Wetzels, C., 89, *99*
Wheeler, B. C., 188, *197*
Whetton, S., 226, *238*
White, D. R., 227, *237*
White, H., 231, *235*
White, M., 48, 49, *57*
Whitten, P., 225, 227, 230, 231, *237, 238, 239*
Whitty, M. T., 47, *57*
Will, P., 84, *100*
Will, R., 263, *279*
Willemain, T. R., 223, 231, *237, 239*
Williamson, S. K., 231, *234*
Wilson, B., 286, *306*
Winett, R. A., 130, *141*
Wittman, C., 231, *234*
Wittson, C. L., 223, *239*
Wizelberg, A., 49, *57*
Wohl, R. R., 186, *196*
Wolf, E. M., 49, *51*
Wolfe, M., 260, 261, 262, *276*
Wolin, L., 154, *160,* 207, 211, *222*
Woo, J., 231, *235, 236*
Wood, A. F., 5, 8, 10, *15*
Woodman, R. W., 266, *278*
Woods, M. L., 188, *197*
Woolard, J. L., 59, *73*
Wooton, R., 230, *239*
Wright, C. R., 128, *141*
Wright, K., 43, 45, 48, 49, *57,* 285, *307*
Wurtzel, A., 129, *141*

Y

Yaezel, A., 231, *235*
Yan, G. G., 184, 192, *198*

Yang, S. C., 191, *197*
Yang,Y., 87, *99*
Yarbrough, E. J., 45, *54*, 285, *305*
Yates, J., 103, 118, *121*
Yelsma, P., 290, *303*
Youn, S., 266, 267, *279*
Young, K. S., 43, 49, *57,* 191, *197*
Yousman, 125, *138*
Yu, P. K., 246, 253, *256*
Yung, Y. F., 215, *220, 221*

Z

Zack, J. S., 45, *55*
Zakon, R. H., 40, *57,* 183, *197*
Zalesny, M. D., 104, *121*
Zaylor, C., 228, *239*
Zelechow, A., 116, *119*
Zeller, T., 150, 151, *163*
Zhu, J. H., 193, *197*
Zillman, D., 185, *197*
Zimmerman, E., 131, *141*
Zmud, R., 287, *302*
Zollo, G., 105, *119*
Zollo, S., 231, *239*
Zuboff, S., 104, 116, *121*
Zuckerman, M., 192, *197*
Zurawik, D., 144, *163*

Subject Index

A

Absorptive capacity (defined), 84, 85, 287
Access, digital divide and knowledge gap, 188
Addictive tendency, 11
Administrative innovations (defined), 83
Advertising, 12, 27, 80, 148, 159–160, 164, 171–172, 177, 186, 203–205, 212, 249, 251–252, 272, 294
  ad pop-ups, 264–265
Affective aggression model, 63
Affordance (defined), 105
Agenda setting, 5, 145–146, 290
Aggressive cues, 64
Al-Jazeera, 23, 27, 146, 284
Americans With Disabilities Act, 81
Analog television, 168–169
  entertainment setting, 291–294
  individual and group setting, 284–286
  legal and regulatory setting, 296–299
  surveillance setting, 289–291
  work and organizational setting, 286–289
Aeropagitica, 245
ARPAnet, 20
Artifacts, 104–105
Artifacts as meta-information, 116–117
Asynchronous connections (defined), 224
AT&T monopoly, 4, 204
Audience fragmentation, 23, 130, 156, 178–179, 292, 296
Audio streaming, 293
Audiocassette tape recorders, 13, 243
Augmentation (defined), 105
Automating, 104, 288

B

Beta, 17, 132, 244, 247, 252, 297
Big brother, 4, 91, 283
BITNET, 40, 183
Blogs (defined), 29, 141, 144–145, 150–151, 158, 160–162, 183–184, 190, 197, 204, 260, 272, 286, 293–294, *see also* Weblogs
Bottom–up influence, 128
Box office, 19, 174
Broadband (defined), 6, 40, 53, 82, 92, 127, 136, 167, 183
Bulletin boards, 111, 125, 134, 189, *see also* Individual setting

C

Cable (defined), 4, 6, 8–10, 12, 20, 24, 32, 124–130, 130–136, 257, *see also* Media technology
  and the 24-hour newscycle; video technology and home entertainment
  Cable Communications Policy Act of 1984, 258, 270–271
Camera phones, 269
CAN-SPAM Act of 2003, 268
CATV, Community Antenna Television (defined), 125
CD (Compac Disk)-ROM, 13, 249
Cell phone, 47, 60, 85, 143, 257
  cell phone privacy, 266, 270–273, 286
Center-based telecommuting (defined), 81

Changing communication infrastructure, 4
Channel repertoire, 175–176, 290, 292, 293
Chat rooms, 39, 41, 42, 44–47, 54–55, 57, 134, 189, 195, 293, *see also* Individual setting
Children's use of media technology, 6, *see also* Computer-mediated technology and children
Children's Online Privacy Protection Act [COPPA] of 1998, 259, 270
Chinese bloggers, 284
Civic involvement, 129–130, 132, 290
Cognitive learning, interactive media and, 285–286, *see also* Effects on cognitive development, 67–68
Cognitive neo-association theory, 62, 63
Communication presence, 85–86
Communication revolution, 79, 91, *see also* Information economy
Communication technology functions, 134
Communication technology and global change, 5, 17–35, 283–284
  Internet communication, 20–21, 27–29
  satellite technology, 19–20, 22–24, 26–27
  VCR technology, 21–22, 26
  videocassette recorders and DVDs, 17–19
Communication technology and social change, 3–15, *see also* Integrated communication technology and social change typology
Community access channels, 125, 129, 134–135
Community and cable TV, 128–130, 133–136
Community functions, 134
Community and the Internet, 128–136
Compatibility, 185, 186, 226
Complexity (defined), 71, 84, 112, 115, 117, 184–185, 292
Compulsive shopping behavior, 209, 214–215
Computer aided software engineering (CASE), 84
Computer games and videogames, 66–67
  computer game addiction, 184
Computers in the classroom, 68, 71–72, 286
Computer-mediated communication (CMC) (defined), 39, *see also* Individual setting
Computer-mediated technology and children, 6, 59–76, 285–286
  computer games and videogames, 66–67, 285
  demographic differences, 60–61
  developmental approach, 65–66
  diffusion of innovations, 64–65
  effects approach, 63–64
  effects on cognitive development, 67–68, 286
  effects on social and emotional development, 68–69
  Internet uses, 59
  macroperspective, 69–70
  microperspective, 70–72
  videogame playing, 61–62
Computer-mediated technology and social interaction, 6, 39–57, 284–285
  e-mail, 40–41, 286
  instant messaging, 41, 286
  motives and uses, 42–43
  online chat rooms, 41–42, 286
  social relations, 44–45, 47–49
  technology attributes, 43–44, 47
  user motivations, 45–47
Consumer addiction, 206
Consumer attributes and behavior, 207–208, 212–213
Consumer innovativeness, 207–208
Consumer motivations, 206–207, 209–212
  psychological gratifications, 206–207
  utilitarian versus hedonic value, 206
Consumer privacy, 205, 216
Consumer setting, 201–240
Consumer shopping dependency, 208–209, 213–214

Contextual cues, 63, 285
Contingency theory, 86, 287
Controlling the Assault of Non-Solicited Pornography and Marketing (CAN-SPAM) Act of 2003, 205
Convergence (defined), 185–186, 283, 292
Cookies spyware, 257, 262, 264–265, 269, 271, 273, 298
Copyright, 13, *see also* Digital media technology and fair use
  Copyright Act of 1976, 244
  Copyright Arbitration Royalty Panel, 250
Corporation for Research and Educational Networking (CREN), 183
Co-workplace, 91
Cross-media competition, 283
CSNET, 183
Cues-filtered-out perspective, 44
Cultivation research, 69, 130
Cultural imperialism, 21–22, 283
Cyber cafes, 21, 135, *see also* Internet cafes

D

Daily me, 156–157
Deficient self-regulation, 50, 215, 216, 294
Demographic differences, 60–61
Dependency and addiction (defined), 184, 191–192, 194, 206, *see also* Internet addiction
Depression, 189, 191, 294, *see also* Internet addiction
Deregulation, 126, 283
Desktop, desktop information, 108–109
  categories of, 112–114
  conceptual analyses, 111–117
  four conceptual dimensions of desktop information artifacts, 112–117
  frequency/importance of use, 111
  personal items, 111, 113
  process-related items, 111, 113
  reminders, 111
  system- and task-related information, 111, 113
  temporary information, 111, 113
  typology of seven desktop artifacts, 111
  unsupported by the system, 111
Desktop conferencing, 43–44
  the desktop and its artifacts, 104–106
Developmental approach, 65–66
Diffusion of innovations theory, 12, 42–44, 62, 64–65, 82–85, 172–173, 185–186, 206, 207, 213, 226–227, 252
Digital broadcasting, 168–170, *see also* Video technology and home entertainment
Digital cable, 126
Digital communication networks, 82, *see also* Work and organizational setting
Digital compression technology, 126
Digital data files, 101–102
Digital divide, 24–25, 30–31, 128, 135, 188, 284
Digital gap, 290
Digital knowledge gap, 157
Digital media culture, 62
Digital media technology and fair use, 12, 243–256
  CD-ROM, 249
  copyrights and fair use, 244–247
  digital media copyrights and fair use, 245
  peer-to-peer sharing, 250–252
  webcasts and music distribution, 249–250
Digital media technology and individual privacy, 12–13, 257–279
  anonymity, 262
  autonomy and information control, 262
  boundaries and balance, 259–260
  concerns over privacy, 264–265
  personal growth and intimacy, 260
  privacy and the law, 267–271
  privacy protection behavior, 266–267
  risk-benefit perceptions of privacy, 265

social safety valve and control, 263–264
solitude and accessibility, 261
surveillance and observability, 261–262
Digital Millennium Copyright Act (DMCA) of 1998, 13, 244, 249–250, 252, 297
Digital news divide, 156
Digital Performance Rights in Sound Recordings Act, 250
Digital television (DTV), 168–170, *see also* Video technology and home entertainment
digital television decoders, 177, 292
Digital video discs (DVDs), 17–19, 171, 173–175, 178, 186, 243, 251, 292, *see also* Peripheral devices
Digital video recorders (DVRs), 9–10, 169, 177, 243, *see also* Peripheral devices
Direct Broadcast Satellites (DBS), 9, 20, 23, 170, 175, *see also* Satellite technology
Direct response TV commercials, 204
Direct subscriber line (DSL), 183
Direct-to-video, 174
Displacement, 290, 292, *see also* Media substitution, Principle of relative constancy
Distance learning, 188, *see also* Education and learning
Domain name system (DNS), 183
"Do not call" list, 273
Dual effects hypothesis, 283–284
Dynamic interactivity model, 187

E

Early adopters (defined), 64
Early majority, 64–65
Economic perspective, video technology, 171–172
Education and edutainment, 184
Education and learning, 199–200
Educational access, 132
Effects approach, 63–64
Effects of cognitive development, 67–68
Effects gap (defined), 127–128
Electronic chats, 185
Electronic Communications Privacy Act (ECPA) of 1986, 258, 269–270
Electronic cottage, 82, 94
Electronic document management (defined), 101–102
electronic document management system, 103
Electronic mail (defined), 40–41, *see also* Computer-mediated technology and social interaction
Emerging information economy, 80, *see also* Information economy
Emotional support, 48–49
Entertainment, 184
Entertainment setting, 165–200, 291–294
Ethnic media, 133
Excitation transfer, 62, 63

F

Facsimile, fax, 8, 85, 102, 106, 108–109, 113–114, 118, 261, 264–265, 286
junk faxes, 264–265
Fair use (defined), 13, *see also* Digital media technology and fair use
Family and child viewing, 178, *see also* Computer-mediated technology and children
Federal Communications Commission, 11, 13, 20, 170, 205, 268, 291, *see also* Legal and regulatory setting
Federal Trade Commission, 11, 205, 217, 268
Fiber optics, 7, 82, 131, *see also* Digital communication
Fifth Estate, 5, *see also* Surveillance setting
Fluid channels, fluidity (defined), 8, 194, 293
Fourth Amendment, U.S. Constitution, 259
Free flow of information, 23
Functional displacement, 152–154, 290
Functional equivalence (defined), 148–149, 290
Functional similarity, 172–173, 175, 186, 194, 286, 292

G

Gambling, 184, 192–193, *see also* Internet addiction
Gender gap, 28, 87
Generation text, 39
Global positioning system (GPS), 12, 257, 264
Glocalization, 23, 284
Google, 127, 284
Gore, Al, 126
Government access, 126, 289
Government surveillance, 257, *see also* Digital media technology and individual privacy

H

Hard news, 144
Hardware market, 61
Health Information Portability and Accountability Act (HIPAA), 233
Hedonic value, 206, 208–209, 215
Hegemony, 22, 24, 284
High definition television (HDTV) (defined), 168–170, *see also* Video technology and home entertainment
Home-based working, 88, 91, 286, *see also* Telework; Information technology and organizational telework
Home shopping, 11, 187, *see also* Interactive media technology and electronic shopping
Home video entertainment content, 170–171
  library building, 173
Home video entertainment technologies, 168–170
Horizontal integration, 4
Human factors, 4, 7–8, *see also* Work and organizational setting
Human interest news, 144–145
Hyperpersonal communication, 44, 284

I

I seek you (ICQ) (defined), 39, *see also* Instant messaging
Identity theft (defined), 257, *see also* Digital media technology and individual privacy
Impulse buys, 11
Individual setting, 37–76, 284–286
Informating, 104, 288
Information age, 79, 156, 274, 290, 295, 297, 298, 301
Information communication technologies (ICTs) (defined), 79, 82, 101, 184
Information economy, 79, 80–82, 300–301
Information gaps, 284, 290, 299, *see also* knowledge gaps
Information and infotainment, 190
  information, infotainment, and blogs, 184
Information society, x, 7, 13, 24, 284, 296, *see also* Information economy
Information superhighway, 126
Information system (defined), 103
Information technology: Analyzing paper and electronic desktop artifacts, 101–121, 288–289
  conceptual analyses, 110–117
  example implications of system change to desktops and work processes, 107–110
  the Desktop and its artifacts, 104–106
  transforming from the physical to the cognitive realm, 103–104
Information technology and organizational telework, 7–8, 79–100, 286–288
  communication presence, 85 86
  defining telework, 80–82
  effects on organizations, 89–90
  effects on teleworkers, 87–89
  organizational implications, 91–92, 286–287
  organizational innovativeness, 83–84
  organizational paradoxes, 91
  social implications, 92–95
  teleworker profile, 86, 87
  worker innovativeness, 85

Information workers, 7, 288, *see also* Work and organizational setting
Informmercials, 210
Infotainment, 10, 14, 183, 185, 187, 190–191, 193, 195, 197, 199
Infringement (defined), 245
Innovators (defined), 64
Instant messaging (IM) (defined), 39–41, 60, 69, 189, 204, 260, 271–273, 286, *see also* Computer-mediated technology and social interaction
Instrumental media use, 150, 203
Integrated communication technology and social change typology, 283–307
Integrated services digital network (ISDN), 85
Intellectual property, 13, *see also* Digital media technology and fair use
Interaction management, 261
Interactive media technology and home shopping, 11, 203–222
  consumer attributes and behaviors, 207–208, 212–213
  consumer motivations, 206–207, 209–212
  consumer shopping dependency, 208–209, 213–215
Interactive media technology and telemedicine, 11–12, 223–239
  interactive video (defined), 224
Interactive TV, 174, 205, 283
Intermedia medium, 186
International news, 143–144, 146–147
International Telecommunications Union (defined), 19
Internet addiction, 51–52, 57, 184, 191–199, 294, 304
  antisocial activities, 184
  computer game addiction, 184
  online gambling, 184, 192–193
  online sex addiction, 184
Internet boom, 167
Internet communication, 20–21, 24–29, *see also* Individual, entertainment and consumer settings
  Internet boom, 167
  Internet (cyber) cafes, 28–29, 32, 132, 135
  Internet use, 59–60
Internet relay chat (IRC) (defined), 39, 41–42, *see also* Computer-mediated technology and social interaction
Internet Spyware Prevention Act of 2005, 269
Intertemporal tiering (defined), 174–175
Introduction, 1
iPod, 179
Iraq war, 5, 144
iTunes, 254

J

Junk faxes, 264–265

K

Knowledge barriers, theory of (defined), 227
Knowledge economy, 80, 102, *see also* Information economy
Knowledge fragmentation, 291–293, 300
Knowledge gap (defined), 127–128, 156, 185, 188, 284, 290, 300
Kohlberg's theories of moral development, 66

L

Laggards, 65
Late majority, 64–65
Legal and regulatory setting, 241–280, 294–299
Lifestyle enclaves, 125
Listservs, 40, 134
Local area networks (LANs), 85, 106
Loneliness and depression, 189–190, *see also* Internet addiction

M

Macroperspective, 69
Maslow's theory of human motivation, 207
Materiality and complexity of information, 112–113

Media consumption: Hours per person, 174
Media dependency theory, 10, 187, 206, 208
Media displacement, 148–149, 290, *see also* Media substitution
Media/cultural imperialism, 21–22, 283–284
Media richness, 43
Media substitution (hypothesis), 172–174, 175–176, 186, 213, 293, *see also* Media displacement, Principle of relative constancy
Media technology and civic life, 8, 125–140, 284
  community and cable TV, 128, 129–130
  community and the Internet, 128–129, 130–131
Media technology and the 24-hour news cycle, 9, 141–163
  agenda setting, 145–146, 150–151
  functional displacement, 152–153
  media displacement, 148–149
  news diffusion, 146–147, 151–152
  news media ownership, 144–145
  the 24-hour news cycle, 144
  uses and gratifications, 149, 154–155
Metamorphosis project, 132–133
MGM v. Grokster, 252
Microperspective, 70–71
Milton, John, 245
Miniature cameras, 269
Minority Report, 13
Mobile work, 82
Mock news, 145
Motion Picture Association of America, 19
Motives and uses, 42–43
MP3 files, 250–251, 254
Multimedia cable, 176
Multiple system operator, 126
Multiple user dungeons (MUDs) (defined), 39, *see also* Online chat rooms
Myspace.com, 286

N

Napster, 250–251, 253–254, 297, 298
Need fulfillment motivation paradigm, 207
Neighborhood identity, 131, 134–135
Net news (defined), 147
Network newscasts, 143–145, 153
Networked interactive media in schools (NIMIS) project, 68
New customer system (NCS) (defined), 106–107
News, 9, 183, *see also* Surveillance setting
News diffusion, 146–147, 151–152
News groups (defined), 155, 293
News media ownership, 144–145
Newton's Third Law, 283
North American Free Trade Agreement (NAFTA), 5
NSFNET, 183

O

Obscenity, 269
Offline teleshopping, 203, 205, 209, 211, 213
Oil crisis, telecommuting and, 82
One way information/media flow, 150, 167, 283
Online chat rooms, 41–42
Online privacy seals, 266–267
Online technology, edutainment and infotainment, 10–11, 183–200
Online teleshopping, 203, 207, 209, 213, *see also* Teleshopping
Optical character recognition (OCR) device (defined), 102
Organizational communication, 7–8, *see also* Work and organizational setting
Organizational implications, 91, 226, 286–289
Organizational innovativeness, 83–84
Organizational learning, 84
Organizational paradoxes, 91–92
Orwell, 4, 283

P

Paper and electronic forms as communication media, 115–116
Paper versus electronic information, 112

Paperless office, 102–103, 107, 119, 121, 288
Parasocial interaction, 186–187, 208–209, 214, 292
Patient-physician relations, 232
Pay-per-view, 167, 172
Peer-to-peer sharing, 250–252
Peripheral devices, 169–170
Personal computer (PC), 6, 40, 60, 92, 101, 106, 143, 177, 243, 250, 296, *see also* Work and organizational setting
Personal digital assistant (PDA), 143, 147
"Personal me," 291
Personal use, 245
Personalogical variables, 64
Phishing, 257, 265
Physical proximity, 148
Piagetian theory, 65–66
Piracy (copyright), 245, *see also* Digital media technology and fair use
Podcasts, 143–144, 184, 204, 286
Pong, 60
Pornography, 69
Portable document format (PDF) (defined), 7, 102, 286
Presence (defined), 85–86, 187
Principle of relative constancy (PRC) (defined), 172–173, 175, 186, 291, *see also* Media displacement, Media substitution
Privacy (defined), 12, 261, *see also* Digital media technology and individual privacy
bodily integrity—intrusive searches/seizures, 267–268
communications privacy—protecting confidentiality, 270–271
decisional privacy—decisions about self/family, 268–269
information privacy—controlling information flow, 269–270
patient Privacy, 233
privacy Rule, 12
Problematic uses of the Internet, 49–50
Propaganda, 17, 22, 284
Psychological impacts, 187–188
Public access channels, 129, 134, 289
Public good nature of information, 246

Q

R

Radio, 3, 12, 60, 80, 118, 125, 143–144, 147–148, 149, 152–153, 160, 173, 186, 203–204, 224–225, 228–229, 236–237, 247, 250, 253, 257, 262, 291, 298
Radio frequency identification (RFID), 12, 257, 262, 264
Radiotherapy, 228
Realism, 86
Recording Industry Association of America, 25, 250
Reinvention, 83, 173, 187, 227
Relative advantage, 82, 173, 185, 213, 287, 292, 296
ReplayTV, 170
Repertoires, 186, 194, *see also* Channel repertoires
Rich media (defined), 83, 86, 204, 291
Ritualized media use, 150

S

S-shaped curve, 65
Sanitized violence (defined), 63
Satellite Home Viewer Implementation Act, 136
Satellite radio, 143
Satellite set top boxes, 169
Satellite technology, 3, 5, 17, 19–20, 22–24, 26, 82, 126, 143, 155, 170, 292, *see* also Communication technology and global change, peripheral devices
SBC, 4
Screen time, 60, 62
Script theory, 60, 63
Security measures, 83
Sensationalism, 144–145
Sexual content, 184, 269, 285, *see also* Computer mediated technology and children
Shopping addiction, 126, 208, 295, 304
Short message service (SMS) (defined), 10, 20, 39–40, 44

Silicon Valley Snake Oil, 286
Smart mobs, 32
Smart Neighborhood Initiative, 135
Social capital, 129, 136, 138, 140, 289, 291
Social cognitive theory, 62, 63, 185, 215, 285
Social embeddedness theory, 227
Social identification model of deindividuation effects (SIDE), 45
Social information processing (SIP) theory, 45
Social interaction, 189–190
Social isolation, well-being, and anxiety, 184
Social presence, 43, 44, 47, 85, 257
Social relations, 44–45, 47–49
Social richness, 86
Soft news, 144–145
Software market, 61, 172
Sonny Bono Copyright Term Extension Act of 1998, 13, 244, 247
Sony Corp. v. Universal City Studios, 247–248, 251–252
Source shifting, 169
SPAM, 11, 12, 205, 254, 261, 264, 267–268, 270–274, 298
Spycams, 264, *see also* Spyware
Spyware, 264, 298, *see also* Digital media technology and individual privacy
Stimulation overload, 263
Store and forward connections (defined), 224
Strong-tie network, 284, 285
Structuration theory, 226
Summary, 281–307
Surveillance setting, 123–164, 289–291
Symbolic interaction, 288
System change to desktops and work processes, example implications of, 107–110

T

Task equivocality, 95, 287
Technical factors, 187
Technologies of freedom, 4, 23
Technology acceptance model (TAM), 226
Technology attributes, 43–44
Technology clusters and media substitution, 172–174, 175–176, 292
Technology determinism, 287
Technology masters, 65
Ted Turner, 126
Telecenters, 130, 134
Telecommuting, 6–7, *see also* Information technology and organizational telework
Teleconsultations, 226, 228, 230–231
Teledermatology, 228
Telegraph, 3, 256
Telehealth, ix, 234, 231
Telehome services, 224, 229, 231
Telemarketing, 11, 204, *see also* Interactive media technology and electronic shopping
Telemarketing and Consumer Fraud and Abuse Prevention Act, 205
Telemarketing Sales Rule, 205
Telematic, 4, 79, 84–85, 91, 92, 186, 283
Telemedicine, 11–12, *see also* Interactive media technolog and telemedicine
  cost effectiveness of, 230
  patient satisfaction with, 230–231
Telenovelas, 21
Telenurses, 232
Telephone, 4–6, 11, 46–47, 52, 60, 85, 87, 92, 108, 117, 152, 154, 203–205, 218–219, 225, 229, 250–251, 253, 257–258, 260–262, 265, 268, 270, 278, 286, 293
Telephone Consumer Protection Act of 1991, 268
Telepsychiatry, 223–225, 228–229
Teleradiology (defined), 224–225, 228–229
Teleshopping, 11, 187, 203–217, 222, 294–295, 305, *see also* Interactive media technology and
  electronic shopping
Telework (defined), 80–81, 286, *see also* Work and organizational setting
  effects on organizations, 89–90
Teleworker profile, 86–87
Teleworkers, 81–82
  effects on, 87–88
Third places, 127, 132

Time shifting, 169, 173, 292
TiVo, 9, 170, 213
Toddlers, 59
Trialability, 185
Transforming from the physical to the cognitive realm, 103–104
TV programming, 170
Twenty-four hour news cycle, 9, 144, 291, *see also* Media technology and the 24-hour news cycle
Two-way TV, 223
Typewriter, 117

U

Umbrella perspective on communication technology, 167–168
USA PATRIOT Act, 259, 268, 273
Usenet, 40, 184
User motivations, 45–47
Uses and gratifications, 42–43, 50, 149–150, 154–155, 173–174, 176, 184–165, 194, 206–210, 292, 293
Utilitarian versus hedonic value, 206, 208, 215

V

Verizon, 4
Versatile technologies (defined), 85
Vertical integration, 4
Video cameras, 269
Video Voyeurism Prevention Act of 2004, 269
Videocassette recorders (VCRs), 9, 10, 17–19, 21–22, 26, 148, 169, 175–176, 243, 251, 292, *see* also Video technology and home entertainment
  Betamax video recorders, 244
  VCR aberration, 172, 186
  VHS format, 17
Videoconferencing (defined), 224, *see also* Interactive video
Videogame playing, 61–62, 66–67
Videophone, 84, 223
Video streams, 213, 293, 296
Video technology and home entertainment, 9–10, 167–182
  economic perspective, 171, 174–175
  economics of distribution, 171–172
  home video entertainment content, 170–171
  home video entertainment technologies, 168–170
  technology clusters and media substitution, 172–174, 175–176
  video Privacy Protection Act of 1988, 259, 271
Violent video games, 6, 66–67, 187, 192, 285
Virtual community, 125
Virtual organization, 82
Virtual public sphere, 133
Virtual workforce, 7, *see also* Information technology and organizational telework
Viruses, 265
Voice recognition, 61

W

Watchdog function, 128
Weak-tie network, 285
Webcasts and music distribution, 10, 213, 249–250
Weblogs (defined), 131, 144, *see also* Blogs
Wide area networks (WANs) (defined), 85
Wireless carriers, 4
Work and organizational setting, 7–8, 77–122, 286–289
Work processes, 109–110
Workarounds (defined), 105
World Trade Organization treaties, 19, 30
World wide web (defined), 126–127, 183, *see also* Individual, entertainment and consumer settings

Y

Yahoo! Local (defined), 131